Regulation and Protectionism Under GATT

Regulation and Protectionism Under GATT

Case Studies in North American Agriculture

EDITED BY

Andrew Schmitz, Garth Coffin, and Kenneth A. Rosaasen

WestviewPress

A Division of HarperCollins*Publishers*

Copyright © 1996 by Westview Press, Inc., A Division of HarperCollins Publishers, Inc.

Published in 1996 in the United States of America by Westview Press, Inc., 5500 Central Avenue, Boulder, Colorado 80301-2877, and in the United Kingdom by Westview Press, 12 Hid's Copse Road, Cumnor Hill, Oxford OX2 9JJ

Library of Congress Cataloging-in-Publication Data
Regulation and protectionism under GATT : case studies in North
    American agriculture / edited by Andrew Schmitz, Garth Coffin,
    Kenneth A. Rosaasen.
        p.   cm.
    Includes bibliographical references and index.
    ISBN 0-8133-8956-9
    1. Tariff on farm produce—North America—Case studies.   2. Tariff
on farm produce—United States—Case studies.   3. Tariff on farm
produce—Canada—Case studies.   4. Protectionism—United States—
Case studies.   5. Protectionism—Canada—Case studies.   6. General
Agreement on Tariffs and Trade (Organization)   I. Schmitz, Andrew.
II. Coffin, Garth.   III. Rosaasen, Kenneth A.
HF2651.F27N77   1996
382'.41'097—dc20   95-43441
                                                                                      CIP

The paper used in this publication meets the requirements of the American National Standard for Permanence of Paper for Printed Library Materials Z39.48-1984.

10   9   8   7   6   5   4   3   2   1

Contents

Acknowledgments

Without the support of dedicated staff and cooperative organizations, this book would not have been possible. In particular, we appreciate the support received from Agriculture and Agri-Food Canada; the Canadian Dairy Commission; the Canadian Chicken Marketing Agency; the Canadian Turkey Marketing Agency; the Canadian Egg Marketing Agency; the Québec Ministry of Agriculture, Fisheries, and Food, the National Farm Products Council, and the Saskatchewan Agricultural Development Fund.

We also want to thank the following people for their expertise and countless hours spent in the production of this book: Kim Box and Carole Schmitz for their competent and persistent editing, Dieudonné Mann for indexing, Roger Hoover for producing the graphics, and Ludovica Weaver for formatting and producing the book.

Andrew Schmitz
Garth Coffin
Kenneth A. Rosaasen

About the Editors and Contributors

J. M. Alston, Professor, Department of Agricultural Economics, University of California–Davis, Davis, California, USA

L. M. Arthur, Adjunct Professor, Department of Agricultural Economics, University of British Columbia, Vancouver, British Columbia, Canada

R. R. Barichello, Associate Professor, Department of Agricultural Economics, University of British Columbia, Vancouver, British Columbia, Canada

M. E. Bohman, Assistant Professor, Department of Agricultural Economics, University of British Columbia, Vancouver, British Columbia, Canada

R. B. Borges, Assistant Professor, Department of Agricultural Economics, Kansas State University, Manhattan, Kansas, USA

D. Christian, Graduate Student Resident, Department of Agricultural and Resource Economics, University of California–Berkeley, Berkeley, California

H. G. Coffin, Associate Professor, Department of Agricultural Economics, McGill University, Macdonald Campus, Ste-Anne-de-Bellevue, Québec, Canada

H. de Gorter, Associate Professor, Department of Agricultural Economics, Cornell University, Ithaca, New York, USA

M. E. Fulton, Professor, Department of Agricultural Economics, University of Saskatchewan, Saskatoon, Saskatchewan, Canada

G. Gartner, Management Consultant, Ottawa, Ontario, Canada

J. A. Janmaat, MBA Student, Simon Fraser University, Burnaby, British Columbia, Canada

M. Katz, Chief, Poultry Unit, Marketing Policy Division, Policy Branch, Agriculture and Agri-Food Canada, Ottawa, Ontario, Canada

R. D. Knutson, Professor and Director, Agricultural and Food Policy Center, Texas A&M University, College Station, Texas, USA

R. Lambert, Professeur Adjoint, Department d' Economie Rurale, Université Laval, Québec City, Québec, Canada

J. S. Lokken, Research Associate, Department of Agricultural Economics, University of Saskatchewan, Saskatoon, Saskatchewan, Canada

K. D. Meilke, Professor, Department of Agricultural Economics and Business, University of Guelph, Guelph, Ontario, Canada

J. L. Outlaw, Assistant Research Scientist, Department of Agricultural Economics, Texas A&M University, College Station, Texas, USA

T. J. Richards, Assistant Professor, School of Agribusiness and Environmental Resources, Arizona State University, Tempe, Arizona, USA

R. F. Romain, Professeur Agrégé, Department d' Economie Rurale, Université Laval, Québec City, Québec, Canada

K. A. Rosaasen, Professor, Department of Agricultural Economics, University of Saskatchewan, Saskatoon, Saskatchewan, Canada

R. R. Rucker, Associate Professor, Department of Agricultural Economics and Economics, Montana State University, Bozeman, Montana, USA

R. Saint-Louis, Professeur, Department d' Economie Rurale, Université Laval, Québec City, Québec, Canada

A. Schmitz, Eminent Scholar, Department of Food and Resource Economics, University of Florida, Gainesville, Florida; Professor Emeritus, Department of Agricultural and Resource Economics, University of California–Berkeley, Berkeley, California; Adjunct Professor, Department of Agricultural Economics, University of Saskatchewan, Saskatoon, Saskatchewan, Canada

T. G. Schmitz, Post-Doctoral Student, Department of Agricultural Economics, North Dakota State University, Fargo, North Dakota, USA

J. J. Skinner, Research Analyst, Canadian Wheat Board, Winnipeg, Manitoba, Canada

G. Skogstad, Professor, Department of Political Science, University of Toronto, Toronto, Ontario, Canada

J. D. Spriggs, Professor, Department of Agricultural Economics, University of Saskatchewan, Saskatoon, Saskatchewan, Canada

B. K. Stennes, Economic Consultant, Agriculture and Agri-Food Canada, Vancouver, British Columbia, Canada

D. A. Sumner, Professor, Department of Agricultural Economics, University of California–Davis, Davis, California, USA

W. N. Thurman, Professor, Department of Agricultural and Resource Economics, North Carolina State University, Raleigh, North Carolina, USA

E. van Duren, Associate Professor, Department of Agricultural Economics and Business, University of Guelph, Guelph, Ontario, Canada

M. M. Veeman, Professor and Chairman, Department of Rural Economy, University of Alberta, Edmonton, Alberta, Canada

J. A. Vercammen, Assistant Professor, Department of Agricultural Economics, University of British Columbia, Vancouver, British Columbia, Canada

Overview of the Effects of GATT

1

Introduction:
Trade and Regulations in Transition

A. Schmitz

Introduction

A considerable amount of confusion exists over what the General Agreement on Tariffs and Trade (GATT) has accomplished. Have a large number of winners and losers emerged from the agreement? As expected, sectors that could potentially incur large losses from free trade fought hard to defeat GATT. In many cases they won, an outcome predicted by several, including Schmitz (1988 and 1994). Statements abounded such as, "GATT is fine for other sectors, but don't let it apply to ours". The fact that many sectors within North American agriculture were able to maintain the protectionist web is not surprising. Significant sums of money were spent on lobbying politicians to defeat GATT and other free trade arrangements, such as the North American Free Trade Agreement (NAFTA) and the Canada–United States Free Trade Agreement (CUSTA).

The overall purpose of this book is to present the major impacts that GATT will have on US and Canadian agriculture; less attention is given to NAFTA and CUSTA. Particular emphasis is placed on US dairy, sugar, peanuts, tobacco and Canadian supply-managed commodities — industries in which protectionism has prevailed in the presence of the new agreements. In spite of the "free trade" rhetoric, trade disputes have arisen between Canada and the United States in all of these industries — disputes tied to GATT, CUSTA, NAFTA and the US Farm Program. In showing why protectionism prevails, we examine in some detail Canadian supply management institutions and political design, along with US sugar policy. Under the economics of rent-seeking, many producers of Canadian supply-managed commodities seem to fare better than they do under open market conditions. This book also discusses whether the supply management cartel will likely persist, at least for the short run, in view of the new rules imposed by GATT.

Background

The Canadian supply management sector, which includes dairy, poultry and eggs, is highly protected from imports from abroad. It evolved in response to low and unstable prices. The system is effective largely because of its complex and unique institutional design, which links the federal government to individual provinces. Because of the institutional design, producers are able to gain and maintain considerable political support. This leads some observers to comment that Stigler's (1971) theory applies to Canadian supply management. In this theory, regulations are put in place to supposedly correct market distortions, but over time the regulators become captured by the regulated, eventually causing the industry to move away from competition. Unfortunately, little has been written about Canadian supply management to test this hypothesis. An understanding of the evolution of the Canadian supply management system and its institutional design is critical to understanding the rent-seeking behaviors of producers. This book shows, given its design, that it is not difficult to explain why GATT is likely to have little impact on Canadian supply management, at least in the short run.

Like Canada's supply management, strong US political support has existed for the protectionism afforded dairy, peanuts, tobacco and sugar. As this book illustrates, this support has not been significantly affected by GATT. Why these sectors have been successful in obtaining political support is not answered in detail by this book with the exception of Chapter 12 in which an expanded version of a political economy model for sugar is provided. In this model, sugar and corn sweetener interests unite in support of the sugar program. When the effects of the sugar program are broken down by region and the supporting votes are regionally tabulated, a basis for strong support of the program is formed. In addition, there are coalitions formed with the peanut, tobacco and dairy sectors. Others like Gardner (1987) and Babcock, Carter and Schmitz (1990) have delved into the political economy dimensions of many farm programs that parallel this book's chapters on Canadian supply management.

Agricultural trade between Canada and the United States is significant, but GATT, NAFTA and CUSTA resulted in an increase in the volume of this trade, however restricted it may have been. Unfortunately, trade disputes have arisen and are continuing for all the major commodities covered in this volume. For example, the US sugar program and NAFTA affect the Canada – US trade of sugar and sugar products. As the book discusses, Canadian sugar refineries are pressing for free trade of sugar and sugar-containing products between Canada and the United States. This action is opposed by US sugar interests. Canada contends that trade is disrupted because of US sugar policy. Also, trade irritants have arisen surrounding Canada – US trade in products that contain peanuts and tobacco, partly because of differing interpretations of the free trade arrangements along with the distortion caused by the US farm program.

Perhaps more important are the trade issues involving Canadian supply management commodities. The United States' complaint is that tariffs on these commodities are set too high by GATT, and they want them lowered. At the heart of the debate are the Canada – US dairy sectors. As this book points out, in resolving some of the conflicts it is important to stress that the US dairy sector deviates significantly from a standard textbook model of competition. The United States contends that reducing Canadian tariffs will lead to significant increases in US dairy product exports to Canada. This book discusses whether this will likely be the case.

For the other supply-managed commodities in Canada there are no US structural counterparts. In the United States the system is very different because eggs and poultry, unlike dairy, are non-US program commodities. The US poultry industry is now vertically integrated and is not governed by production controls. From a trade perspective, the industry contends that reducing tariff barriers will significantly increase US exports of chicken and other products to Canada. As we show, given the high level of chicken tariffs agreed to under GATT, this expanded trade will take a long time to materialize. However, if NAFTA and CUSTA apply and tariffs are lowered, then the outcome will obviously be different.

Program Costs, Structural Change, and Coalitions

Program Costs

The programs underlying the commodities covered in this book have been under attack because, according to their opponents, they impose significant costs to society. The following gives some flavor to the debate. In 1993, the US General Accounting Office (GAO) reported on the cost of the US sugar program:

> Sweetener users bear the cost of supporting sweetener producers. Some studies have estimated a high cost by comparing the supported domestic price of sugar with the prevailing world price, resulting in a cost to domestic users of over $3 billion annually. However, recognizing that the world price would go up significantly in the long run if the United States and other countries purchased more sugar on the world market, GAO chose a more conservative approach. Using a long-run world price for refined sugar, and including program-related High Fructose Corn Syrup (HFCS) user costs, GAO estimates that the program costs sweetener users approximately $1.4 billion annually. This is an average based on 1989, 1990, and 1991 cost estimates.

Since the sugar program keeps domestic sugar prices higher than they would otherwise be, manufacturers of sugar's main competitor — HFCS —

can keep their prices higher as well. GAO has estimated that manufacturers of HFCS receive an additional $548 million annually because of the sugar program. HFCS manufacturer benefits are also concentrated: Four HFCS firms accounted for 87% of domestic production in 1990. This concentration of benefits occurred largely because of the substantial investment required to produce HFCS. This investment has made it difficult for new firms to enter the market.

For Canadian supply management, the estimated net societal costs are reported in Table 1.1 (Barichello, 1982). Note that for dairy alone the cost of the program is roughly Cdn $215 million annually. Table 1.2 gives the results from various studies for eggs and broilers; Table 1.3 gives the results from various studies of turkeys. According to these results, the largest costs are associated with dairy.

TABLE 1.1   Some Economic Effects of Selected Canadian Agricultural Policies

Major Commodities	Economy Gain	Producer Gain	Consumer Gain	Taxpayer Gain
		Million $		
Wheat	*	+470	0	-307
Barley	-3	+246	0	-160
Rapeseed	*	+70	0	-49
Eggs	-19	+55	-74	0
Broilers	-13	+57	-73	0
Dairy	-214	+955	-980	-303

*Less than $1 million
Source: Barichello (1982)

As the results in Table 1.3 show, there is considerable disagreement about the costs of the program. For example, Van Kooten and Spriggs (1984) contend that for eggs, broilers and turkeys, the efficiency losses are small while Veeman (1982) contends they are sizable. Also, Schmitz and Christian (1993) contend that for sugar the costs of the program are overstated because key elements such as supply response to risk aversion are ignored.

TABLE 1.2   Economic Effects of Poultry Industry Regulation in Canada, Farm Gate Level, Selected Years

	Barichello *1980*	*Arcus* *1979*	*Veeman* *1979*	*Harling &* *Thompson* *1975–77*
	Million $			
Eggs:				
Economy Gain	-19	—	-0.4	-5
Producer Gain	+55	+45	+38	+74
Consumer Gain	-74	-56	-39	-80
Broilers:				
Economy Gain	-13	—	-5	-11
Producer Gain	+57	+71	+71	+94
Consumer Gain	-73	-77	-76	-121
Importer Gain	+4	—	—	—

Sources: Barichello (1982); Arcus (1981); Veeman (1982); Harling and Thompson (1983)

This book does not further explore the costs and benefits of programs associated with the commodities covered. However, work in this area along the lines of Schmitz and Schmitz (1994) is required. Some examples of questions that still need to be answered are: 1) Does protectionism lead to industry inefficiency, or does it lead to rapid pursuit and adoption of new technologies? 2) Is regional comparative advantage blocked by supply management, and does it lead to resource inefficiencies? 3) How costly would the programs be if the border prices used to calculate costs were estimated free trade prices rather than actual prices that are the outcome of worldwide distortionary government programs? These and many other elements must be included in the new agenda regarding the measurement of farm program costs.

TABLE 1.3   Annual Income Transfers and Efficiency Losses Due to Supply
Restrictions for Eggs, Broilers and Turkeys in Canada, 1979

		Veeman (1982a)	Van Kooten & Spriggs (A)	Van Kooten & Spriggs (B)
			Million $	
Eggs	- rent	na	105.7	105.7
	- income transfer	na	98.1	105.7
	- efficiency loss	107.7	1.9	2.2
Broilers	- rent	na	139.6	139.6
	- income transfer	na	100.4	139.6
	- efficiency loss	148.1	6.1	11.8
Turkeys	- rent	na	59.3	59.3
	- income transfer	na	35.9	59.3
	- efficiency loss	67.2	4.8	13.0

Sources: Van Kooten and Spriggs (1984); Veeman (1982)

Structural Change and Market Niches

Farm programs, regulations and institutional designs affect the structure
of the agricultural sector. For example, supply management in Canada
has a major impact on the degree of vertical integration of the entire
chicken industry. Canada's highly regulated industry is much less
vertically integrated than the US chicken industry, largely because of
different regulations. Also, farm processing cooperatives are common in
supply-managed sectors but are less apparent in the nonsupply-managed
sector of the United States.

This book explores the relationship between supply management and
processing cooperatives in the Canadian dairy industry. It asks such
questions as: 1) Are cooperatives more likely to evolve in a supply
management system than in an open market arrangement? 2) How is
competition at the processing and retailing levels affected by producer
regulations? 3) Can noncompetitive behavior exist even in the presence
of regulations? 4) An issue often raised is that regulation stifles an
industry's progressiveness because it causes the industry to be unable to
compete in niche markets. This book highlights the problems of niche
marketing in Canadian supply-managed sectors. Because of economies
of scale in processing, the US Chicken industry, for example, is able to

market many more product lines than the Canadian industry is able to market. This raises the question: 1) If Canada is to compete in the diverse market abroad, is further processing required? 2) If so, what are the implications of processing plant consolidation for achieving the economies of scale needed to produce many product lines?

Coalitions

Protectionism for a given industry is made easier if several commodity subsectors form a coalition supporting a common goal. For example, in Canada not only are producers affected by import quotas and tariffs but so are processors, importers and retailers. Generally, if processors and producers together seek protection, politicians are more likely to respond positively than they would if the groups were in opposition to one another. We model the type of implied coalitions in the Canadian scene in order for there to exist a small level of imports into Canada and a high level of tariffs in the presence of GATT. We show that importers likely had very little influence on the GATT outcome and look at how the outcome would change if the political influence of various groups changed. For example, we address what would happen if the government, instead of the private sector, collected import quota revenue.

The Political Economy of Reform

The pressure to lower governmental support for agriculture is apparent in both Canada and the United States. In the new 1995 US Farm Bill, changes will be made that will affect the future direction of agriculture. The amount of support that will be withdrawn from sectors such as sugar, peanuts, tobacco and dairy is open to speculation. The passing of the 1995 Farm Bill will provide data on just how effective these sectors are at lobbying politicians for support. In Canada, significant changes have already been made toward removing governmental involvement in agriculture, but these changes are outside of supply management. For example, two programs that have significantly affected the grains and livestock sectors have been the railway transportation subsidy (CROW) and the Gross Revenue Insurance Program (GRIP). Both are to be eliminated. In the future, unless new safety net programs are put into place, the grain economy in Canada will receive little governmental support. If the United States follows Canada's move, the 1995 Farm Bill will lower support to US grain producers.

Policymakers, especially in Canada, face something of a dilemma. The supply-managed sector discussed in this book receives income support through its ability to control production and limit imports. The resulting, effective rate of protection in this sector far exceeds the support given to other sectors. In future policy discussions, this imbalance will have to be addressed. Given the political support for supply-managed commodities, especially from Ontario and Québec that together have over one-half of the Canadian population (when coupled with the Québec Separatist Movement), those trying to lower support for producers of supply-managed commodities will have to tread lightly. As Table 1.4 shows, the production of supply-managed commodities has a much more relative economic significance in Ontario and Québec than in the Prairie Provinces. In terms of production, Ontario and Québec together produce 78.5% of the industrial milk and 54% of the eggs. In Ontario and Québec the greatest percentage of farm income comes from supply management, whereas in the Prairies the largest share comes from grain, oilseeds, pork and beef.

From a related perspective, Table 1.5 shows cash receipts from farming in Saskatchewan, Ontario and Québec. Less than 5% of Saskatchewan's farm income is derived from supply management production whereas Québec has roughly 40%.

Not only has the level of support for Prairie agriculture dropped relative to many of the other agricultural areas in Canada, but the high tariffs on imports of supply-managed commodities have also significantly affected Canada's grain exports, especially those to the United States. For example, in the 1994 Canada-US durum wheat dispute, in which US tariffs were imposed, the issue about tariffs on supply-managed goods was raised (Duncan and Koo, 1995). "If the Canadians can have high tariffs so can the Americans," one observer commented. Likewise, in the sugar trade dispute, the United States made the claim that the US sugar program is no more protectionist than the Canadian supply management program, thus arguing that protection from Canadian exports of refined sugar and sugar-containing products is justified. On the other side, Canadian sugar refiners are probing for free trade, obviously in opposition to the policies of Canadian supply management.

The policies in the United States are very different from those in Canada especially with regard to the poultry producing sector. This poses a policy dilemma: How does one level the playing field in trade when policies between the two countries are so different? Also, what about differences between trading institutions? Take, for example, wheat. In Canada, wheat is traded through a marketing board called the Canadian Wheat Board (CWB); the United States uses an open market system. Goodwin and Smith (1995) attacked the central desk-selling of the CWB, contending that its price discrimination policies give Canadian wheat producers an unfair advantage over their US counterparts. Ironically, in Canada there is growing opposition to the CWB.

TABLE 1.4   Regional Production Shares of Supply-Managed Commodities in Canada, 1991

Region or Province	Industrial Milk	Eggs	Turkey	Chicken
	(% of National Total)			
Atlantic Provinces	4.5	8.9	4.8	8.0
Québec	47.4	16.5	23.0	30.3
Ontario	31.1	37.9	42.0	35.1
Manitoba	3.9	11.3	7.2	3.6
Saskatchewan	2.6	4.7	3.7	2.7
Alberta	6.6	8.7	8.7	8.2
British Columbia	4.0	12.0	10.6	11.9
Canada (%)	100	100	100	100
Total Production	161[a]	443[b]	132[c]	551[d]
Change from 1971	(-20%)	(-7%)	(+28%)	(+107%)

[a]Million kgs of butterfat (Canadian Dairy Commission quota)
[b]Million dozen (Canadian Egg Marketing Agency quota)
[c]Million kgs eviscerated (Canadian Turkey Marketing Agency quota)
[d]Million kgs eviscerated (Canadian Chicken Marketing Agency quota)
Source: Statistics Canada: Quarterly Farm Cash Receipts

TABLE 1.5   Cash Receipts from Farming Operations, Selected Canadian Provinces, 1994

	Saskatchewan	Ontario	Québec
Total Crop Receipts	3,630,535	2,489,122	936,200
Total Supply Management, Livestock Receipts*	150,830	1,807,340	1,686,770
Total Livestock Receipts	968,576	3,425,298	2,782,646
Total Payments	250,766	144,010	489,726
Total Receipts	4,849,877	6,058,430	4,208,572

*Includes eggs, dairy, hens and chickens and turkeys
Source: Statistics Canada: Quarterly Farm Cash Receipts

Trade disputes will continue between Canada and the United States. As this book discusses, significant linkages exist between tariff and nontariff barriers in supply-managed commodities and the grain, oilseed and livestock sectors. The length of time that US farm programs will remain in effect is open to question. However, Canadian supply management is likely to be around for a long time. As this book shows, the system is designed to be almost immune from major tampering by politicians. However, there is a possibility that the supply management system could self-destruct. This is made possible because GATT, while assuring high levels of external tariffs, does not insist on internal production controls.

The above message could be wrong, however, depending on whether or not it is NAFTA or GATT that rules on tariffs. The United States contends that Canada is obligated under NAFTA to eliminate the tariffs imposed by GATT. The United States contends that under Article 302 of NAFTA that no member shall impose or increase any existing import tariff. Canada argues that Article 302 does not apply to tariffs for volumes of product above the minimum access rule since these tariffs did not exist prior to GATT. The outcome of this dispute will have significant impact on the degree of protection that Canada will receive. This debate was not resolved in a meeting held on March 1, 1995. Further meetings were scheduled for the future (Brown, 1995).

Road Map

This book is divided into three sections: Section I covers chapters 2–5; Section II deals with chapters 6–12; and Section III covers Chapters 13–19.

Section I contains a broad overview of the effects of GATT and other trading arrangements. It also details the effect of GATT on the Canadian supply management sectors.

Individual chapters in section II examine the effect that GATT had on the US dairy industry, the tobacco sector, the US peanut market and the US sugar industry. These sectors have remained highly protected even in the presence of the various free trade arrangements. In addition, the US dairy industry is compared to its Canadian counterpart. In this context, the role of cooperatives is also discussed.

One of the most fascinating research topics is the origin, structure and consequences of the Canadian supply management industries. Unfortunately, little has been written on Canadian supply management. Section III brings to light a new body of literature on the topic of regulation, rent-seeking and the economic outcomes of highly regulated industries.

Section I

Knutson sets a broad stage for GATT in Chapter 2 by highlighting its effects and the effects of other free-trading arrangements on major commodity sectors. It also draws out the implications of the free trade agreements with respect to Canada – US agricultural trade with particular emphasis on grains and dairy.

Chapter 3 by Schmitz, de Gorter and Schmitz, discusses the impact of GATT on the Canadian supply management sectors. If GATT rules are to be adhered to then tariffication will have little effect on the Canadian supply management sectors, at least in the short run. Tariffs under GATT were set high relative to other agricultural sectors in Canada.

Alston and Spriggs point out in Chapter 4 that previous studies of trade policy reform have treated domestic quotas as exogenous. In their model, quotas on imports are chosen jointly with domestic production quotas to optimize on total welfare and its distribution among producers, consumers, importers and taxpayers. GATT and NAFTA impose minimum import access requirements. This chapter shows the unintended consequences of GATT and NAFTA regulations that arise from the optimizing response of the Canadian government subject to the constraints of the minimum access requirements. Chapters by Schmitz and Vercammen, which concludes this section, answers the question of why imports into Canada have remained low. They show by using a game theoretic approach why GATT had little effect on supply management. The outcome suggests that producers are able to effectively lobby politicians. The political cost of transferring rents to importers at the cost of producers is too high. In this context one would expect high tariffs and little change in the level of imports.

Section II

Chapter 6 by Outlaw and Knutson deals with a case study of the US dairy industry. They demonstrate how it has been affected by GATT and how the industry is continuously changing in the presence of government policy and rapid changing technology. They show how these changes could affect the Canadian dairy industry if trade barriers are lowered.

Chapter 7 by Barichello, Lambert, Richards, Romain and Stennes, is a summary of three independent studies that compare the Canadian – US dairy industries by examining, in some detail, how costs vary between countries and regions. They find that certain regions in Canada are much more competitive than are others. This is also true in the United States. For example, the production costs in California are lower than those in Wisconsin which partly explains why the relative growth in the dairy sector has shifted to California. The first study compares interspatial and intertemporal productivity differences between Alberta and Wisconsin to show that the Wisconsin dairy sector holds a distinct and growing advantage over Alberta. The second uses a multilateral cost of production (COP) comparison to determine interregional and international advantages to illustrate that California has the lowest COP in North America, followed by Wisconsin, New York, Alberta, Ontario and Québec. The third study explains COP variations within the Ontario and Québec dairy industries to demonstrate that there are not appreciable economies of scale in either the Ontario or Québec dairy industries.

Chapter 8 by Bohmann and Janmaat deals with the role of cooperatives under a supply-managed, vertically coordinated system. They show that cooperatives may in fact play a lesser role when regulations are present under a supply management system. However, cases are also developed where the role of cooperatives would actually increase.

A great deal of discussion now centers around adding value to a product before it is sold to consumers. Chapter 9 by van Duren and Meilke explores the potential for niche marketing of Canadian supply management products. Since Canada's supply-managed industries are relatively small by world standards and also have a relatively high cost structure as a result of product quotas, the most appropriate value-adding strategy for firms in the sector is a niche strategy. Presumably, choosing the appropriate niche would, in turn, generate exports. This notion has considerable popular appeal and support among marketing boards, governments and some analysts. But is it feasible? The authors examine this question by first exploring the concepts of value, value added and niche strategies from economic and strategic management perspectives. They then discuss changes caused by multilateral trade liberalization and other factors that are likely to occur in the business environment facing the Canadian supply-managed industries as they move into the twenty-first century. They also assess the Canadian supply-managed industries' capabilities to compete using a niche strategy.

Chapters 10, 11 and 12 focus on the effects of GATT and other free-trading arrangements on the tobacco, US peanut and US sugar sectors, respectively. In addition, political economy dimensions are added. These industries were not significantly effected by GATT. Even so, there are significant trade disputes between Canada and the United States concerning these specific commodities that will have to be resolved. The extent to which GATT and other free trade arrangements have fueled the debate between the two countries is not clear. Very few people even realize that there is significant trade in tobacco, peanut and sugar products between Canada and the United States.

Specifically, Sumner in Chapter 10 shows that a supply management scheme has limited US tobacco output, without significant import protection, for more than half a century. The US tobacco program used production quotas throughout the period during which the United States was the world's largest importer and exporter. Ironically, in 1994 at the same time the United States was instituting a new policy of import protection, Australia, under pressure from the ban on nontariff import barriers contained in the Uruguay Round Agricultural trade agreement (URA), was in the process of eliminating its own long-standing import barriers. Australia converted its domestic content rule to a tariff-rate quota for leaf tobacco and reformed its domestic supply management system. This chapter describes the supply management policies of the United States and Australia with respect to the tobacco industries. Particular emphasis is placed on how current policy changes relate to the URA in order to provide insights that may be applied in the context of Canadian supply management schemes.

In the United States, peanut sales for edible use are restricted by a federal quota system, and imports are virtually banned. Rucker, Thurman, and Borges in Chapter 11 analyze the probable effects of GATT and related agreements on trade in peanut butter between Canada and the United States, on US peanut trade and on the welfare of US peanut producers. A side-agreement to GATT that halts growth in US imports of Canadian peanut butter will increase the demand for US-grown peanuts and decrease any treasury costs associated with the US peanut program. Rucker, Thurman and Borges conclude that the primary effect of GATT on US markets will be an increase in raw peanut imports, which will reduce the demand for US-grown peanuts. The net effect of such increased imports on growers will depend upon how US policymakers respond. If there is no response in the aggregate quota or support price, this aspect of GATT will serve mainly to increase costs to the US Treasury. If there are policy responses as seems likely, then they forecast that the annual loss in peanut producer surplus will be in $10 million–$18 million range, which translates into less than one cent per pound of quota peanuts produced. Further, possible expansion in foreign demand for US peanuts, as a result of secondary GATT effects, may reduce (or even reverse) these losses.

The US sugar program began in 1934 and has been the subject of numerous debates. Recently, for example, it has been under attack by Canadian sugar refiners. Schmitz and Christian in Chapter 12 show that the program was largely unaffected by GATT, NAFTA and CUSTA. Historically, the sugar program received strong political support. This support can be attributed to many factors, including the wide geographical dispersion of various components of both the sugar and corn sweetener industries. A strong tie between these two industries also contributes to strong political support.

Section III

Coffin, Rosaasen and Saint Louis trace the origins and evolution of Canadian supply management in Chapter 13. The regulatory systems used to govern supplies and influence prices in Canada's dairy and poultry sectors are examined in terms of their organizational and jurisdictional features. The conditions prevailing in these sectors prior to the implementation of supply management are explored as factors contributing to the formation of such a system. This chapter also outlines the implications for provincial cost-price relationships and opportunistic behavior under the price-leveling and regulatory aspects of national marketing plans.

Chapter 14 by Skogstad deals with the political economy perspective of supply management. This chapter appraises the viability of Canadian supply management in the altered GATT environment by focusing upon reform initiatives underway in the poultry and dairy sectors. It argues that the composition of key decision-making structures in poultry supply management systems and the rules under which they have functioned have stymied the adjustment and undermined the legitimacy of supply management. The future survival and adaptation of poultry supply management rests upon reforming its institutions and rules. However, the reformist strategy in poultry supply management is flawed, and the private and public stakeholders in the sector are limited in terms of their ability to build a long-lasting consensus. In contrast, the existing institutional framework and the ongoing reform process in dairy supply management are more likely to enable consensus-building among private and public stakeholders which is necessary for its adaptability and survival.

Rosaasen, Lokken and Richards in Chapter 15 discuss provincialism and the problems that the regulators have in managing an industry when provincial interests are at stake. They illustrate the many problems surrounding supply management and suggest avenues for improvement. This chapter provides examples of allocation problems in the "regulatory playing field." It dispels the myth that current quota values and industry size and distribution accurately reflect relative regional profitability, and it outlines the problems with regulations, including the capture of regulators by the industry. Divisive issues such as import quota rights, interprovincial quota trading and provincial quota allocation are also discussed. This chapter leads to the conclusions that there is a need for the supply-managed system to define goals and priorities and that regulators need to adjust the system to make it more consistent with economic criteria. This may prevent system collapse and high adjustment costs. A move to include more economic principles in the regulation of supply management is a prerequisite for system survival.

In Chapter 16, Gartner highlights this design by comparing the dairy sector with the poultry industry. He highlights the governing boards that exist under each program and discusses which of the models is more efficient from a resource usage standpoint. This chapter also describes and analyzes the political-economic dynamics of institutional change within the Canadian supply management systems for poultry and eggs, using the Canadian egg and chicken marketing models as case studies.

Consumers in Canada have never had a loud voice in opposing supply management in Canada. Many have contended that the prices that consumers pay for Canadian products are considerably higher than those in the United States. This has been a subject of considerable public debate. It is not explored in detail in either the chapters by Veeman or by Gartner.

Chapter 17 by Veeman and Arthur is concerned with observations and conclusions based on the authors' representation of consumers' interests on the national dairy and poultry task forces. As part of a broad policy review for agriculture and food, these bodies and the subsequent Deputy Ministers' Supply Management Steering Committee sought consensus of major stakeholders on changes to improve the costs, inflexibilities and lack of transparency of Canada's national supply management programs. In contrast to naive public-interest theories of regulation, as in hypotheses that supply management has enabled producers to offset the market power of processors and distributors, the operations of these particular programs, operated at the expense of Canadian consumers, involve many areas of joint interest and benefits to producers, processors, wholesalers and importers of supply-managed products. Many of the major problems identified in the task force process continue to persist, a situation that reflects the political influence of supply-side interests in the current supply management system.

As in many agricultural industries, a complex interrelationship exists between the producers, processors, wholesalers and retailers. Very few models deal with these linkages. Most work focuses on producers and/or retailers thus ignoring two important sectors — the processors and the wholesalers. In Chapter 18 Fulton develops a conceptual model of how prices and output are determined under supply management. This model explicitly incorporates the role of retailing, processing and farm sectors and the bargaining or market power that the players in these sectors possess. The model also incorporates the regional (or horizontal) structure of the Canadian market. Since the mid-1980s, there has been a substantial increase in the interprovincial chicken trade because of new products, lower transportation costs and new market arrangements. Although marketing boards have historically been accustomed to setting price and quantity separately, these changes mean the separation of price and output decisions is no longer possible. The result is greater provincial rivalry among marketing boards and among processors as different companies and regions seek to expand their production base and increase pressure for vertical integration while producers attempt to provide themselves with production and price guarantees. Considerable strain and antagonism are also apparent in the political decisions regarding the direction that supply management should take in the future.

Chapter 19, the final chapter in this volume, analyzes the economic implications of a major production increase by Ontario chicken producers on the Canadian broiler industry to illustrate issues surrounding the topic of "Will the cartel stand?" An economic spatial oligopoly model of Canada which estimates trade flows and the economic benefits to provincial producers as well as to consumers, is developed. The model shows that a major increase in the supply of broilers from Ontario would cause economic losses to accrue to producers in all provinces if the current supply management system remains intact. On the other hand, this policy could force retaliatory action on the part of other provinces. This action would cause a breakup of the national supply management cartel. Under a new regime in which marketing boards from each individual province compete with each other for a share of the Canadian Market, Ontario producers could accrue additional economic benefits if they could act as a leader in the industry. In this scenario, producers in the prairie provinces would also accrue economic benefits while producers in British Columbia, Québec and maritime provinces would suffer losses.

References

Arcus, P. 1981. *Broilers and Eggs*. Economic Council of Canada Technical Report No. E13, Ottawa, Ontario, Canada.

Babcock, B., C. A. Carter, and A. Schmitz. 1990. "The Political Economy of US Wheat Legislation." *Economic Inquiry* 18: 335–353.

Barichello, R. R. 1982. "Government Policies in Support of Canadian Agriculture: Their Costs." Paper presented at the United States/Canada Agricultural Trade Research Consortium Meeting, Airlie House, VA (16–18 December).

Brown, R. J. 1995. "US – Canada dispute over poultry, dairy pushed into May." *Feedstuffs, the Weekly Newspaper for Agribusiness* (27 March).

Duncan, M., and W. Koo. 1995. "The United States/Canada Durum Wheat War." *Choices* 1ˢᵗ Quarter: 30–34.

Gardner, B. 1987. "Causes of US Farm Commodity Programs." *Journal of Political Economy* 95: 290–?10.

Goodwin, B. K., and V. H. Smith. 1995. *Price Discrimination in International Wheat Markets*. Report prepared for US Wheat Associates, Washington, DC.

Harling, K. F., and R. L. Thompson. 1983. "The Economic Effects of Intervention in Canadian Agriculture." *Canadian Journal of Agricultural Economics* 31: 153–176.

Schmitz, A. 1988. "GATT and Agriculture: The Role of Special Interest Groups." *American Journal of Agricultural Economics* 70: 994–1005.

_____ . 1994. "Special Interests and the GATT Outcome." Paper presented at the

International Conference of Agricultural Economists, Harare, Zimbabwe, Africa (August).

Schmitz, A., and D. Christian. 1993. "The Economics and Politics of US Sugar Policy," in S. V. Marks and K. E. Maskus, eds., *The Economics and Politics of World Sugar Policies*. Ann Arbor, MI: University of Michigan Press.

Schmitz, A., and T. G. Schmitz. 1994. "Supply Management: The Past and Future." *Canadian Journal of Agricultural Economics* 42: 125–148.

Stigler, G. 1971. "The Theory of Economic Regulation." *Bell Journal of Economics and Management Science* 2: 3–21.

US General Accounting Office (GAO). 1993. *Sugar Program Under Changing Conditions*. P. 4. GAO/RCED 93–84, Washington, DC.

Van Kooten, G. C., and J. Spriggs. 1984. "A Comparative Static Analysis of the Welfare Impacts of Supply-Restricting Marketing Boards." *Canadian Journal of Agricultural Economics* 32: 221–230.

Veeman, M. M. 1982. "Social Cost of Supply-Restricting Marketing Boards." *Canadian Journal of Agricultural Economics* 30: 21–36.

2

Post-GATT Assessment of the World Marketplace

R. D. Knutson

Abstract

From a US perspective, the General Agreement on Tariffs and Trade (GATT) led to a clear set of winners including primarily the wheat, feed grain, oilseed and livestock sectors. The benefits to these sectors result from income enhancement, particularly in developing country sectors. For example, the wheat sector benefits because of anticipated reductions in European Union (EU) subsidies. Dairy, cotton, rice, tobacco, peanuts and sugar stand to be the primary losers. However, trade irritants between the United States and Canada may well have been increased as a result of the North American Free Trade Agreement (NAFTA), Canada–United States Trade Agreement (CUSTA), and GATT. In the long run, continued movements in the direction of free trade will require adjustments in marketing board and supply management policies. Movements toward regional trading blocs could be anticipated as an alternative to generally freer trade.

Introduction

This chapter provides an assessment of the impact that GATT has had on some of the major traded commodities. There are winners and losers, but certain large commodity sectors, such as US peanuts, remain highly protected. The same is true, however, for supply-managed commodities, such as the chicken and dairy industries in Canada.

Making a post-GATT assessment of the world marketplace is not an easy task. Virtually every government has made its own assessment of the impacts of GATT, each with a spin designed to put its best light on the agreement. Perhaps not surprisingly, the United States Department of Agriculture (USDA) study found no negative impacts on US agriculture (March, 1994) — a result that could have been intended to increase the

chances of GATT's approval by the US Congress. Once GATT was approved by Congress, questions arose as to what the impacts on world markets for farm and food products would likely be. An assessment of the post-GATT marketplace is made more difficult by the following:
- the existence of CUSTA and NAFTA
- concerns about the effectiveness of the newly created World Trade Organization (WTO) (Sparks Commodities, 1994)[1]

Moreover, no one has ventured into an analysis of the impact of GATT on competitive and bargaining relationships within the world marketplace. The changes in these relationships, in addition to policy effects, directly impact the relative position of producers. For example, what is the impact of GATT on CUSTA and NAFTA? This is a particularly important issue from the perspective of US dairy farmers who look for GATT to favor a loosening of Canadian dairy import controls, thereby opening markets for a more internationally oriented US dairy industry.

An assessment of GATT's impacts on the world marketplace has three dimensions:
- determination of the winners and the losers from the combination of GATT, CUSTA and NAFTA
- evaluatation of the implications for competitive and bargaining relationships in the world marketplace
- evaluatation of implications for producers and, more specifically, for programs that are of a supply management nature

Winners and Losers

An analysis of GATT's winners and losers is dependent upon one's timetable. The short-run effects are considerably more clear than the long-run impacts. Here, short-run refers to the period covered by GATT, 1995 to 2000. The short-run analysis assumes that the provisions of GATT are carried out as specified in the agreement. It also assumes that the type of programs existing in the 1990 Farm Bill, the EU's Common Agricultural Policy (CAP) and the Canadian Wheat Board (CWB), is continued. The long-run effects depend upon how the provisions of GATT, NAFTA and CUSTA are implemented, the precedence set by these agreements for future trade negotiations and farm policies of the key trading countries. In the absence of perfect long-run foresight, three alternative scenarios are set forth.

Short-Run Winners and Losers

This short-run analysis of GATT's winners and losers is based on an independent, third-party study, that was commissioned by the US Congress and completed by a consortium of the Food and Agricultural Policy Research

Institute (FAPRI) and the Agricultural and Food Policy Center (AFPC). FAPRI has locations at Iowa State University and the University of Missouri. It maintains a set of international and US domestic agriculture sector models that are used to project the effects of changes in supply and demand conditions on market prices and government costs. AFPC is located at Texas A&M University. It maintains a set of about 75 representative farms ranging in size from very small to extremely large and dealing with an assortment of commodities. Utilizing FAPRI's projected prices, these farms are simulated over a five- or ten- year period for the purpose of providing policymakers with analyses of the impacts of proposed policy changes as compared with the status quo.

The results of the GATT study are reported in three publications:

- The international market impacts are reported in FAPRI from Iowa State University.
- The US agriculture impacts are reported by Womack, Young, Adams and Brown (1994) from the University of Missouri.
- The US farm-level impacts are reported by Richardson, et al. (1994) from Texas A&M University.

This study indicates a decisive set of winners and losers as measured from a US perspective (Table 2.1). These winners and losers have important implications for commodities produced in Canada and particularly for dairy and poultry.

TABLE 2.1   GATT's Winners and Losers from a US Perspective

Commodity	Change in Net Exports[a]	Change in Price[a]	Farm-Level Returns[b]
	Percent		
Winners			
Corn	3.9	3.2	7.9
Barley	1.0	0.8	9.2
Wheat	2.3	2.5	5.6
Soybeans	0.3	3.1	4.9
Broilers	15.8	2.2	n a
Pork	80.1	2.3	23.7
Beef	3.3	1.6	5.0
Losers			
Nonfat Dry Milk (Dairy)	-27.0	-6.2	-1.4
Cotton	2.7	1.7	-4.4
Rice	4.0	7.8	-14.1

Sources: [a]Womack, et al.; [b]Richardson, et al.

Winners

Corn. The clearest GATT winner from a short-run US perspective is corn with a 3.9% increase in US net exports, a 3.2% price increase and a 7.9% increase in net cash income to the 1,575-acre Nebraska corn farm that receives 97% of its receipts from corn. This highly favorable result is a reflection of US dominance of international corn trade — roughly a two-thirds export market share. Increased competition for feed grains has significance for Canada because of the correlation between the price of corn and barley and because of the importance of feed grains as major inputs for the production of broilers, eggs, hogs, fed beef and milk. As will be discussed later, the implications of freer trade in corn are even greater for Mexico under NAFTA.

Barley. The FAPRI analysis projects a 1% increase in net barley exports — a price increase of 0.8% compared to 3.2% for corn. The 4,000-acre North Dakota representative barley farm realizes a 9.2% increase in net cash income with only one-fourth of its receipts generated from barley (and the remaining three-fourths generated from wheat and oilseeds).

From a feed grain perspective, barley prices follow corn prices while the brewing business gives malting-quality barley a life of its own. Barley imports have become a source of conflict between the United States and Canada. This irritant may not only be of short-run significance to the United States. While from a US perspective, a short-term fix may have been provided by the 1994–95 voluntary restraint agreement, the long-run competitive advantage for barley production probably lies with Canada. This reality makes barley imports a long-run irritant for US barley producers.

The nature of the barley conflict between the United States and Canada is influenced by policy decisions of both countries. The debate in Canada over a continental barley market and the role of the CWB in barley marketing would have a substantial effect on trading relationships (Carter, 1993; Schmitz, Gray and Ulrich, 1993).

Oilseeds. Soybeans are the only major US crop that have not become farm-program dependent. As a result, it should not be surprising that soybean farmers are a major beneficiary of GATT. This is the case, despite only a projected 0.3% increase in US soybean exports. However, with restraints exercised on the EU, soybean production prices have increased by 3.1%. As a result, the 1,250-acre Missouri representative farm, with 55% of its receipts generated from soybeans, 14% generated from wheat and 31% generated from corn, has increased its net cash income by 4.9%.

The future of soybeans has special significance for Canada, which imports significant quantities of US beans and soybean meal to complement its poultry, milk and hog production. However, implications of a relatively strong soybean market on rapeseed and canola prices are of at least equal importance. Relative to rapeseed and canola, which are produced primarily for the value of their oil, the protein component of soybeans is much more important than its oil component. Soybean oil is, to a greater degree than either rapeseed or canola, a by-product of soybean production. Roughly 20% of the value of a tonne of soybeans is attributable to oil, and 80% is attributable to soybean oil meal. Comparable relationships for both rapeseed and canola are 40% oil and 60% meal. With meat and poultry production increasing in response to higher consumer incomes stimulated by increased trade resulting from GATT, soybeans stand to benefit to a relatively greater extent. Soybeans receive a strong demand push from the soybean oil meal market that, in turn, generate soybean oil as a by-product. Canola benefits relatively more because its oil is believed to be of higher quality. The effects of these two forces are difficult to sort out but need to be recognized as factors affecting the post-GATT marketplace.

Broilers. Broilers represent the epitome of industrialized agriculture in the United States. The four largest US integrators control 40% of the 22 billion pounds of US broiler production. The US broiler industry is a major beneficiary of GATT and has had a 15.8% increase in net exports. The 12-city-average broiler price, as used in this analysis, has risen by 2.2%. Restrictions on broiler imports by Canada, designed as a means of protecting its supply management programs and maintaining the structure of its industry, will continue to be a significant point of contention between the United States and Canada (USDA, April, 1993). At the same time, US integrators welcome the absence of Canada as a competitive factor in the world export market for broilers.

Hogs. Clearly, US pork producers are a major beneficiary of GATT, realizing a projected 80% increase in net exports and a 2.3% increase in price. The integrated 12,400-sow hog farm that receives 100% of its receipts from pork production has increased its net cash income by 23.7%. The 450-sow Illinois farm that receives 82% of its receipts from hogs has increased its net cash income by 6.6%. The clear trend in the United States is toward an integrated hog industry. Economists differ over whether the structural form of this industry will be the same as broilers or whether the slaughter/packer function will be separated by ownership from hog production. There also exists an interesting US trend for production to move in the direction of the Great Plains where fewer environmental problems are encountered because of a more sparse population and lower rainfall. The Canadian hog industry can be expected to experience the same structural pressures as the US industry. That is, without a supply management system in Canada, large integrated production systems are likely. Moreover, the Canadian industry could become concentrated in a very small number of production areas (provinces) to match the efficiencies of large-scale packing plants.

Beef. Despite having been protected by a countercyclical import quota, the beef industry was able to increase net exports by 3.3%. As a result, prices rose by a projected 1.6%. The result is a 5.0% increase in net cash income for the 400-cow representative Montana cattle ranch. Yet, our representative ranches have had problems turning a profit. The cow-calf segment of the US beef industry tends to be heavily influenced by smaller, part-time operators who have less than 50 cows. These weekend or hobby ranchers subsidize their operations from off-farm jobs, professions or investments. Off-farm income makes them less sensitive to price and profit changes. Because of this lack of sensitivity to market conditions, the post-GATT marketplace for cattle ranching is not likely to improve greatly. However, cattle feeders may realize greater benefits from GATT than do ranchers. Large US feedlots are becoming more closely tied to packing operations and can be expected to become more market-oriented — producing the type of beef desired by domestic and international markets. Market orientation is a relatively new phenomenon for the beef industry.

Wheat. In contrast with corn, the United States is just one of five major players in the world wheat trade — the others are Canada, Australia, Argentina and the European Union. With the European Union being the major loser in GATT, the remaining four countries share the spoils. The resulting projection from a US perspective is a 2.3% increase in net exports and a 2.5% increase in wheat prices. The 2,800-acre representative Kansas wheat farm, with 86 % of its receipts from wheat, increases its net cash income by 5.6%. Like barley, wheat has become a source of conflict between the United States and Canada. This conflict extends to Mexico and, therefore, permeates NAFTA, CUSTA and GATT. Sources of conflict lie in the combination of policy changes precipitated by NAFTA and GATT and by the comparative advantage of Canada in quality wheat production. The following are some examples of contentious issues between the United States and Canada: 1) Canadians view the Export Enhancement Program (EEP) as an unfair subsidy, particularly with US producers being assured prices that are well above world levels.[2] 2) To the dismay of many US producers, doing away with the Canadian railway transportation (the Crow Rate) subsidy, will put greater pressure in 1995 on wheat flowing through the United States. The conditions under which this flow takes place and the role of the CWB are critical issues. 3) The growth of Canadian durum wheat exports to the United States has been slowed by the US imposition of tariffs in 1994.

In view of the above, it is critical that farmers in both the United States and Canada understand each other's policies. What would happen if free trade, including the elimination of the CWB and US farm subsidies, suddenly broke out under the CUSTA/NAFTA accord? Analyses of relative costs and comparative advantages are seriously lacking. Moreover, these are issues that cannot be analyzed on a single-commodity basis. They must be evaluated in a general equilibrium world market context. Reliable data and models are critical for an accurate assessment of the post-GATT marketplace.

Losers

Dairy. From an historical perspective, dairy has been the most protected of US agricultural industries.[3] However, the price-support level has been reduced sufficiently over the past decade such that the price paid for milk used for manufacturing (the Minnesota-Wisconsin price) has been above the support level. Butter is the only milk component that generally has been resting on the support price. The result has been considerable pressure for the adjustment of production relative to commercial demand. Yet, in 1993 and 1994, the Dairy Export Enhancement Program (DEEP) was extensively utilized to promote exports of nonfat dry milk in competition with the European Union and New Zealand. It is anticipated that penetrating the market for Mexican imports of nonfat dry milk could become a major target of US export subsidies in the future. However, GATT provisions for reducing DEEP subsidies could stifle this objective. Moreover, tariffication and the reduction of tariff levels hold the potential for increased manufactured dairy-product imports.

The result of reduced exports and increased imports is a 27% decrease in net exports of nonfat dry milk and a 6.5% reduction in its price. A potent, domestic cheese market, bolstered by increasing domestic consumption caused by expanded consumer income, results in a 1% increase in the price of cheese despite a 13% increase in US imports of cheese. Since cheese has the greatest impact on the level of milk prices received by US dairy farmers, the price of milk increases by 0.2%. However, because of higher feed costs, net cash income to our 225-cow Central New York dairy declines by 1.4%. The USDA analysis found a net benefit of NAFTA for dairy in projected, lower Commodity Credit Corporation (CCC) purchases that triggered an increase in the support price for milk (March, 1994). There are many dairy experts who feel that both of these analyses are overly optimistic.

Because of these negatives, the United States is particularly concerned about the issue of access to higher-priced Canadian dairy product markets. This access issue involves the resolution of conflicts between CUSTA, GATT and at least the spirit of NAFTA. The root causes of these conflicts are the domestic production control policies of Canada and the price support policies of the United States. From the perspective of both countries, while CUSTA found it politically expedient to exclude dairy, poultry, eggs and sugar until January, 1998 (Goodloe and Simone, 1992; Waverman, 1993), GATT calls for comprehensive tariffication with a minimum 15% reduction in tariff levels, and the spirit of NAFTA requires the elimination of tariffs (USDA, April, 1993; USDA, May, 1994).

Cotton. From the perspective of merchants, cotton has benefited from a 2.7% increase in exports and a 1.7% increase in price. However, cotton producers are losers in GATT because of the uniqueness of the operation of the marketing-loan program that involves direct farmer payment of the difference between the loan rate and the world market price. This payment is in addition to the domestic price received by producers and the deficiency payment. Therefore, when the world market price rises, farmers' gross receipts per unit decline within a range. Therefore, within this range, the increase in cotton net exports and the higher cotton prices ironically are converted into lower producer returns. As a result, the 3,310-acre Texas South Plains cotton farm's net cash income declines by 4.4%.

Yet, the post-GATT marketplace for cotton is very difficult to analyze. Future cotton production patterns in the former Soviet Republics are uncertain. This region has been the world's largest cotton production area. Likewise, China's production is difficult to assess. In addition to uncertainty regarding Chinese supply and demand data, important questions arise regarding the allocation of lands between cotton and food production. This is an uncertainty that also affects the world demand for wheat and rice.

Rice. Opening the Japanese and Korean rice markets to imports is viewed by the United States as one of the major accomplishments of GATT (USDA, March, 1994). As a result, rice exports are projected to increase by 4%, and market prices are projected to rise by 7.8%. However, because of reduced-marketing, loan-deficiency payments, the 1,500-acre Texas all-rice farm realizes a 14.1% lower net cash income.

Other Commodities. While not studied explicitly by FAPRI and AFPC, at least three other highly protected commodities (sugar, tobacco and peanuts) warrant comment as potential GATT losers although USDA fails to acknowledge them as such. Under GATT, protection will be afforded these industries by a tariff-rate quota to be reduced over time. These reductions will make it more difficult to maintain the favorable sugar and peanut price levels that have existed in the past.

From time to time, sugar and peanuts have been a source of conflict between the United States and Canada. These industries are the topics of later chapters.

Aggregate Short-Run Impact

From a US perspective, while GATT has both winners and losers, the absolute size of the gains are far larger than the losses. Cash receipts have increased by an average of about $4 billion (1.7%), net cash income has increased by 1.9%, net farm income has risen by 3.2% and government costs have fallen by 9.5%. With this magnitude of benefits, it is not surprising that most farmer-oriented congresspersons or senators voted in favor of both NAFTA and GATT. Yet many US farmers and some politicians question the merits of GATT, NAFTA and CUSTA. While some of this opposition reflects the interests of the commodity losers from GATT negotiations, other interests simply result from a lack of understanding of policies and their effects or, perhaps, a basic mistrust of politics.

The magnitude of these aggregate benefits, relative to the costs, will vary depending on the particular country situation. To a degree, US farmer benefits ironically depend upon reductions in the level of government support provided to farmers during the past decade. The realization of benefits also depends on expanding food demand that results from increases in income and a growing world population. Each country's situation depends on the combination of its endowment of resources and the agricultural policies it employs. Those countries/commodities/farmers that have had the highest levels of protection can be expected to have the greatest losses. Those having had the least protection realize the greatest gains.

Long-Run Evaluation

Recall that the long-run, as used here, extends beyond the implementation of GATT to future rounds of GATT. Long-run implications depend upon how GATT is implemented and the outcome of future negotiations. While it would be ideal to have a single, projected outcome like the short-run analysis presented above, such a projection is far too dependent on the political forces that develop in each of the major trading countries (European Union, United States, Japan, Canada, Australia and Mexico) and on the effectiveness of the new WTO. As a result, three alternative scenarios are presented and evaluated for their implications on the world marketplace:
- ◆ continued movements toward free trade
- ◆ status quo and/or new subsidy forms
- ◆ trading blocs

Any one of these three scenarios is equally probable, yet each has decidedly different implications for the world marketplace. Moreover, the options arguably are mutually exclusive, and mixtures are unlikely. If this is the case, producers' choices regarding the preferred option should be reasonably clear-cut.

Continued Movements Toward Free Trade

The free trade option implies progressive reduction of trade barriers in future rounds of GATT negotiations. This option implies the complete dismantling of all domestic farm programs affecting production. This crucial point extends to the practical workability of the NAFTA provisions. That is NAFTA was negotiated with the provision that each country could maintain the structure of its domestic farm program (Congressional Budget Office, 1993; Grennes, 1993). This verbiage was, and is, a political deception. Free trade cannot exist within the framework of existing farm programs. From a US perspective, the provision that each country could maintain its existing domestic programs was required in order to obtain the votes needed for NAFTA's approval.

Free trade between Canada, Mexico and the United States in the context of existing domestic farm programs is an oxymoron. US target prices, marketing loans, nonrecourse loans and price support provisions cannot be maintained if free trade is the objective of NAFTA. Free trade, in a broader context of major exporting and importing countries, would require the dismantling of the EU's CAP and a complete opening of major Pacific Rim markets.

The status of various types of marketing boards under NAFTA and, more generally under GATT, has received relatively little attention. Because of a wide variation in the operations of marketing boards, each probably needs to be evaluated independently to determine the extent to which it distorts market forces. The following represents some initial thoughts from such an analysis:

♦   Clearly, board policies that manage supplies and restrict market access violate at least the spirit of GATT. This will likely be a principle contention of the US government with regard to Canadian dairy and poultry policy.

♦   Boards having an exclusive, state-authorized, export-marketing function might be evaluated differently than those having a combination domestic and export function. The domestic/export combination effectively authorizes various forms of two-price strategies that are obvious trade barriers. However, the power of an exclusive, state-authorized, export agency to practice two-price strategy is implied.

♦   Boards having exclusive export authority would be inconsistent with the spirit of GATT if they operated in conjunction with domestic subsidy policies such as lucrative disaster, crop or income-insurance programs.

♦   Boards that operate solely as an export marketing agency without other forms of government assistance, such as the New Zealand Dairy Board, would need to be evaluated based on their ability to exercise monopolistic market influence.

In the current CUSTA/NAFTA/GATT configuration, the agricultural economy of Mexico is clearly more adversely affected than either Canada or the United States. While the provisions of NAFTA were designed to provide for a 15-year phaseout of tariffs and import quotas on corn and soybeans, Mexico's previously protected pricing structure for these staple commodities began to deteriorate as early as 1989 when import licenses were removed (Harvey, 1994). In 1994, the Mexican PROCAMPO system of direct payments was implemented to replace corn and soybean price supports. However, concern now exists that small producers will be effectively excluded from the market by imports (Harvey, 1994).

The rural, small-farm structure of Mexican agriculture relating to the production of corn (a staple in the Mexican diet and the basis of its rural "ejido" culture) is being completely restructured. Hinojosa and Robinson (1991) estimate that full liberalization of Mexican imports of food and agricultural products could displace 800,000 of 3.7 million subsistence farmers. This restructuring is felt to be a potential source of substantial political instability, and has already been confirmed in Mexico's largest corn-producing state, Chiapas (Harvey, 1994). Restructuring and political instability have been accelerated by the devaluation of the peso. The resulting financial and social instability only serves to emphasize the fact that Mexico is the biggest, short-run loser in NAFTA. Ironically, in the long run, after structural adjustment has taken place, Mexico will likely receive the biggest gains.

The degree of instability created by GATT and NAFTA is an important issue. Johnson (1991) asserts that free trade would result in more stable world market prices. His basis is that domestic farm programs and related trade barriers foster world price instability. In a free market, all production and consumption decisions would be directly impacted and adjusted to world market price changes. While Johnson's utopian world market perspective has merit, farmers would be subject to more instability in receipts than exists under the current domestic programs. That is, while it is granted that world free market prices could be more stable than they are today because of the highly inelastic nature of supply and demand, farm prices would be highly unstable. Domestic programs shield farmers from the effects of this instability even though they cause increased world market price instability. Therefore, the marginal revenues on which farmers make decisions are more stable with current domestic programs than under free trade conditions.

Farmers, who are knowledgeable about utilizing the futures markets or who are members of cooperatives that pool receipts, are in the best position to deal with the consequences of the increased price instability that would exist in a free market.[4] One could anticipate increased cooperative activity in a free market environment as farmers seek means of reducing risk. This can be accomplished either through cooperatives that offer producers improved market information combined with various means of hedging or through marketing agreements with pooling and/or bargaining associations.

Clearly, the institutional gainers from free trade would be the market intermediaries such as multinational traders, futures brokers and marketing consultants. They benefit from the larger production volume, the increased need for futures/options activity and the requirements for improved market information. Even in the presence of domestic programs, market intermediaries, particularly multinational traders, are dominant forces in the market with the ability to manufacture instability (Schmitz, McCalla, Mitchell, and Carter, 1981).

Status Quo and/or New Subsidies

The second long-run alternative impacting the world marketplace is the status quo — a continuation of subsidies. This could happen in either of two ways: 1) GATT could fall apart with an ineffective WTO 2) New forms of subsidies could be substituted for the current ones.

The new WTO faces an herculean task in implementing the provisions of GATT in a manner that will subsequently lead to free trade. The European Union is already arguing that a target-price system is legal under GATT, even if it is not decoupled. Moreover, there are disagreements over what constitutes decoupling. Questions exist over whether, and on what conditions, Japan and Korea will follow through on minimum access to their rice markets. US producers of highly-protected commodities (rice, cotton, sugar, peanuts, milk and tobacco) have not yet had their final word on the status of safeguards against imports. In some cases, tariff levels, established as a result of the tariffication by the United States and Canada, resulted in even higher levels of protection on politically sensitive commodities, such as Canadian dairy and US sugar. In future multinational trade negotiation rounds, there could be even more heated debates over the characteristics of marketing boards that constitute barriers to trade.

New forms of subsidies are already in the developmental stages in the United States under the aegis of income assurance, green payments and self-help. Income assurance would provide farmers a guarantee of a certain percentage of past receipts (70% has been most frequently discussed) or would employ a combination of price insurance (potentially utilizing the options market) and yield replacement insurance. The cost of either, in terms of government outlays or the restructuring of agriculture, has not been studied. Moreover, one has to be pessimistic about the ability of the US government to run an income assurance program. In past years, USDA has experienced a succession of failures in managing its crop insurance program. The Canadian experience with insurance has not been significantly better although performance varies considerably among the provinces.

Green payments are subsidies made by the government in return for farmers engaging in environmentally sound, best management practices. Green payments are proposed by environmentalists as an alternative to current subsidies. Environmental advocates have explicitly suggested that savings from the release of Conservation Reserve Program (CRP) land ought to be converted into green payments. While these payments are to be made for environmental purposes, they hold potential for replacing current subsidies. Presumably, these green subsidies would go to those areas of agriculture that have the greatest environmental problems. One could speculate that a higher proportion of subsidies would go to the livestock and poultry industries and their treatment of animal waste, to lands subject to erosion, to riparian areas and for programs such as integrated pest management. One proposal calls for green milk payments!

Both income assurance and green payments will face the challenge of competing for a share of a downsized federal budget. The prospects for green payments were substantially reduced by the 1994 election of a new Republican majority in the US Congress. Efforts to reduce spending make new and potentially costly programs unlikely. Moreover, the election of the new majority reflects a wave of opinion against government regulation. Invasion of property rights by environmental regulations, such as those applied to endangered species, has become a focal point of this concern.

Milk producers in the United States have tested the political water for acceptance of a marketing board mechanism designed to limit government cost exposure. The new board would be empowered to assess producer levies to cover the costs of conducting export activities at less than domestic prices and controlling production. While there are GATT legality questions surrounding this self-help concept, the House Subcommittee on Livestock in the 103rd Congress reported this proposal to the full Committee on Agriculture. The proposal did not make it out of the full committee. Aside from GATT issues, this proposal's biggest barrier was the conflict between farmers who have a desire to expand production versus those who are satisfied with current production levels. Yet there is substantial interest in the 104th new majority Congress for policies designed to expand trade, which could involve some type of marketing board or order that could implement some type of two-price plan.

These various forms of extension of status quo policies are not likely to impact competitive relationships in the world marketplace as do current subsidies. The advantage continues with those countries that can afford to subsidize their agriculture.

Trading Blocs

The trading bloc option would carve the world up into groups of countries for the purpose of pursuing commercial and political activity. Such coordination may, at the extreme, involve sharing a common currency among countries. A trading bloc's member countries would have common trade policies and would resolve to move in the direction of common domestic policies that affect trade. As noted previously, the need for common domestic policies is a perspective that deserves further discussion and debate in light of NAFTA.

If the long-run result of GATT is to substantially reduce domestic subsidies, it is likely that producers will be looking for other means of fending for themselves in the marketplace. Trading blocs with export-oriented marketing boards could be considered one of several options in such an environment.

Other options include a common set of domestic subsidies across countries with some uniform formula for allocating costs.

In certain respects, this trading bloc scenario has structural similarities to one of the initial EU proposals in the GATT negotiations. That EU proposal involved the formation of country blocs with understandings regarding the appropriate division of world market trade shares among blocs or countries.

A division of the world market is but one of many options for trading relationships among countries. A pro-trade alternative would eliminate protectionist policies, such as the division of world market shares or subsidies within the trading blocs. In other words, there would be free trade among blocs. In this case, the purpose of the bloc would be to foster an economic, technological and social environment that is conducive to growth by maximizing the comparative advantage of each member of the bloc. That is an optimistic perspective on what the leaders of NAFTA countries could accomplish in working with the European Union and Pacific Rim leaders.

The world marketplace with trading blocs would most likely closely resemble an economic structure of bilateral oligopoly. In the long-run, the major oligopolists might include: 1) the countries of the Western Hemisphere, 2) an expanded European Union covering its proposed new members plus most of Eastern Europe and the former Soviet Republics and 3) the Pacific Rim countries including China.

To add emphasis to trading bloc possibilities, four free trade areas already exist in our hemisphere — NAFTA, The Caribbean Common Market (CARICOM), the Central American Common Market (CACM) and the Southern Cone Common Market (MERCOSUR). The interest in trading blocs appears to be expanding. In early 1995 there were at least nine trading bloc groups of countries. Within these blocs there are ongoing discussions about bringing additional countries into the blocs. The European Union plans to include Eastern European countries, and NAFTA plans to include Chile.

It is interesting to note that some of these trading bloc structures hold the potential for encompassing a substantial share of production and trade in particular commodities. For example, a Western Hemisphere bloc would include a clearly dominant share of world production of feed grains and oilseeds. A key to success would be designing strategies that utilize the resulting superior market position for the long-run advantage of the bloc. Visionary agricultural leadership would be essential.

How the multinational traders would fare in a trading bloc structure depends, in part, on the strategies employed by the blocs. Initially, trading company expansion may be favored within the bloc because these entities consolidate their market position in their new, broader based market. In the long run, however, there would be incentives to expand their scope of operation, trade and influence across blocs. This would take advantage of different economic policies, the comparative benefits of the participating countries and the substantial economies of size associated with market intelligence and increased business volume.

Producers would have the potential for an improved market position under a trading bloc structure relative to the free trade option but would be disadvantaged relative to the status quo. They would find it more difficult to influence the course of agricultural policy within the blocs — just as Canadian poultry and milk producers are having more difficulty getting their way in the CUSTA/NAFTA/GATT configuration. In a trading bloc context it would be important that producers be organized across the participating countries as a balancing force in the economic and political marketplace.

Implications for Producers

Structurally, agriculture is highly diverse. The diversity of the Agricultural and Food Policy Center's (AFPC) representative farms staggers the imagination of most farmers — dairies ranging in size from 55 to 2,150 cows, hog operations from 75 to 12,400 sows, feed grain farms from 760 to 4,500 acres and wheat farms from 1,175 to 4,250 acres. These farms by no means reflect the extremes of US farm sizes and structures because they are designed to reflect moderately sized commercial farms. For example, they certainly do not reflect the integrated broiler industry where the largest firm produced a ready-to-cook (RTC) equivalent of 2.1 million metric tonnes with 6,000 contract growers (350,000 RTC kg/grower) in 1993 (Tyson, 1993).

Part of that diversity is brought on by government programs. Research utilizing representative farms consistently indicates that moderately sized farms are the most vulnerable to a dropping of subsidies in the United States (Smith et al, 1995). Structural adjustment to no subsidies will be the greatest in those commodities that are the most farm program dependent. In the United States, that is wheat, rice and cotton, among the major crops, and sugar, peanuts and tobacco, among the minor crops. In livestock, it is dairy, wool and mohair. In these commodities the impacts of GATT-imposed movements to freer trade will cut deeper.

Of these commodities, the United States and Canada share a common concern over wheat. Simultaneous changes in Canadian and US policies will probably serve to further inflame border issues.

The dropping of the Canadian Crow transportation subsidy has the potential for intensifying competitive relationships between the United States and Canada in wheat and barley. The dropping of this subsidy alone could mean that the overall level of US agricultural protectionism will exceed the level of Canadian protectionism. However, such a conclusion must await consideration of the level of subsidies contained in the 1995 Farm Bill — again illustrating how rapidly change is occurring.

Conclusion

Regardless of what happens in GATT and the new World Trade Organization, producers and governments will seek a tolerable degree of shelter from the instability that characterizes agriculture. While both Canada and the United States do not have to worry about food security since they produce in excess of their domestic needs, they do have to worry about trade security. That is they need sufficient stable supplies so that importing countries can safely rely on Canadian and US farmers as their supply sources. Without this level of security on the part of governments and peoples in other countries, the incentives for uneconomic self-sufficiency policies will prevail.

Will these US and Canadian producer and government needs be satisfied by marketing boards, trading blocs, storage programs, income assurance, price supports, income supports or some type of farmer savings program? Time will answer this question. Of all the options available, truly free trade could be the most improbable.

Notes

1. *Food and Fiber Letter 2*, May 23, 1994. Sparks Commodities.

2. In reality, the target price and deficiency payments do not apply to all acres. Farmers do not receive deficiency payments on 15% of their acres (flex acres) and may be required to set aside additional acreage.

3. Sugar and peanuts are a close second and third.

4. One might question why cooperatives raise no free trade issues while it was previously indicated that marketing boards may constitute trade barriers too. The answer lies in the voluntary nature of the cooperative form of business organization. At a minimum, farmers need voluntary forms of market intervention whereby they can offset the power position of multinational market intermediaries.

References

Carter, C. A. 1993. *An Economic Analysis of a Single North American Barley Market*. Agriculture Canada, Ottawa, Ontario, Canada.

Congressional Budget Office. 1993. *A Budgetary and Economic Analysis of the NAFTA*. US Congress, Washington, DC.

FAPRI. 1994. *Implications of the Uruguay Round for Agriculture*. Iowa State University, Ames, IA (June).

Goodloe, M., and M. Simone. 1992. *A North American Free Trade Area for Agriculture: The Role of Canada and the US–Canada Agreement*. Agriculture Information Bulletin No. 644, ERS/USDA, Washington, DC.

Grennes, T. 1993. "Toward a More Open Agriculture in North America," in Steven Globerman and Michael Walker, eds., *Assessing NAFTA: A Trilateral Analysis*. The Frazer Institute, Vancouver, British Columbia, Canada.

Harvey, N. 1994. "Rebellion in Chiapas: Rural Reform, Champesino Radicalism and the Limits to Salinismo." Transformation of Rural Mexico No. 5, Center for US-Mexico Studies, University of California, San Diego, CA.

Hinojosa, R., and S. Robinson. 1991. "Alternative Sources of US-Mexico Integration: A Computable General Equilibrium Approach." Giannini Foundation Working Paper No. 609, University of California, Berkeley, CA.

Johnson, D. G. 1991. *World Agriculture in Disarray.* 2d ed. New York, NY: St. Martin's Press.

Richardson, J. W., P. T. Zimmel, R. D. Knutson, D. P. Anderson, A. W. Gray, E. G. Smith, and J. L. Outlaw. 1994. "Impacts of GATT on Representative Farms in Major Production Areas of the United States." *Agricultural and Food Policy Center Briefing Series* 94–3. Department of Agriculture Economics, College Station, TX.

Richardson, J. W., J. L. Outlaw, D. P. Anderson, A. W. Gray, P. T. Zimmel, J. W. Miller, B. T. Young, E. G. Smith, and R. D. Knutson. 1993. "Implications of the 1990 Farm Bill and FAPRI November 1993 Baseline on Representative Farms." Agricultural and Food Policy Center Working Paper 93–6. Department of Agricultural Economics, College Station, TX.

Sacheti, S., A. Schmitz, and G. Winters. 1994. "Government Response and the Gains From R&D: The Case of Bovine Somatotropin." Giannini Foundation Working Paper 729: 642–1212, University of California, Berkeley, CA.

Schmitz, A., R. Gray, and A. Ulrich. 1993. "A Continental Barley Market: Where Are The Gains?" Department of Agricultural Economics, University of Saskatchewan, Saskatoon, Saskatchewan, Canada.

Schmitz, A., A. F. McCalla, D. O. Mitchell, and C. A. Carter. 1981. *Grain Export Cartels.* Cambridge, MA: Ballinger Publishing Co.

Smith, E. G., J. W. Richardson, A. W. Gray, R. D. Knutson, J. L. Outlaw, and J. Penson. 1995. "Macroeconomic and Respresentative Farm Impacts Resulting From Extending the 1990 Farm Bill, Program Elimination, and Marketing Loan Only Scenarios." Agricultural Food Policy Center Working Paper No. 95–16, Department of Agricultural Economics, Texas A&M University, College Station, TX (May).

Sparks Commodities. 1994. *Food and Fiber Letter* 2 (23 May).

Tyson Foods, Inc. 1993. Annual Report. Springdale, AR.

USDA. 1993. *Agriculture in the US–Canadian Free Trade Agreement.* FAS/USDA, Washington, DC (19 April).

_____. 1994. *Issue: What is the Status of the On-going Agricultural Negotiations Between Canada and the United States.* FAS/USDA, Washington, DC (24 May).

_____. 1994. *Effects of the Uruguay Round Agreement on US Agricultural Commodities.* Office of Economics, ERS/USDA, GATT-1, Washington, DC (March).

Waverman, L. 1993. "The NAFTA Agreement: A Canadian Perspective," in Steven Globerman and Michael Walker, eds., *Assessing NAFTA: A Trilateral Analysis.* Frazer Institute, Vancouver, British Columbia, Canada.

Womack, A. W., R. E. Young, III, G. M. Adams, and D. S. Brown. 1994. "Implications of the Uruguay Round for US Agriculture." University of Missouri, Columbia, MO.

3

Consequences of Tariffication

A. Schmitz, H. de Gorter, and T. G. Schmitz

Abstract

With the passing of the General Agreement on Tariffs and Trade (GATT) in 1994, both import quotas via minimum access requirements and tariffs now apply to the Canadian supply-managed sectors. Minimum access commitments were increased slightly; however, tariffs on imports beyond these commitments appear to be prohibitive in many cases — at least in the short run. The cartel will not be as easily held together today as it was under Article XI of GATT.

Introduction

Trade between the United States and Canada in supply-managed commodities, such as chickens, has always been a contentious issue. When supply management was instituted in Canada, three pillars were established: 1) cost of production formula, 2) base and over-base production quotas and 3) import quotas. The latter, which significantly restricted the exportation of US goods into Canada, was allowed under Article XI of GATT because import quotas were used in combination with domestic production controls. The supply management system, because of its direct use of production controls and import quotas, generated significant rents for producers and importers (Schmitz and Schmitz, 1994) but was costly for consumers (Veeman, 1982). As a result, any attempts to lower trade barriers were opposed by special interest groups. This was certainly true in the Uruguay Round of GATT when Canada finally agreed to the tariffication of its border protection in the dairy, poultry and egg sectors. Throughout the negotiations, Canada argued against tariffication of border measures, preferring that Article XI be strengthened rather than weakened. A major impetus for this position was the concern by producers that protection afforded them through production and import quotas would be diminished.

This chapter evaluates the economic impact of the tariffication of supply-managed sectors in Canada as finalized in the recently concluded GATT. We find that: 1) The level of protection afforded producers has not changed in light of the proposed reduction of tariff levels that will take place over the next six years. 2) Farmers may improve their welfare, particularly if they expand exports. 3) Tariffication increases the possibility that internal forces may cause the supply management cartel to crumble.

Tariffication Defined

It is important to understand that both import quotas (to implement minimum access commitments) and production quotas are maintained under GATT. Thus, the term tariffication is misleading because, even after GATT, import quotas and tariffs remain in effect. Canada is obligated to a minimum level of imports, regardless of market conditions. This minimum access commitment is to be allocated through import quotas. Imports falling under the minimum access requirement have a zero tariff. Because world prices are well below domestic prices, imports reach the minimum allowable (with import rents going to import quota holders in Canada) and often exceed the minimum when they fit the requirements of "supplementary import quotas."

Under GATT, changes were made to Canada's minimum access requirements, but these requirements varied according to product classification. For example, requirements for chicken did not change as the minimum access requirement remained at 7.5% of domestic production. In the dairy sector, only the access commitments for butter and ice cream increased. For example, Canada will provide access for 1,964 tonnes of butter in 1995 and will increase access to 3,274 tonnes in the year 2000 (The year 2000 access represents only 3.9% of Canada's 1993 level of butter consumption). Turkey and eggs imports will also increase. Import access at FTA levels initially applied but will be replaced by GATT quota levels of 5,588 tonnes and 21.37 million dozen, respectively, by the year 2000. Tariffs are applied to all potential imports above minimum access requirements. These tariffs will be prohibitive if they do not induce further imports above the minimum quota.

The tariffs (for imports beyond the minimum) agreed to in GATT are summarized in Table 3.1. The base period is 1995, and all tariffs are to be reduced by 15% over six years. These extremely high tariffs are calculated as the 1986–88 average of the domestic and world price gap. If the primary objective of supply management is to improve farmer welfare, then the Canadian government has an incentive to calculate a high baseline tariff. Some flexibility is given to Canada about the choice of domestic and world prices, exchange rates, transportation and handling costs. The possibility that Canada overstated baseline tariffs is supported by comparing implicit tariff levels calculated by Moschini and Meilke (1991) for chicken. They estimated that tariff levels for chicken averaged 31% for 1986–88, far above the 280% reported in Table 3.1 and agreed to by GATT. Another comprehensive assessment, of empirical issues involved in measuring baseline tariffs for the supply-managed sectors in Canada, is provided by Cymbal and Veeman (1994). They conclude that specified tariffication schedules embody appreciable potential increases in the level of protection afforded these sectors.

TABLE 3.1   Canadian Food Tariffs Under GATT

Item	1995	2001
Butter	351 %	299 %
Cheese	289 %	246 %
Milk	283 %	246 %
Chicken	280 %	238 %
Skim milk powder	237 %	202 %
Eggs	192 %	164 %
Turkey	182 %	155 %

Source: Agriculture Canada, 1994

To illustrate the protective nature of the tariff, consider the year 1989 when there was a price gap of roughly Cdn $93¢/kg between Canadian and US prices (including transport costs). If one were to apply the 280% tariff in Table 3.1, the price gap would have to widen to Cdn $2.93/kg for the tariff to be equivalent. If the US price remains the same, then the Canadian price would have to be Cdn $4.56/kg for the tariff to be equivalent. This implies a Canadian price increase of over 75%. In addition, consider the implications of a falling Canadian dollar versus the US dollar. The Canadian dollar has dropped significantly since 1989. Suppose the Canadian dollar dropped by 15%. In this case, given a tariff of 280%, the implicit Canadian chicken price would exceed Cdn $5.00/kg — more than twice the Cdn $2.48/kg figure. Of course, a strengthening of the Canadian dollar has the opposite effect.

In recent years it appears that the price gap between the United States and Canada has narrowed for most commodities under supply management. This is especially true for eggs and has made the threat of free trade less of a concern. More importantly, this narrowed price gap implies that many established tariffs are even more prohibitive than initially thought (Moschini and Meilke, 1991), and they will remain so through the year 2001.

In review of the above discussion, suppose that the preliminary tariffs are reduced by 25 to 30%. Even in this case, the level of protection via the tier-two tariff is high, and generally, the increase in imports is likely to be minimal. One is left with the notion that tariffication will have minimal impact throughout the life of the Uruguay Round of GATT.

Canada's supply management sector was left highly protected as a result of GATT as were the Canadian dairy sector and Japanese and Korean rice industries. Canada negotiated the highest tariffs for ice cream and butter but granted greater access through the minimum access rule. Japan and Korea did not agree to tariffication, rather they allowed greater access via import quotas. Given the high levels of tariffs for Canadian butter and ice cream imports, the outcome was similar to that in Japan and Korea. It is also interesting to consider the degree to which other countries moved to freer trade via tariffication. Consider, for example, wheat in the European Union (EU). Josling and Tangermann (1994) stated that the European Union was successful in choosing high tariffs for the wheat sector. They write,

> The internal price for Common Wheat is taken as ECU 241/ton, calculated from the Intervention Price, plus 10%, plus seasonal increments. The external price is taken as ECU 93/ton, specified to be the f.o.b. Argentina price, plus transport cost, minus a quality adjustment. The difference between these two prices gives an initial bound duty of ECU 149/ton. This method, though technically in keeping with the modalities document, certainly builds in "water" into the tariff. The EU market price is often below the intervention price for wheat. The external price is also a little low. The c.i.f. Rotterdam price used by the European community in calculating its variable levies was higher during the base period, and when suggesting a "trigger" price for the special safeguard mechanism, the EU used a figure of ECU 143/ton as an external price.

As further examples, little happened to US sugar tariffs and quotas as a result of GATT (Schmitz and Christian, 1996). This was also true for the highly protected US tobacco and peanut industries (Sumner, 1996; Rucker, Thurman and Borges, 1994).

The Economic Effects of Tariffication

The economic effects of tariffication with an import and production quota scheme are depicted in Figure 3.1. The underlying supply curve in Canada is given by S while the domestic demand schedule is depicted by D. The minimum access commitment requires a specified level of imports, thereby shifting the domestic demand curve facing farmers to D'. The price is determined by the intersection of the supply curve and the marginal revenue of D', resulting in a domestic price, P*, a production quota, Q*, and an import quota equal to the distance ab.

Consider a world price, P_w, (assumed to be unaffected by changes in production and consumption in Canada) below the domestic support price, P*, such that imports under free trade would be exactly equal to the negotiated import quota (imports under the minimum access agreement). The world price under such conditions must coincide with the intersection of the supply and import quota adjusted demand curves (at point A). Quota rents to importers are the area, abcd, because the tier-one tariff is zero.

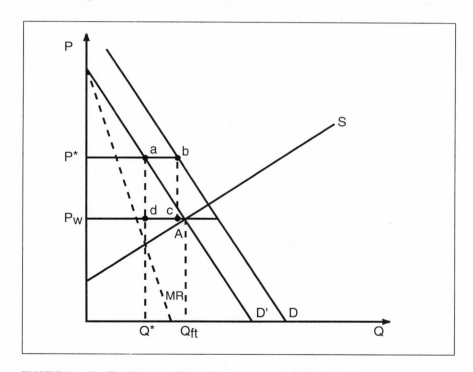

FIGURE 3.1   Tariff with Import Quota Equal to Free Trade Imports

What are the effects of the tier-two tariff as negotiated in GATT and summarized in Table 3.1? It is very possible that the tariffs are greater than the gap between P^* and P_w even after the 15% reduction in tariffs by the year 2001. In this case, the effect of tariffication is zero (except for perhaps some minor adjustments to the minimum access commitments negotiated in GATT).

A more interesting scenario is the possibility that tariff reductions will impact domestic prices. Consider a baseline tariff, T_o, equal to P^*-P_w in Figure 3.2. A partial reduction in T_o over time will result in lower prices to domestic consumers and an increase in consumption. Profit-maximizing farmers will increase their production quotas rather than lose their domestic market share to imports. A reduction in the tariff to T_1 in this case results in import quota holders' loss of the shaded area while farmers lose the hatched area minus the crosshatched area. Interestingly, the government never receives any tariff revenues. The production response to changes in tariffs is given by the function F_1 in Figure 3.3 and is everywhere decreasing under the assumptions depicted in Figure 3.1 and 3.2. Indeed, the function, F_1, is decreasing for all cases where the world price is such that baseline imports under the quota are equal to or greater than imports under free trade. (The world price line is at or above point A in Figures 3.1 and 3.2). It should be noted that import tariff levels exist for which production does not respond to changes in tariffs.

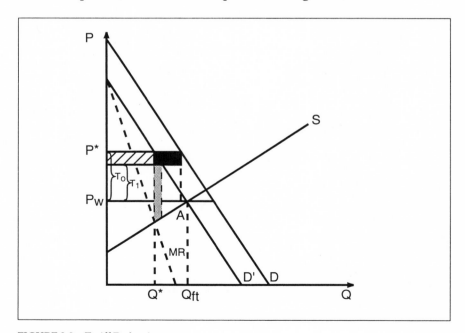

FIGURE 3.2   Tariff Reduction

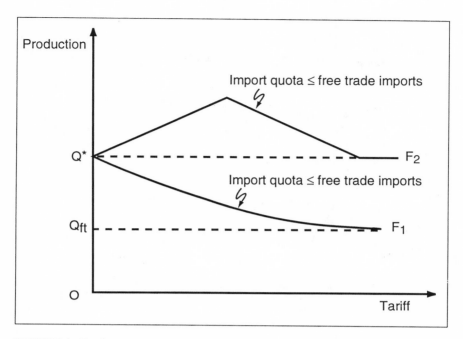

FIGURE 3.3   Production Response to Tariffs

Consider the case where the world price generates a free trade import level greater than that of the baseline import quota. Such a scenario is depicted in Figure 3.4. The world price, P_w, is set below point A of Figures 3.1 and 3.2 and, for expository purposes, is set at the intersection of the marginal revenue and supply curves (point B in Figure 3.4). A tariff that is greater than T_o results in no change in the baseline imports (production equals demand on D'). Nevertheless, a tariff that is greater than T_o can still be low enough to reduce producer and consumer prices below that of P^*. In this case, farmers and import quota holders lose and consumers gain (as in the cases analyzed in Figures 3.1 and 3.2 above).

A tariff set at or below that of T_o will generate an increase in imports from its baseline levels. This can be shown by considering a zero tariff. By construction, free trade and the monopoly production levels coincide. Any small tariff below T_o will result in farmers increasing production until point A is reached (at tariff, T_o). All levels of tariffs above T_o will result in farmers beginning to reduce production again. This relationship between production and tariffs is given by the function, F_2, in Figure 3.3. This means imports exceed minimum requirements if the tariff is less than or equal to T_o and that the government begins to collect tariff revenues for a positive tariff. Quota holders will still collect rents, albeit smaller than if the tier-two tariff was higher. Figure 3.5 depicts a tariff, T_1, that generates government tariff revenues and is represented by the shaded area. Import quota rents are represented by the crosshatched area.

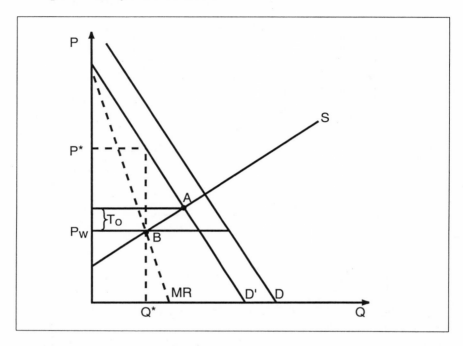

FIGURE 3.4   Tariff with Import Quota Less Than Free Trade Imports

A remaining question is: What scenario best depicts the realities of the Canadian supply-managed commodity sectors? One can infer from the empirical results of the Moschini and Meilke (1991) study on chicken that Figures 3.4 and 3.5 and the function, F_1, in Figure 3.3 are the more relevant. Moschini and Meilke (1991) determined that the actual tariff was three times that required (still positive) to preserve the status quo (minimum access commitment) import level. This means that a positive tariff is required to reach point A in the aforementioned figures, necessarily implying that the world price falls below point A. Whether this is the case in other sectors can only be determined by further, empirical research. It should be noted that Canada must increase the minimum access for some products and hence increase the level of import quotas. Although chicken imports are set at the appropriate level of 7.5% of domestic production, imports of several milk products are required to increase in the next few years. The resulting economic effect of this is that farmers lose welfare while import quota holders gain. Consumer prices are unaffected.

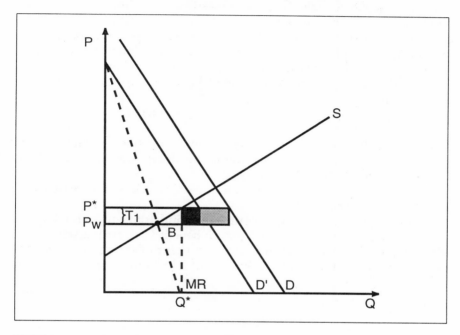

FIGURE 3.5   Tariff that Generates Government Revenues

Optimal Producer Responses to Tariffication

Thus far the analysis has only examined the economics of tariffication when farmers adjust production to changes in domestic consumption. Several other countries continue to have domestic price support schemes in agriculture that employ import control schemes, and yet, they export at the same time. A classic example is the European Union where grain imports are controlled under the domestic price support scheme, and its exports are financed by export restitution payments. The US wheat program is a similar case. Would it therefore be possible for Canada to continue controlling imports while expanding production so that extra production could be exported? To overcome dumping laws in GATT, the program might have to be implemented via a producer marketing board where differential prices can be charged on world markets (like the United States and Europe do now in their wheat programs).

The hatched area in Figure 3.6 depicts the economic gains to Canadian farmers if production is expanded beyond current production quotas and sold on world markets at P$_w$. The only alteration required in the current program is the transformation of current production schemes to domestic marketing quotas for sales into the domestic market. Import quotas and high domestic prices are maintained. However, a producer levy that is equal to the difference between domestic and world prices is charged for overquota production (as in Europe). This will ensure that total production is maintained at the free trade levels, Q$_{ft}$, as shown in Figure 3.6.

As evident from a close inspection of Figure 3.6, the potential gains from adopting this scheme are higher with:
*   a more narrow gap between P* and P$_w$ that will occur over time as tariffs decline.
*   a tariff that is higher than the current gap between domestic and world prices as indicated in studies by Moschini and Meilke (1991) and Cymbal and Veeman (1994).
*   a higher minimum access commitment, which will increase over time for some products.
*   lower production costs over time resulting from new technologies.

The only case in which Canada does not benefit from such a scheme is when the world price is at or below the point of intersection of the marginal revenue and supply curves (point B in Figure 3.6).

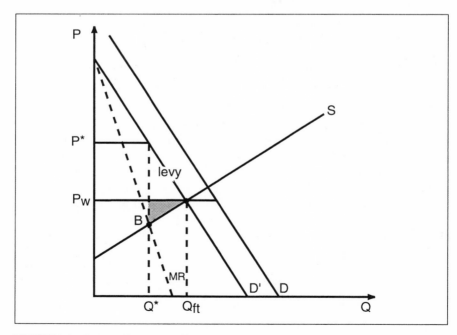

FIGURE 3.6   Gains to Farmers with Exports

Elimination of supply management cannot be ruled out, particularly if import quotas increase over time and domestic prices decrease toward world price levels. There may come a point in the future where the hatched area in Figure 3.6 is greater than the benefits from higher prices for domestic consumption. A lesson may be learned from the New Zealand experience. Previously, New Zealand charged much higher prices for domestic sales than for export sales. Liberalization of New Zealand's farm policy in the 1980s changed all of that. Nevertheless, there is a noticeable degree of confidence in New Zealand's dairy sector today. Dairy production should increase substantially in the next year (Mulrennan, 1993) with land prices being the only, major limiting factor. New Zealand has no price discrimination or quota with the farmer-operated dairy board that markets its processed dairy products. Farm prices are substantially below those received by Canadian farmers, and yet, New Zealand accounts for close to 25% of the world trade in dairy products, even though it only has 2% of world dairy production. While most countries' dairy markets are governed by subsidies, protectionism and export surpluses, New Zealand's expansion relies only on world market competition. Even though Canada's current situation is in sharp contrast with New Zealand's, it is still within reach of becoming an important world market if major reforms are undertaken.

Other Issues

Distribution of Import Quota Rents

The debate continues as to who should obtain the import quota rents. One option is to return them to the government or to producers. Regardless, the allocation of these rents affects the location of production, value added and pricing patterns among provinces.

Value Added

Supply management in Canada has to increase its emphasis on further processing. The type and degree of pricing in certain cases results in increased imports from the United States, that is Canadian processors cannot access cheap inputs as can their American counterparts, thereby accelerating final product imports into Canada. Price structures will have to be changed, but they should be made consistent with trading rules. Also, economies of scale in processing have to be present in order to increase product differentiation.

Will the Cartel Stand?

Even with the high levels of protection granted under GATT, the supply management cartel will not be held together automatically. Confrontations will continue over such issues as quotas and market access. Tariffication may make it more difficult for the cartel to stand, given its existing members, than was the case under Article XI. For example, recently Ontario is trying to break the broiler cartel by producing beyond its quota allocation. This type of response would have been in violation of Article XI. However, with tariffs in place, the story is different. In this case, Ontario, for example, could attempt to drive out competition by expanding production and lowering prices. Once the competition had been driven out, it could return to the initial industry position but with fewer players in the industry. The tariff scenario, with minimum access requirements in place, ensures that producers are not forced to the free trade price or to the free trade import levels. We, therefore, speculate that the cartel will remain in place. However, games are not possible where the cartel is in a constant state of flux.

Trade-offs

GATT set the tariff levels for all commodities, not just those under supply management. The regional implications for Canadian agriculture that have resulted from GATT are quite clear. Western Canadian agriculture, where grains dominate, did not receive the gains for which they had hoped because tariff and nontariff barriers on grains were not significantly reduced worldwide. Also, tariff levels on imports into Canada were not nearly as high for grains as they were for supply-managed commodities (Table 3.2).

TABLE 3.2   Tariff Equivalents for Canadian Grain Imports

Product	Base Tariff	2000 Tariff
Durum Wheat	57.7 %	49.0 %
Wheat, Other	90.0 %	76.5 %
Barley, Malting	111.4 %	94.7 %
Barley, Feed	25.1 %	21.3 %

Source: Agriculture Canada, 1994

From a broader perspective, what is the interrelationship between the supply-managed sectors and grains and livestock? The US ruling in August, 1994, on Canadian wheat exports to the United States was a setback, especially for durum wheat. (Cdn $23 tariff/tonne for volumes of imports between 300,000 tonnes and 450,000 tonnes and Cdn $50/tonne for 450,000 tonnes and above.) Were there trade-offs between commodities such as durum wheat and chicken? For example, for Canada to expand and/or maintain 1994 wheat export levels to the United States, do tariffs have to be lowered on supply-managed commodities? We don't know the answer to this, but we hypothesize that there is a connection when it comes to negotiating time. This issue has been raised often by Western grain growers, but no satisfactory answers have been forthcoming.

In terms of political influence, it appears as if supply management is much better at rent-seeking than are many of the other commodity groups. Theory suggests that supply management is a structure that is conducive to getting what it wants from Ottawa. Rent-seeking is the name of the game — more power to a sector that can control and direct its destiny. Some sectors receive a handsome dividend from investing in lobbying and rent-seeking while others do not, so the outcome is not guaranteed.

Conclusion

Supply management will face different choices in the future as it will have to deal with issues such as regional allocation of production, quota transferability, processing capacity and consolidation and value added. However, this will all be simplified by the fact that, at least for the next several years, supply management will be a sector that is highly protected from imports from abroad. This will be true even when the preliminary tariffs become final.

As a caveat, in a recent dispute over poultry and dairy products (Brown 1995), the United States contends that Canada is obligated to phase out its tariff rate quotas on the disputed products, according to Article 302 of NAFTA. However, Canada maintains that NAFTA applies only to the within-quota tariffs that fall under the GATT minimum access rule since these were the only tariffs in place under NAFTA. Canada argues that Article 302 does not apply to the (prohibitive) tariffs for volumes of product above the minimum access rule since these tariffs did not exist prior to the GATT agreement. This debate was not resolved in a meeting between Canada and the United States on March 1, 1995. The debate has been pushed forward until sometime in May. This chapter focuses on the long-term effects of tariffication under GATT though clearly the current dispute over NAFTA adds another layer of complexity to the discussion presented here.

Now that tariffs are in place and visible, pressure for tariff reduction will intensify. This, however, may present an opportunity. By increasing efficiency within the industry in the presence of high tariffs, the future impact of reduced tariffs could greatly be mitigated. The pressure to lower tariffs will largely come from the United States where many feel that the new tariffication under GATT will have a significant negative impact on US exports to Canada.

References

Agriculture Canada. 1994. GATT/WTO Fact Sheets, Pp. 1–25. Ottawa, Ontario, Canada (April).

Brown, R. H. 1995. "US-Canada dispute over poultry, dairy pushed into May." *Feedstuffs, the Weekly Newspaper for Agribusiness* (27 March).

Cymbal, W., and M. Veeman. 1994. "Canadian Agriculture and Article XI. An Economic Analysis of Tariffication for Poultry Products." Department of Rural Economy, University of Alberta, Edmonton, Alberta.

Josling, T., and S. Tangermann. 1994. "The Significance of Tariffication in the Uruguay Round Agreement on Agriculture." Paper presented to the North American Agricultural Policy Research Consortium Workshop, Vancouver, Canada (14 May).

Moschini, G., and K. D. Meilke. 1991. "Tariffication with Supply Management: The Case of the US-Canada Chicken Trade." *Canadian Journal of Agricultural Economics* 39: 55–68.

Mulrennan, F. 1993. "Cheap Milk Producers Take on the World." *Farming Independent.* New Zealand (June–August).

Rucker, R., W. Thurman, and R. Borges. 1994. "The Effects of GATT on US Present Market." Draft Paper, North Carolina State University, Raleigh, NC (May).

Schmitz, A., and D. Christian. 1996. "The US Sugar Industry." This volume.

Schmitz, A., and T. Schmitz. 1994. "Supply Management: The Past and Future." *Canadian Journal of Agricultural Economics* 42: 125–148.

Sumner, D. 1996. "Tobacco Supply Management: Examples from the United States and Australia." This volume.

Veeman, M. 1982. "Social Costs of Supply Restricting Marketing Boards." *Canadian Journal of Agricultural Economics* 30: 21–36.

4

Supply Management Under Minimum Import Access Requirements

J. M. Alston and J. D. Spriggs

Abstract

Previous studies of trade policy reform have treated domestic quotas as exogenous. In our more realistic model, quotas on imports are chosen jointly with domestic production quotas to "optimize" total welfare and its distribution among producers, consumers, importers and taxpayers. The General Agreement on Tariffs and Trade (GATT) and the North American Free Trade Agreement (NAFTA) agreements impose minimum import access requirements. This chapter shows the unintended consequences of GATT and NAFTA regulations that arise from the optimizing response of the Canadian government subject to the constraints of the minimum access requirements.

Introduction

In Canada, supply management programs currently exist for eggs, broiler chickens, turkeys, and milk. In these programs, domestic supply is regulated by controlling both domestic production and imports. Trade barriers that prevent international arbitrage from undermining domestic prices, are an essential feature of this system. In the negotiations leading up to the Canada–United States Free Trade Agreement (CUSTA), the NAFTA, and the Uruguay Round GATT agreements it was argued that the import quotas involved in supply management programs were trade-distorting.[1] Consequently, these agreements included provisions that will modify supply management in Canada in the post-GATT era. The main features of these agreements, as they relate to supply management, are tariffication and minimum access requirements. Tariffication refers to the agreement to replace import quotas on supply-managed commodities with tariffs, which are to be gradually phased down. It is widely acknowledged, however, that there is a great deal of "water in the tariff" line agreed under the GATT/World Trade Organization (WTO) agreement. Thus, tariffication will not reduce the rate of import protection afforded Canada's supply-managed industries; rather, in the near term, the rate of protection will increase. In effect, the new two-part tariff amounts to a continuation of the import quotas. The important changes, then, relate to market access.

Minimum access requirements (MARs) were established under both the CUSTA/NAFTA and GATT/WTO agreements for a range of agricultural commodities, including those under supply management. In the GATT/WTO agreement, specific minimum import quantity commitments have been established for each product in each year. Under the NAFTA (and CUSTA) agreements, MARs were also established for poultry and eggs. As is currently true in the case of chicken, for some products in some years, the NAFTA rather than the GATT access commitment may become binding. The NAFTA commitment is different from the GATT/WTO commitment in that it is expressed as a fraction of the previous year's production rather than as a specific quantity. This difference is significant. The NAFTA commitment allows Canada to vary import quantities by varying domestic production, a feature not shared by the GATT/WTO commitment.

In this chapter, we develop a pressure-group model in which four interest groups have a direct financial stake in supply management: producers (who are also the owners of production quotas), middlemen (who are also the owners of import quotas), consumers, and taxpayers (for simplicity, in most of the analysis we aggregate consumers and taxpayers). We analyze how the "optimal" settings of production and import quotas vary in response to changes in market parameters, conditions of supply and demand, and "welfare weights" reflecting the differential political influence of the different groups. In addition, we show how the introduction of external constraints (that is adopting minimum import access requirements and replacing with a tariff the "efficient" instrument given by the unconstrained choice of import quotas) influences the outcome in terms of production quotas, trade, and deadweight losses from supply management.

Theoretical Model

We begin with a simple model of a hypothetical supply-managed industry. We adopt similar assumptions to those used by Vercammen and Schmitz (1992): we assume linear supply and demand equations, competitive domestic supply and demand conditions, and a small country in trade, with a "perfect" freely transferrable production quota applied to output and another "perfect" freely transferrable quota applied to imports.

Figure 4.1 represents the market equilibrium situation. In Figure 4.1, SS' represents the unregulated supply function that would prevail in the absence of supply management, DD' is the domestic demand function, and P_W is the exogenous price of imports. In the absence of intervention, we would see the free trade values of domestic consumption (C_F) and production (Q_F), with imports (I_F) making up the shortfall. The government chooses both a domestic production quota (Q) and an import quota (I) which together determine the total domestic availability ($C = Q + I$) which, in turn, determines the domestic price (P).[2] Import quota owners receive rent equal to ($P - P_W)I$ = area (**H+J**), which is a pure gain to those to whom the import quotas are given. Producers who are given production quotas earn quota rents equal to ($P - MC)Q$ = area (**G+E**), but they lose producer surplus equal to area (**E+F**) so their net gain is area (**G-F**). Consumers lose consumer surplus of area (**G+H+J+K**). Summing these effects, the net social cost of the policy is equal to the sum of the Harberger triangle of distortion in consumption, area (**K**) plus the triangle of distortion in production, area (**F**).

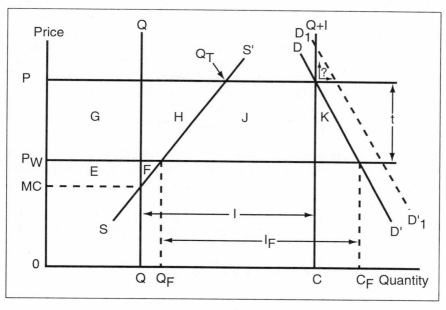

FIGURE 4.1   Price, Quantity and Economic Welfare Effects of Supply Management Assuming Exogenous Policies

In this static context, tariffication with the tariff set to maintain the rate of protection would change nothing except that the import quota rents would become tariff revenues (see Alston and Spriggs, 1995 for more details). A conventional view would appear to be that imposing a minimum access requirement (requiring imports greater than I) would result in increased imports, increased total consumption, lower domestic price, and greater total domestic welfare (that is smaller deadweight losses). Indeed, if the domestic quota remained fixed at Q, increased imports would ensure lower prices and greater domestic welfare. However, this would be partly at the expense of domestic producers.

The key question is whether domestic production would be caused to change. The existence of supply management indicates that producers are relatively powerful politically and it seems unlikely that the domestic production quota would remain unchanged in the face of changes in imports and prices. It is conceivable, for instance, that domestic production might fall by an amount greater than the required increase in imports, leading to opposite effects on price and consumer welfare. Thus, in order to understand the consequences of the MAR, we require an explicit model of policy choice.

The following algebraic model is developed to show the choice of the quota on imports, I, and the quota on domestic production, Q, with and without a minimum import access restriction. We assume policy is chosen as if the goal is to maximize a weighted sum of consumer surplus (CS), domestic producer surplus (PS), and importer surplus accruing to holders of import quotas (MS) subject to a minimum access requirement.[3] Further, suppose the MAR is established as a specific fraction, v, of the previous (and, in a static model, current) year's production, as under the CUSTA/NAFTA commitment (the binding MAR in the case of Canadian chicken). Hence, assuming a binding MAR, the import quota satisfies

$$I = vQ = rC$$

where $r = v/(1 + v)$ and v is the minimum fraction of production (and r is the minimum fraction of consumption) that must be supplied as imports. Consequently, the relationship between production and consumption is that $C = (1 + v)Q$ and $Q = (1 - r)C$.

Assuming a linear welfare trade-off and treating producer surplus as a numeraire with a weight of 1, the maximization problem can be stated mathematically as:

(1)
$$Max\ W = W(PS, CS, MS)$$
$$= PS + \theta_1 CS + \theta_2 MS,$$
$$subject\ to\ Q \leq (1 - r)C.$$

Producer and consumer surpluses can be defined in terms of the price-dependent domestic demand function, $D(C) = P$, and supply (or marginal cost) function, $S(Q) = MC$, as follows:

(2)
$$CS = \int_0^C D(C)\,dC - P.C,$$

(3)
$$PS = P.Q - \int_0^Q S(Q)dQ,$$

(4)
$$MS = (P - P_W)I = (P - P_W)(C - Q).$$

Combining equations (1) through (4) gives the objective function:

(5)
$$Max\ L = P.Q - \int_0^Q S(Q)dQ + \theta_1\left[\int_0^C D(C)dC - P.C\right]$$
$$+ \theta_2(P - P_W)(C - Q) + \lambda[(1 - r)C - Q].$$

The first-order conditions for a maximum are:

(6)
$$\frac{\partial L}{\partial C} = Q\frac{\partial P}{\partial C} - \theta_1 C\frac{\partial P}{\partial C} + \theta_2\left[(C - Q)\frac{\partial P}{\partial C} + P - P_W\right] + (1 - r)\lambda = 0,$$
$$\frac{\partial L}{\partial Q} = (P - S(Q)) - \theta_2(P - P_W) - \lambda = 0,$$
$$\frac{\partial L}{\partial \lambda} = (1 - r)C - Q \geq 0.$$

We assume that an imposed MAR will be binding on the policy choice, and the algebraic solution to equation (6), in the case when the MAR is set high enough not to be binding, can be used to represent the pre-MAR situation. By comparing the model solutions when the constraint is and is not binding, we can evaluate the impacts of the changes introduced under GATT and NAFTA. We therefore consider two situations: (a) when the unrestricted maximization of equation (1) through the joint choice of domestic production quotas and import quotas (that is the pre-MAR scenario); and (b) when the constraint is binding (the likely post-MAR scenario).

No Minimum Access Restriction

If the MAR does not apply (or is not binding), $\lambda = 0$ in equation (5), and the first-order conditions are:

(7)
$$\frac{\partial W}{\partial C} = \frac{\partial P}{\partial C}[Q - \theta_1 C + \theta_2(C - Q)] + \theta_2(P - P_W) = 0,$$

(8)
$$\frac{\partial W}{\partial Q} = P - S(Q) - \theta_2(P - P_W) = 0.$$

Combining these conditions,

(9) $$\frac{\partial W}{\partial C} = \frac{\partial P}{\partial C}[Q - \theta_1 C + \theta_2(C - Q)] + P - S(Q) = 0.$$

Now, assume linear supply and demand relationships:

$$S(Q) = MC = \beta_0 + \beta_1 Q \text{, and } D(C) = P = \delta_0 - \delta_1 C,$$

where β_1 and δ_1 are strictly nonnegative. Note that the free trade quantities are defined by:

$$MC = P_W = \beta_0 + \beta_1 Q_F \text{, and } P_W = \delta_0 - \delta_1 C_F,$$

and we can denote the free trade import share as $r_F = (C_F - Q_F)/C_F$. Substituting these linear supply and demand equations into (9) gives:

(10) $$Q = \frac{\delta_0 - \beta_0}{(1 - \theta_2)\delta_1 + \beta_1} + \frac{\delta_1(\theta_1 - \theta_2 - 1)}{(1 - \theta_2)\delta_1 + \beta_1} C.$$

Solving equations (7) and (10) for C yields:

(11) $$C = \frac{\delta_1(1 - \theta_2)(\delta_0 - \beta_0) - \theta_2(\delta_0 - P_W)[\beta_1 + (1 - \theta_1)\delta_1]}{\delta_1[\beta_1 + (1 - \theta_1)\delta_1][(\theta_1 - 2\theta_2)] - \delta_1^2(1 - \theta_2)(\theta_1 - \theta_2 - 1)}.$$

Or, using the supply and demand equations above,

(12) $$C = \frac{(1 - \theta_2)(\delta_1 C_F + \beta_1 Q_F) - \theta_2[\beta_1 + (1 - \theta_2)\delta_1]C_F}{[\beta_1 + (1 - \theta_2)\delta_1][(\theta_1 - 2\theta_2)] - \delta_1(1 - \theta_2)(\theta_1 - \theta_2 - 1)}.$$

Substituting this equation into (10) yields Q and substituting into the demand equation yields an equation for the price, P.

A Binding Minimum Access Requirement
When the MAR constraint is binding, λ, the shadow value of the MAR is greater than zero. Solving the first order conditions with $(1 - r)C = Q$ yields:

(13) $$[(1 - r)(1 - \theta_2) + (\theta_2 - \theta_1)]C\frac{\partial P}{\partial C} + (1 - r)(P - MC) + r\theta_2(P - P_W) = 0.$$

Substituting the supply and demand equations into (13) and solving for C, we obtain:

(14)
$$C = \frac{\delta_1 C_F[(1-r)+r\theta_2]+(1-r)\beta_1 Q_F}{2\delta_1[(1-r)+r\theta_2]-\delta_1\theta_1+(1-r)^2\beta_1}.$$

Differentiating equation (14) with respect to r and expressing the result in elasticity form,

$$\epsilon_{C,r} = \frac{\partial C}{\partial r}\frac{r}{C} = \frac{\delta_1(\theta_2-1)r-\beta_1(1-r_F)r}{\delta_1(1-r)+r\theta_2(1-r)(1-r_F)\beta}$$

(15)
$$-\frac{2r[\delta_1(\theta_2-1)-\beta_1(1-r)]+r(1-r)\beta_1 Q_F}{2\delta_1[(1-r)+r\theta_2]-\delta_1\theta_1+(1-r)^2\beta_1}.$$

Recall, under a binding MAR, imports are $I = rC$ and domestic production must satisfy $Q = (1 - r)C$. Thus we can define elasticities for price, production and imports with respect to r as: $\epsilon_{P,r} = \epsilon_{C,r}/\eta$ (where η is the demand elasticity); $\epsilon_{Q,r} = \epsilon_{C,r} - r/(1-r)$, and $\epsilon_{I,r} = \epsilon_{C,r} + 1$.

Equation (15) shows that changing a binding MAR has an ambiguous effect on quantity produced and consumed, and therefore on price, depending on the values of welfare weights and supply and demand parameters. This ambiguity of effects is likely to carry over into welfare measures, as well. These aspects are difficult to resolve algebraically or graphically, so we now turn to numerical simulations.

Welfare Impacts of a Binding MAR

It can be seen that the introduction of a binding MAR can have implications for the optimal settings of endogenous policies such as the domestic quota and total consumption. These changes must involve a reduction in the value of the weighted welfare function W, given in equation (1), since it would be maximized (by the definition of a binding MAR) with a smaller fraction of imports. In this sense, the restrictions derived from the GATT/WTO agreement or NAFTA involve a loss of efficiency in the importing country. In pursuing a "second-best" optimum, subject to the constraint of the MAR, the government may change its policies so as to increase or reduce domestic production and increase or reduce domestic consumption. The directions of change in the two quotas will depend, we suggest, on the supply and demand elasticities, the welfare weights, and the value of the MAR parameter, r. In turn, these adjustments will determine whether the binding MAR restriction involves an increase or decrease in unweighted domestic and global welfare ($W^* = PS+CS+MS = W_G$).

It is of interest to evaluate the implications of the various parameters for the MAR-induced changes in the value of both the weighted and unweighted welfare measures (W and W^*, respectively). To do this, as an illustrative example, we conducted numerical simulations using the equations of the model laid out above, with the equations parameterized using data from Moschini and Meilke (1991) for the Canadian chicken meat industry. The key parameters are the demand elasticity (-0.5), the supply elasticity (1.0), the 1987– 89 average values for the quantity produced (521 thousand metric tonnes) and consumed (561 thousand metric tonnes), the world price ($1.58 per kilogram), and the domestic price ($2.23 per kilogram). In addition, the marginal cost was assumed to be 70% of the domestic price at the distorted equilibrium. These values are sufficient to parameterize the model and solve for the effects of interest using a range of values for the welfare weights. The model was specified and solutions were obtained using the *Solver* option in *EXCEL 5.0*.

Figure 4.2 shows plots of the unweighted economic welfare (that is W^*) obtained by maximizing the weighted welfare (W) subject to the constraint of the MAR for various combinations of relative political weights for consumers and middlemen (given a weight of 1 for producer welfare). For example, the uppermost plot in Figure 4.2 shows the values for domestic welfare when producers, consumers and middlemen all have equal weights (the same plot applies also when the middleman weight is reduced from 1.0 to 0.5 or 0). When the MAR is zero with equal weights, we have the competitive market solution (corresponding to C_F and Q_F in Figure 4.1) and the value of total domestic welfare is maximized at $2,059 million per year. The MAR becomes binding at 18%, the competitive fraction imported. Up to that value, welfare is unaffected by the MAR, but for MARs of 20% and beyond, welfare progressively declines as output becomes restricted below the competitive quantities.

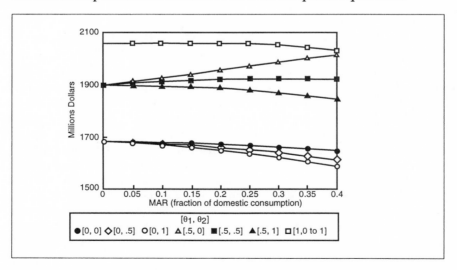

FIGURE 4.2 Economic Welfare Values (W^*) for Different Relative Political Weights (Producers=1, Consumers=θ_1, Importers=θ_2)

The middle set of three plots, beginning at $W^* = \$1,900$ million apply when the weight on consumer welfare is $\theta_1 = 0.5$, with a range of values for the weight on middlemen welfare ($\theta_2 = 0, 0.5, 1.0$). All three curves begin at the same point, reflecting the deadweight loss arising from the departure from the competitive solution as a consequence of having relatively powerful producers. This deadweight loss amounts to $159 million per year ($2,059 – $1,900 million per year). In each of these three cases, regardless of the weight on middlemen, the unconstrained optimum is autarky. The explanation is that, even when middlemen and producers have equal weight, it is still cheaper to source the entire consumption from domestic production. When there is zero weight on middlemen, the increasingly binding requirement to import, as the MAR rises, causes welfare progressively to rise. Production declines but consumption rises and the gains to consumers outweigh the losses to producers. This is the exceptional case, and a weight of zero on importer (middleman) welfare seems unlikely. In contrast, when there is some weight on middlemen ($\theta_2 = 0.5$ or 1.0), production declines faster and consumption increases more slowly so that the losses to producers outweigh the gains to consumers, and welfare progressively falls when the MAR becomes progressively larger.

The lower set of three plots in Figure 4.2, beginning at $W^* = \$1,682$ million per year, represents the case where consumer welfare receives no weight at all and the weight on middlemen varies ($\theta_2 = 0, 0.5,$ and 1.0). Once again all three curves start at the same point since, once again, regardless of the weight on middlemen, the least-cost way to source the optimal total quantity in the unconstrained setting is from domestic production. So long as the weight on consumer welfare is zero, the monopoly solution prevails with deadweight cost equal to $377 million per year ($2,059 – $1,682 million per year). The imposition of a MAR causes the deadweight cost to rise due to the resulting adjustments in total consumption and domestic production. The increase in deadweight cost is greater when more weight is attached to middleman welfare, as can be seen by comparing the three plots. In a situation where importers are already relatively favored, a binding MAR is likely to lead to greater deadweight losses.

The most significant feature of Figure 4.2 is that, with one exception, every one of the plots shows a monotonic decline in W^* as the MAR percentage increases from zero: that is the enforcement of "freer trade" under the new GATT or NAFTA, using a MAR, leads to a reduction in global welfare and unweighted domestic welfare in the importing country, Canada.[4] The exception arises when a consumer welfare weight of 0.5 is combined with an importer welfare weight of zero — but this is a case where import quotas would be a most unlikely policy anyway.

Auctioned Import Quotas or Tariffs

Suppose the import quotas are owned by the government and auctioned so that the rents go to taxpayers rather than middlemen or foreigners. In terms of the implications for the distribution of welfare, this is equivalent to replacing import quotas with tariffs — a "pure" tariffication. This case could be represented in Figure 4.2, if we could assume that taxpayers and middlemen had equal welfare weights, so that it would make no difference whether the import quota rents go to middlemen or taxpayers. Alternatively, it is common to aggregate general taxpayers with consumers in models of agricultural policy (for example, Gardner 1983, 1987), which amounts to assigning them equal weights (that is setting $\theta_1 = \theta_2$). Under this weighting scheme, with our other (static) model assumptions, the optimum import and domestic quantities are the same as would be obtained by optimizing a tariff. Figure 4.3 shows the total domestic (and global) welfare from optimizing a trade-off of producer welfare (with a weight of 1) against welfare of taxpayers-cum-consumers (with weights of $\theta_1 = \theta_2 = 0, 0.25, 0.50, 0.75$ and 1.0) given values of the MAR ranging from 0 to 40%. As can be seen in Figure 4.3, welfare effects of a binding MAR with auctioned import rights are ambiguous, depending on the value for the relative weight of taxpayer-cum-consumer welfare ($\theta_1 = \theta_2$) and on the setting of the MAR.

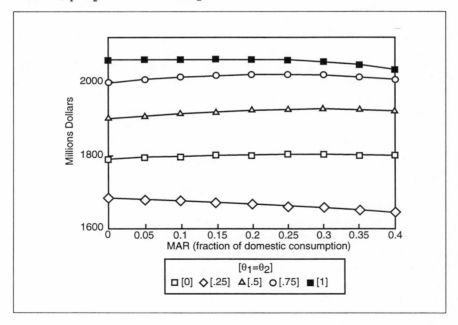

FIGURE 4.3   Global Welfare Values with Equal Welfare Weights for Consumers/Taxpayers (θ_1) and Importers (θ_2)

Conclusion

The objective of pursuing trade agreements under the GATT and NAFTA was to promote freer trade among nations and, thereby, generate benefits both globally and to the individual contracting parties. Canada (among others) reluctantly agreed to meet the minimum import access requirements. In the CUSTA and NAFTA, Canada agreed to introduce MARs for poultry and eggs. In the new GATT/WTO agreement, Canada agreed to MARs and will replace the import quotas for supply-managed commodities with tariff-rate quotas, beyond which prohibitive tariffs will apply.

At first blush, a restriction to increase the fraction of consumption supplied by imports seems to assure an increase in imports and a reduction in the distortions of domestic production and consumption as well as trade. This presumes total domestic production remains fixed while domestic consumption rises to accommodate the restriction; but domestic production may rise or fall while domestic consumption may rise by more or less than the increase in imports. The nature of the domestic policy response to the external constraint depends on the relative political influence of interest groups affected by the policy. Where producer welfare is relatively important (as evidenced by the existence of the policy in the first place) the adjustment to the constraint of the MAR is likely to be in a direction that reduces losses to producers and increases the burden on consumers. This could involve a loss of overall welfare. Whether it is a proportional rather than a specific MAR may be important in that context, since reducing total consumption reduces the amount that must be imported to satisfy a percentage MAR.

A theoretical analysis of the consequences of imposing a MAR shows that the effects on consumption are ambiguous, depending on parameters such as the welfare weights on consumers relative to producers and middlemen. An empirical illustration using data on the Canadian broiler industry shows that global and domestic deadweight losses due to supply management could well increase when an import MAR is imposed. The result applies regardless of whether the import quota rents go to middlemen (as at present, in the pre-MAR setting) or to domestic taxpayers (as tariff-quota rents under a revised import policy).

Imposing an import MAR can result in reducing domestic and global welfare and may seem counterintuitive at first. It arises because, in our analysis, the domestic production quota and the import quota are endogenous. The Canadian government, choosing these quotas in response to a MAR constraint imposed by NAFTA or the GATT (with their objective of freer trade), acts as if it is maximizing a different objective (the weighted welfare of domestic interest groups). It should not be surprising to economists to learn that, in pursuing a "second-best optimum" under the constraint of the MAR, the government may choose policies that involve a reduction in some other objective (that is freer trade, unweighted domestic welfare, or global welfare). In fact, such an outcome seems likely in many cases, as our empirical examples illustrate.

The key general lesson from this analysis is that, when designing trade agreements that restrict policies of individual governments, care should be taken to recognize the endogenous nature of policy determination, and things should not be taken as given when they are not. In the context of multinational agreements, rules should be chosen with due regard to incentive compatibility if the objective is to force policies to adjust towards freer trade and enhanced domestic and global social welfare. In a GATT agreement that imposes few restrictions on domestic policy use, and partial restrictions on instruments used at the border, there is ample scope for perverse outcomes and good reason to expect them to proliferate.

Notes

1. The CUSTA took effect on January 1, 1989. The United States, Mexico, and Canada signed NAFTA on December 17, 1992, and it took effect on January 1, 1994. The Uruguay Round Agreements established the new WTO which replaced GATT. Negotiations for the new GATT/WTO Agreement were concluded in Marrakesh on April 15, 1994 and the agreement is to be implemented in 1995.

2. In practice, what seems to happen in some cases is that the domestic production quota is fixed and a price is established using a cost-of-production formula, with the import quota being chosen aiming to clear the market at the predetermined price. At this stage of the analysis, the question of how the import and domestic quotas are chosen is put aside and, regardless of how they are chosen, it is the total availability that determines price.

3. The same outcome could be derived from a self-willed government (SWG) model, in which the government actively maximizes an objective function, or from a clearinghouse government (CHG) model in which the government is passive and the policy is driven by competition among groups (for example, see Alston and Carter 1991).

4. With some other combinations of welfare weights, not shown in the figure, domestic (and global) welfare rises initially but eventually falls as the MAR percentage progressively increases.

References

Alston, J. M., and C. A. Carter. 1991. "Causes and Consequences of Farm Policy." *Contemporary Policy Issues* IX: 107–121.

Alston, J. M., and J. Spriggs. 1994. "Endogenous Policy and Supply Management in a Post-GATT World." Invited paper presented at the conference on "Supply Management in Transition Towards the 21st Century," Macdonald Campus of McGill University, Ste-Anne-de-Bellevue, Québec, Canada (revised January, 1995).

Gardner, B. L. 1983. "Efficient Redistribution through Commodity Markets." *American Journal of Agricultural Economics* 65: 225–34.

_____ . 1987. *The Economics of Agricultural Policies.* New York, NY: MacMillan.

Moschini, G., and K. Meilke. 1991. "Tariffication with Supply Management: The Case of the US–Canada Chicken Trade." *Canadian Journal of Agricultural Economics* 39: 55–68.

Vercammen, J., and A. Schmitz. 1992. "Supply Management and Import Concessions." *Canadian Journal of Economics* 25: 957–971.

5

Imports into Canada: Why Have They Remained Low?

J. A. Vercammen and A. Schmitz

Abstract

In this chapter we use a game theoretic framework to help explain why imports of supply-managed commodities remain low. This outcome can be explained where producers are more politically powerful relative to other interest groups, such as importers. The political cost of transferring rents to importers at the cost of producers is too high.

Introduction

In Canadian supply-managed industries, production control schemes exist in conjunction with import quotas. Empirical studies have found that these arrangements are noncompetitive in nature and often result in a large transfer of rents from consumers to producers (Schmitz, 1983; Veeman, 1987; Barichello, 1981). However, sizable rents also accrue to those holding the license to import the restricted commodity because, in many cases, the wedge between the Canadian domestic price and the world price is large. Even under the new General Agreement on Tariffs and Trade (GATT) struck in December, 1993, tariffication does not eliminate import quota rents because of Canadian minimum access commitments (Schmitz, de Gorter and Schmitz, 1994).

Imports have never made up a large share of Canada's domestic consumption of supply-managed commodities.[1] Moreover, any increases in these shares have typically been in response to external pressures, for example, the Canada–United States Free Trade Agreement (CUSTA) and GATT, rather than domestic pressures. Of interest in this chapter is understanding why Canadian importers have not been more successful in increasing their share of the lucrative Canadian market. If both parties had equal and perfectly effective say in the political process, then the level of domestic production and imports would reflect the outcome of a quantity-setting game between Canadian producers and importers. Could the equilibrium level of imports in such a game be consistent with the low level of imports that are actually observed? On the other hand, if one group, for example, producers, had relatively more power in the political process, then the quantity-setting game would be constrained and biased in favor of the more powerful group. Does this scenario better explain the relatively low and stable level of imports into Canadian supply-managed industries?

A model is constructed below which shows the equilibrium outcome of a single-period game between a monopolistic producer and a monopsonistic importer operating within a "small" country such as Canada. The extent to which importers behave as a collective unit in Canadian supply-managed industries is not clear. Nevertheless, the mere fact that there exists an Association of Regulated Importers within Canada lends support to this assumption. Two alternative behavioral specifications, Cournot-Nash and Stackleberg Leader-Follower, are assumed when the equilibrium is derived. The sensitivity of the results to the assumed supply and demand elasticities are explored as well.

For convenient interpretation of the results, the equilibrium outcome is expressed as percent deviations from the free trade, competitive outcome. The actual competitive outcome is unknown in Canadian supply-managed industries since the market has not witnessed a competitive equilibrium for many years. Nevertheless, useful comparisons between equilibrium import levels in the theoretically specified game and actual import levels in today's supply-managed industries can still be made. For example, one could assume that the import quota level is set equal to the competitive level of imports because imports were "frozen" at historical levels when the supply management regime was put into place.

The extent to which the importer can obtain a reasonable share of the market within the game is unclear a priori. On one hand, the importer is small, that is he would have a small share of the Canadian market under competition, in comparison to the producer. Therefore, one would expect a small share for the importer relative to the competitive outcome. On the other hand, because of the small country assumption, the importer faces a constant marginal cost for obtaining its product whereas the producer faces a rising marginal cost. This cost advantage should tend to enhance the market share of the importer. In any event, examining the range of possible outcomes for the equilibrium level of imports under different parameter and behavioral assumptions and comparing those outcomes to actual import levels is a useful way to determine the extent to which the political process is biased in favor of a particular group.

Model

The industry in question consists of a single producer, a single importer and a large number of consumers of a homogeneous commodity. (The government does not intervene). The importing firm purchases the commodity at a constant world price, P_0. For simplicity, the importing firm is assumed to incur no transaction costs when importing nor does it pay for the right to import.[2]

Let the country's inverse domestic demand schedule and the producing firm's marginal cost schedule be given by:

(1) $P_d = a - b(Q+I)$ and $MC = \alpha + \beta Q$

where Q, I, P_d and MC denote domestic production, imports, the demand price and marginal cost, respectively, and a, b, α and β are positive constants. Also, let $y \equiv Q_0/(Q_0+I_0)$ be the competitive market share of the producing firm where Q_0 is domestic production under competition and I_0 is imports under competition.

When the two firms behave noncompetitively, the respective quantities chosen by these firms can be expressed as percent deviations from their competitive values. Let δ_Q be the percentage difference between production with and without competition, and δ_I be the percentage difference between imports with and without competition, that is:

(2) $\delta_Q \equiv (Q_0 - Q_r)/Q_0$ and $\delta_I \equiv (I_0 - I_r)/I_0$

where Q_r and I_r are noncompetitive domestic production and imports, respectively.

The demand and marginal cost parameters can be expressed equivalently in terms of the absolute value of the elasticity of demand η (measured at the world price, P_0, and competitive consumption level Q_0+I_0) and the elasticity of supply ϵ (measured at the world price, P_0, and the competitive production level, Q_0). In particular:

(3) $a = (1 + 1/\eta)P_0,$ $\alpha = (1 - 1/\epsilon)P_0,$

            $b = (1/\eta)P_0/(Q_0+I_0)$ and $\beta = (1/\epsilon)P_0/Q_0.$

Rents for the producing firm with import quotas in place (denoted PR_r) are given by the area below the domestic price line and above its marginal cost schedule up to its choice of production, that is:[3]

(4) $PR_r = (P_d - \alpha - \beta Q_r/2)Q_r.$

Next, we specify a rent deflator and consider a relative measure of producer rents rather than an absolute measure. A convenient deflator is the rents earned by a competitive producer denoted by PR_0. Hence, an expression for the rents of the producing firm, given production, Q_r, and imports, I_r, relative to competitive producer rents can be derived by dividing equation (4) by $PR_0 = (1/2)\beta Q_0^2$, substituting the demand equation from (1) for P_d, rearranging terms and then substituting into the resulting expression the parameters defined earlier:

$$(5) \qquad PR_r / PR_0 = 2\left[1 + r(\gamma\delta_Q + \overline{\gamma}\delta_I) - (1/2)\overline{\delta}_Q\right]\overline{\delta}_Q$$

where $r \equiv \epsilon/\eta$ is the ratio of the competitive supply elasticity and the absolute value of the competitive demand elasticity. A bar over a variable denotes 1 minus the value of that variable, that is $\overline{y} = 1-y$.

Equation (5) shows that relative producer rents increase at a decreasing rate for larger values of δ_I and increase but eventually decrease for larger values of δ_Q. In other words, the producer always prefers fewer imports and benefits from restricted imports until the monopoly solution has been attained. Import quota rents (denoted by QR_r) can also be specified as a function of domestic production and imports and can be expressed relative to the rents earned by the domestic producer under perfect competition. These rents are calculated as the difference between the domestic price, P_d, and world price, P_0, then multiplied by the restricted volume of trade, I_r. To derive the appropriate expression, subtract $P_0 = a-b(Q_0+I_0)$ from $P_d = a-b(Q_r+I_r)$, multiply the resulting expression by I_r, divide that expression by $PR_0 = (1/2)\beta Q_0^2$, rearrange terms and then substitute in the appropriate parameter expressions:

$$(6) \qquad QR_r / PR_0 = 2r\overline{\gamma}\left[\overline{\delta}_I - \gamma\overline{\delta}_Q\overline{\delta}_I - \overline{\gamma}\overline{\delta}_I^2\right]/\gamma.$$

For a given level of domestic production, importer quota rents are concave in δ_I and reach a maximum at $\delta_I = (1/2)[1-r(2y-1)y/\overline{y}]/[1+r\gamma]$. Import restrictions beyond δ_I decrease import quota rents because the loss in revenues from the lower volume of imports more than offsets the gain in revenue from the higher domestic price. The opposite is true for import restrictions prior to reaching δ_I. Because the expression for δ_I is negative (positive) for sufficiently large (small) values of y, maximum importer rents may occur when imports are either larger or smaller than the competitive level.

Figure 5.1 plots equations (5) and (6) for $y = 0.7$ and $r = 2$. For illustrative purposes, rents are plotted as a function of imports, holding the level of production fixed at the competitive level and at the monopoly level (with respect to competitive imports). The top half of Figure 5.1 shows that the producing firm always prefers fewer imports and always prefers to produce at the monopoly level rather than the competitive level. The bottom half of Figure 5.1 shows that the importer prefers imports in excess of (less than) the competitive level when production is at the monopoly (competitive) level and that the importer is always better off with a lower volume of domestic production. A conflict arises because the producer prefers the monopoly level of production and fewer imports, but the former is inconsistent with the latter if the importing firm controls the level of imports.[4] Below, we assume that the equilibrium level of production and imports emerges endogenously in a noncooperative game between the producer and importer.

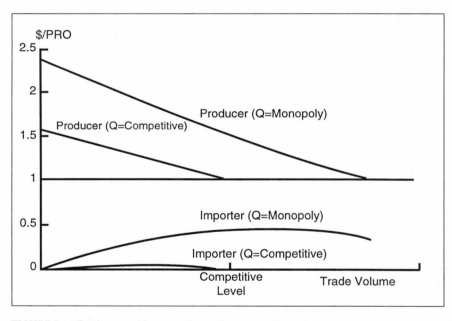

FIGURE 5.1   Producer and Importer Rents. (gamma=0.7, r=2)

Imports With and Without Competition

Cournot-Nash

In a Cournot-Nash (CN) equilibrium, the producing firm takes the level of imports as given when choosing output, and the importing firm takes the level of domestic production as given when choosing imports. The first-order conditions for the producing firm and importing firm can be derived from equations (5) and (6) and expressed as[5]

$$
(7) \qquad \begin{bmatrix} (1+2r\gamma) & r\bar{y} \\ \gamma & 2\bar{y} \end{bmatrix} \begin{bmatrix} \delta_Q^{CN} \\ \delta_I^{CN} \end{bmatrix} = \begin{bmatrix} r\gamma \\ \bar{y} \end{bmatrix}.
$$

Using Cramer's rule, solution values for the firms' choice variables can be derived and written as

$$
(8) \qquad \delta_Q^{CN} = \frac{(3\gamma-1)r}{3r\gamma+2} \quad and \quad \delta_I^{CN} = \frac{1-r(3\gamma-2)\dfrac{y}{\bar{y}}}{3r\gamma+2}.
$$

We can now identify the set of parameters consistent with $\delta_I^{CN}<0$, that is fewer imports entering with competition than without. Using the second expression in equation (8), note that $\partial\delta_I^{CN}/\partial\gamma=(3r\gamma^3-6r\gamma^2-2\gamma+1)/(3r\gamma+2)^2<0$. This result implies that the competitive import share must be sufficiently low to ensure $\delta_I^{CN}<0$. In particular, $\delta_I^{CN}<0$ if $r(3\gamma-2)y/\bar{y}>1$ or, equivalently, $1-\gamma < 1-(2r-1)/6r-([2r-1]^2+12r)^5/6r$.

For example, with $\epsilon=1$ and $\eta=0.5$ implying that $r=2$, competitive imports are less than noncompetitive imports for $\gamma \geq 0.73$ (or, equivalently, a competitive import share less than 0.27). For $r=1$, the critical value of γ is slightly higher (0.77 rather than 0.73), and for $r=4$, the critical value of γ is slightly lower (0.70 rather than 0.73). In general, higher values of r imply a lower critical value of γ, or alternatively, a wider range of competitive import shares are consistent with the result that competition entails fewer imports. It is straightforward to establish that for a relatively wide range of demand and supply elasticities, a competitive import share of less than 20% results in fewer imports entering the country with competition than without.

Figures 5.2 illustrates the CN equilibrium. Domestic demand and the producer's marginal cost are labeled D and S, respectively. The importer's best response function has half of the slope of the demand schedule and emanates upward from the point where the world price line, P_0, intersects demand. In the CN equilibrium, imports equaling I_{CN} will enter the country. To see this, note that imports, I_{CN}, imply a residual demand, D', and marginal revenue, MR', for the producer. The producing firm will produce at Q_{CN}, which is where its marginal cost equals marginal revenue. This output will sell at a price, P_{CN}. At this point of production, the importer has no incentive to change its level of imports implying that I_{CN} is indeed the CN equilibrium level of imports. Notice that more imports enter the country in the CN equilibrium than under competition.

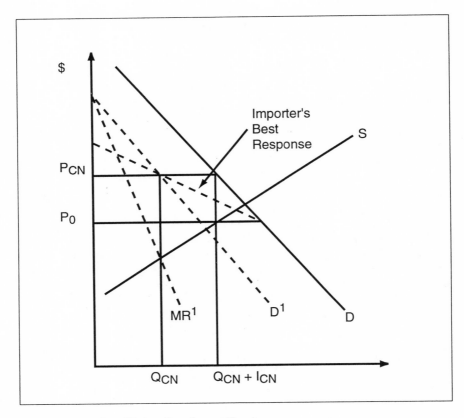

FIGURE 5.2   CN Equilibrium (Low Import Share)

In Figure 5.3, demand and marginal cost are the same as in Figure 5.2, however, the world price is lower implying a relatively larger market share for the importer under competition. Notice that, as a result of this larger market share, imports decline when moving from competition to the CN equilibrium. The reason why a sufficiently small, competitive import share is necessary for our results to hold is because with a small import share, the importer is able to "free ride" in the sense that when the producer reduces output in order to raise the domestic price, the importer takes advantage of this higher price by expanding imports. If an importer with a relatively high share of the market attempts to "free ride," the domestic price remains suboptimally low.

Finally, consider why a wider range of competitive import shares is consistent with our result. Recall that r equals the ratio of the supply and

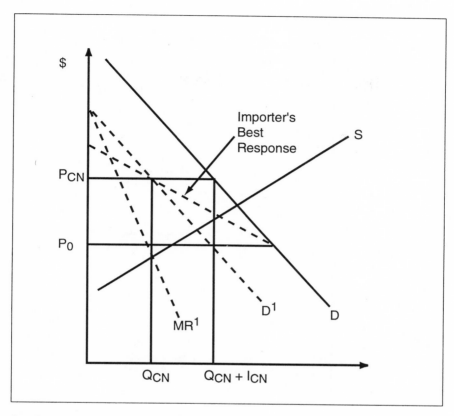

FIGURE 5.3   CN Equilibrium (High Import Share)

demand elasticities measured at competitive prices and quantities. A flatter marginal cost schedule implies that the producing firm decreases production to a relatively greater extent when moving from the competitive to the noncompetitive equilibrium. Hence, the domestic price markup is relatively larger in the noncompetitive equilibrium, making it relatively more attractive for the importing firm to increase imports beyond the competitive level. A steeper demand schedule also implies that the domestic price markup is relatively larger in the noncompetitive equilibrium, again making it relatively more attractive for the importing firm to increase imports.

Stackleberg

If the results in which we are interested arise for relatively high values of γ, then a Stackleberg follower-leader equilibrium may be a more appropriate behavioral assumption. When the producing firm behaves as a Stackleberg leader and the importing firm behaves as a Stackleberg follower, the following first-order conditions emerge from equations (5) and (6):

$$(9) \qquad \begin{bmatrix} (1+3/2r\gamma) & r\overline{\gamma} \\ \gamma & 2\overline{\gamma} \end{bmatrix} \begin{bmatrix} \delta_Q^{SL} \\ \delta_Q^{SF} \end{bmatrix} = \begin{bmatrix} 1/2r\gamma \\ \overline{\gamma} \end{bmatrix}.$$

Solution values for production and imports under the Stackleberg specification can be derived from equation (9) using Cramer's rule:

$$(10) \qquad \delta_Q^{SL} = \frac{(2\gamma-1)r}{2r\gamma+2} \quad and \quad \delta_I^{SF} = \frac{1-r(3/2\gamma-1)\dfrac{\gamma}{\overline{\gamma}}}{2r\gamma+2}.$$

As above, we are interested in identifying the parameter values that are consistent with fewer imports entering the country with competition than without. Because δ_I^{SF} is increasing in γ (just as in the CN case), it follows from the second expression in equation (10) that $\delta_I^{SF} < 0$ (which establishes our result) if $\gamma > (r-1)/3r+(r^2+4r-1)^{.5}/3r$.

Recall that in the CN case, 0.73 was the critical value of γ when r=2. In the Stackleberg case, 0.77 is the corresponding critical value of γ. Other critical values are γ=0.82 when r=1 and γ=0.73 when r=4. Thus, in the Stackleberg equilibrium, a slightly smaller range of competitive import shares gives rise to our result as compared to the CN equilibrium. However, our result, that noncompetitive imports exceed competitive imports when the competitive import share is less than 20%, still holds for a wide range of supply and demand elasticities.

Figure 5.4 illustrates the Stackleberg equilibrium. The difference between Figure 5.4 and Figure 5.2 is that, in the former, the importer's best response function also serves as the residual demand facing the producing firm. Hence, the producing firm produces at Q_s, which is where the marginal revenue from its residual demand equals its marginal cost. The equilibrium level of imports is therefore I_s and the domestic price is P_s. Domestic production is higher in the Stackleberg case than the CN case (because the producing firm's marginal revenue curve has a smaller intercept and is flatter) implying a relatively lower domestic price and thus relatively fewer imports.

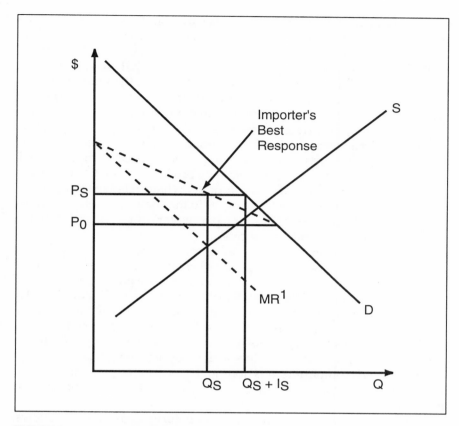

FIGURE 5.4   Stackleberg Equilibrium

Political Market

The results of the previous section indicate that for a wide range of supply and demand elasticities, for competitive import shares of less than 20% and for the two alternative behavioral assumptions, Cournot-Nash and Stackleberg Leader-Follower, the share of imports is higher without competition (the status quo) than with competition (free trade). This finding suggests that if producers and importers both had equal and effective say in the political process, then import levels under supply management would be higher than competitive import levels. Clearly, this does not appear to describe the case of Canadian supply management where it is highly unlikely that the current level of imports are above the competitive level of imports. Indeed, if this were the case, one would expect to see fewer imports entering Canada as the tariffs under the new GATT agreement take effect.

If the outcome of a game between a monopolistic producer and a monopsonistic importer does not describe the situation in Canadian supply-managed industries, then the alternative hypothesis gains support. In this case, the alternative hypothesis implies that producers have proportionately more power in the political process than importers.[6] In other words, at the current level of imports the government can transfer an additional dollar to importers by changing the production and/or import quota appropriately, and this will result in x dollars lost by producers. For a given distribution of "political power" across producers and importers, the larger is x, the less likely the transfer to importers will take place because of opposition voiced by producers. As well, for a given value of x, the more political power producers have relative to importers, the less likely the transfer will take place. In the analysis to follow, a value for x is derived for a feasible set of parameter values to facilitate the discussion about the political process.

Computation of the value of x, as described above, is equivalent to calculating the slope of the surplus transformation curve evaluated at the current level of trade. Gardner (1983) analyzes surplus transformation curves in the context of efficient redistribution in commodity markets. He suggests that governments are most interested in policies which have high-transfer efficiency, that is a dollar loss from one group should translate as closely as possible into a dollar gain for another group. Gardner's suggestion is incorporated into this analysis by assuming that when rents are transferred to importers, it is done in the most efficient manner so that x is minimized. In this formulation of transfer efficiency, consumer surplus is ignored.

The slope of the surplus transformation curve is derived as follows. Let $S_Q \equiv PR_r/PR_0$ and $S_I \equiv QR_r/PR_0$. S_Q and S_I are the surplus measures of the producer and importer, respectively, expressed relative to the surplus of the producer under competition. Now solve (6) for δ_Q to obtain

$$(11) \qquad \delta_Q = 1 - \frac{1}{\gamma}\left(1 - \overline{\gamma\delta}_I - \frac{\gamma S_I}{2r\overline{\gamma\delta}_I^2}\right).$$

Treating δ_Q as a function of δ_I for a particular value of S_I, the derivative of equation (11) with respect to δ_I can be expressed as

$$(12) \qquad \frac{\partial \delta_Q}{\partial \delta_I} = \frac{1}{\gamma}\left(\overline{\gamma} - \frac{\gamma S_I}{2r\overline{\gamma\delta}_I^2}\right).$$

Equation (5) can now be maximized subject to equations (11) and (12). If the total derivative of (5) with respect to δ_I is set equal to zero, the following expression results:

$$(13) \quad \frac{\partial S_Q}{\partial \delta_I} = \left[r\left(\gamma \frac{\partial \delta_Q}{\partial \delta_I} + \bar{\gamma} \right) - \frac{1}{2} \frac{\partial \bar{\delta}_Q}{\partial \delta_I} \right] \bar{\delta}_Q + \left[1 + r(\gamma \delta_Q + \bar{\gamma}\delta_I) - \frac{1}{2} \bar{\delta}_Q \right] \frac{\partial \bar{\delta}_Q}{\partial \delta_I} = 0$$

Equations (11) and (12) can be substituted into equation (13), and the resulting expression can be solved for the value of δ_I, which maximizes the producer's surplus subject to the importer earning rents, S_I and δ_Q. If this derived expression for δ_I is substituted into (5) along with equations (11) and (12), an expression for the surplus transformation curve is obtained.

As discussed above, we want to determine the value of the slope of the surplus transformation curve at the current level of trade since this is the derived value for x. As in the previous section, our best guess for the current value of δ_I is zero. That is the current level of imports is believed to equal the competitive level of imports. Assuming once again that r=2 (demand is twice as elastic as supply) and γ=0.8 (the domestic producer would have 80 % market share in a competitive outcome), the following values were derived using the above formulas. With S_I=0.6, the optimal values for δ_Q and δ_I are 0.242 and 0.160, respectively. With this policy configuration, S_Q=1.538. With S_I=0.7, the optimal values for δ_Q and δ_I are 0.206 and -0.162, respectively, and S_Q=1.379. Notice that the slope of the surplus transformation curve is indeed being evaluated near δ_I=0.

A simple calculation using the above data reveals that the slope of the transformation curve, dS_Q/dS_I, is x=-1.59. For every dollar transferred to importers, producers give up $1.59. One of the reasons why this number is comparatively large is because consumers also enjoy a higher level of surplus when rents are transferred to importers. To see this, note the expressions for δ_Q and δ_I can be combined and written as $Q_r+I_r=(Q_0+I_0)(\gamma \bar{\delta}_Q+\bar{\gamma}\bar{\delta}_I)$. In the context of the example, when S_I= 0.6, then $\gamma \bar{\delta}_Q+\bar{\gamma}\bar{\delta}_I$ = 0.8*0.794+0.2*0.840=0.773. When S_I= 0.7, the analogous value is 0.868. Hence, Q_r+I_r is larger when importers earn a higher level of surplus, implying that consumer surplus increases when a transfer to importers is made.

How should x=-1.59 be interpreted? As discussed earlier, if producers and importers have equal weight in the policy-setting arena, then it makes sense that importer rents are not increased above current levels because producers lose $1.59 for every dollar gained by importers. In this case, the net "political benefit" from transferring a dollar to importers probably does not offset the net "political cost" from transferring $1.59 away from producers. If producers have more "political strength" than importers, then the above argument is further strengthened, and it is even less likely that a transfer to importers will take place for political reasons. In fact, the above results suggest that if it were not for Canadian-guaranteed access commitments, which ensure that a minimum amount of imports will always be brought in under supply management, the import share would erode since a dollar taken away from importers would translate into $1.59 for producers.

Conclusions

This chapter proposes two alternative hypotheses as to why import shares in Canadian supply-managed industries have remained relatively low. The first is that producers and importers have equal and effective say in the political process, implying that production and import levels that are currently observed reflect the equilibrium outcome of an unconstrained quantity-setting game between producers and importers. Using both the Cournot-Nash and Stackleberg behavioral specifications and alternative assumptions about the competitive supply and demand elasticities, it is shown that for relatively low competitive import shares of less than 20%, the equilibrium level of imports without competition would exceed the competitive level of imports. Since this outcome appears highly unlikely for the case of Canadian supply management, this hypothesis is rejected.

The second hypothesis is that import levels have remained low because the political cost of transferring rents to importers at the expense of producers is too high. This hypothesis seems much more credible because it is calculated that a dollar of rents transferred to importers by increasing the import quota results in a $1.59 loss for producers. Clearly, one has to treat this result with caution because a number of rather strong assumptions were made during the derivation of this number. Nevertheless, the basic idea that producers stand to lose considerably at the margin when import quotas are increased is probably a reasonable explanation as to why importers have not been more successful at achieving higher rents in the Canadian supply-managed industries.

Notes

1. Chicken imports have a guaranteed share of approximately 7.5%, and the share for butter is less than 5% (Schmitz, de Gorter and Schmitz, 1994).

2. For a model of supply management, see Vercammen and Schmitz (1992). "Supply Management and Import Concessions." *Canadian Journal of Economics* 25: 957–71.

3. In supply-managed industries, a number of factors including rising input costs may be responsible for an upward-sloping, industry supply curve, implying that the area above this curve may not correspond to the rents earned by producers. We ignore this distinction in the current analysis.

4. Although not illustrated, an analogous conflict arises over the optimal level of

domestic production. A noncompetitive importer always prefers less than the competitive level of production while the producer may prefer more or less depending on the level of imports.

5. In the linear case, the best response function for the producer and importer is both linear and downward-sloping. Hence, according to Dixit (1986) the Cournot-Nash equilibrium will be both stable and unique.

6. Gardner (1987) and Carter, Faminow, Loyns and Peters (1990) use characteristics of the political environment and interest groups to explain levels of intervention in agriculture in the United States and Canada. Our hypothesis fits nicely within their framework.

References

Barichello, R. R. 1981. "The Economics of Canadian Dairy Industry Regulation." *Economic Council of Canada Regulation Reference and the Institute for Research on Public Policy*. Technical Report E/12, Ottawa, Ontario, Canada.

Carter, C., M. Faminow, R. Loyns, and E. Peters. 1990. "Causes of Intervention in Canadian Agriculture." *Canadian Journal of Agricultural Economics* 38: 785–796.

Dixit, A. K. 1986. "Comparative Statics for Oligopoly." *International Economic Review* 27: 107–122.

Gardner, B. 1983. "Efficient Redistribution through Commodity Markets." *American Journal of Agricultural Economics* 65: 225–234.

_____. 1987. "Causes of US Farm Commodity Programs." *Journal of Political Economy* 95: 290–310.

Schmitz, A. 1983. "Supply Management in Canadian Agriculture: An Assessment of the Economic Effects." *Canadian Journal of Agriculture Economics* 30: 135–152.

Schmitz, A., H. de Gorter, and T. Schmitz. 1994. *Consequences of Tariffication of Supply Management in Canadian Agriculture*. Proceedings of "Supply Management in Transition Towards the 21st Century," Macdonald Campus of McGill University, St. Anne de Bellevue, Québec, Canada (28–30 June).

Veeman, M. M. 1987. "Marketing Boards: The Canadian Experience." *American Journal of Agriculture Economics* 69: 992–1000.

Vercammen, J. A., and A. Schmitz. 1992. "Supply Management and Import Concessions." *Canadian Journal of Economics* 25: 957–71.

Case Studies of GATT's Effects

6

Regulation — The US Dairy Industry

J. L. Outlaw and R. D. Knutson

Abstract

In the spirit of the General Agreement on Tariffs and Trade (GATT), the US dairy industry is developing a new mindset toward becoming internationally competitive. While the 1995 Farm Bill may not eliminate US dairy programs, substantial restructuring is possible. Reduced levels of government support are an integral part of that strategy. The result is a massive restructuring of US milk production patterns into a more efficient configuration. This contrasts directly with the Canadian strategy of supply management with high supports. Implications of the US strategy for the Canadian dairy industry are explored.

Introduction

Insulated from the world market for decades, the US dairy existed as two quite separate, structural segments:

- A large farm, drylot dairying sector (500 cows and up) located primarily in the West, Southwest and Florida.
- A small and moderately sized farm sector (less than 70 cows) centered in the Northeast quadrant bounded by Minnesota, Missouri, Kentucky and Maryland.

Production in the drylot sector progressively increased, but few recognized the efficiency advantages held by these farms. Instead, there was extensive denial of the changing economic dimensions of milk production and its potential implications. This was possibly because the structure of the moderately sized farm sector was insulated to a degree from structural change due to a relatively high and stable milk price supported by the federal government.

The high level of support collapsed in the mid 1980s as a victim of overproduction and high government costs. Rejecting the option of controlling production, the US government dramatically lowered the price support level by 25%. By doing so, they exerted substantial pressure for structural adjustment on the dairy industry. Milk production in major, traditional, milk-producing states such as Minnesota, Missouri, Kentucky and Vermont fell as dairies that had fewer than 70 cows found themselves unable to compete. Due to differences in structural adjustment, California passed Wisconsin as the largest US dairy state. Defying conventional wisdom, large farm, drylot dairies began to develop outside of the West, Southwest and Florida — for example, in Western New York, Central Texas and Idaho.

Despite lower milk price supports and rapid structural change, there was little decline in overall production, except in particularly disadvantaged states. This lack of decline in production brought about the unpopular Dairy Termination Program (DTP) that was caused by an increase in cow numbers in the West and Southwest. The dynamics of structural change could be one possible explanation for this increase. Those farms that adjusted to this structural change by expanding the scale of their operations realized lower costs (Reimund, Stucker and Brooks, 1987). They frequently purchased the younger cows and heifers from those farms that were going out of business and, through improved management (feeding, health care, breeding and housing), substantially increased output per cow.

A more efficient dairy industry with lower milk prices led to questions concerning the ability of the US dairy industry to compete internationally in the presence of a more level trading field. Because of the increased efficiency of the industry, the need for US dairy price supports and federal milk order programs was questioned. The US General Accounting Office (GAO) called for deregulation policies designed to change the mindset of the dairy industry from protectionism to market orientation. (GAO, 1993).

The purpose of this chapter is to explain the changes that occurred in US dairy policy and structure. These changes raise a number of issues that have implications for structural and policy change in both Canada and the United States. These issues are highlighted, although not necessarily answered.

US Dairy Policy

The dairy industry is one of the most highly-regulated, agricultural sectors in the United States (Manchester, 1983). In addition to extensive sanitation regulations, there are two basic, economic, regulatory mechanisms:

- the dairy price support program (DPSP)
- the federal milk marketing order program (FMMOP)

These two programs directly affect the pricing of all milk produced in the United States. However, California, the largest dairy state, has its own state marketing order program that operates independently from the federal order program (Cropp 1995).

The Dairy Price Support Program

The DPSP undergirds the price structure of all milk produced in the United States. The Agricultural Act of 1949 provided permanent legislation for the program. Its original objective was to support the farm price of milk at 75%–90% of parity. Parity prices gave a hundredweight (cwt) of milk the same purchasing power in the current period as it had in the 1910–14 base period. Parity was not effective because no adjustments in parity prices were made to reflect technological changes that would effectively lower the support price level (Knutson, Penn and Boehm, 1995). Thus, the parity concept was abandoned in 1981. Each farm bill enacted by Congress since 1981 has amended the 1949 act by legislating the level of price support to be provided (Fallert, Blayney and Miller, 1990). The result was a gradual decrease in the price support to the current level of $10.10/cwt.

The Commodity Credit Corporation (CCC) of the United States Department of Agriculture (USDA) stood ready to buy butter, cheese and nonfat dry milk at product prices that would allow processors to return the support price to the farmer (Manchester, 1983). In March, 1995, purchase prices for butter, nonfat dry milk and cheese were $0.65, $1.034 and $1.12/lb, respectively. At these purchase prices, processors theoretically could sell their commodity to the CCC and return the support price of $10.10/cwt to the farmer.

Over the past 15 years there have been dramatic changes in the dairy price-support philosophy. The support price was increased to a peak of $13.49/cwt in 1981 (USDA, October, 1992). As a result of the USDA maintaining a high support price, milk prices have been at or very near the support level since 1949. Thus, the CCC has acquired substantial, but variable, stocks of cheese, butter and nonfat dry milk. The reduction of the support price level to $10.10/cwt in 1990 meant that government wanted to decrease the amount of money spent on the dairy program. Accordingly, CCC cheese purchases under the support program were all but eliminated. Periodically, small quantities of butter were purchased because of seasonal production overdemand. Over the past six years, butter is the only one of the three products that the CCC has consistently purchased in large quantities. However, in recent years, butter purchases have declined because of increased consumer demand that resulted from lower butter prices and bad publicity about margarine.

On the threshold of the 1995 Farm Bill debate, economists disagree over the consequences of eliminating the milk price support program. One school of thought holds that, since the price of milk used for manufactured products (butter, nonfat dry milk and cheese) has consistently been above the support level, there should be no adverse effect. The other school holds that, since butter has been continuously purchased by the CCC, a reduction in the price support for butter should lower the price of milk. In addition, those who subscribe to this school of thought believe that without the government as an alternative market, prices will be much more unstable (Gruebele, 1978; USDA, 1985).

Federal Milk Marketing Orders

Authorized by the Agricultural Marketing Agreement Act of 1937, the FMMOP sets minimum Grade A milk prices that processors would pay to dairy farmers or their cooperatives. Currently, there are 37 separate, but coordinated, federal orders covering different geographical areas of the United States. The only major, milk-producing area that is not covered by federal orders is California, which operates its own state program.

Federal orders price about 70% of US milk production on the basis of use. This is called classified pricing. There are typically four classes of milk in each marketing order.

- ◆ Class I — milk used directly for fluid consumption as whole, low-fat or skim milk.
- ◆ Class II — milk used as fluid cream or used in soft dairy products, such as cottage cheese and frozen desserts.
- ◆ Class III — milk manufactured into cheese and butter.
- ◆ Class IIIA — milk manufactured into nonfat dry milk.

Processors are charged higher prices for Class I milk than for Class II, III or IIIA milk. Milk producers in a marketing order area are paid a blend or average price based on the amount of milk used in the order for Class I, II, III, and IIIA. Milk utilization refers to the percentage of milk used by a marketing order in each of the four milk classes. For example, the 1994 Class I utilization for all milk produced under milk marketing orders in the United States was about 40%. Because processors paid more for Class I milk, orders with higher Class I utilization rates paid producers a higher blend price.

The difference between the Class I price and the Class III price is referred to as the Class I differential and is made up of two components:

- ◆ The Grade A differential, which was originally established to encourage farmers to upgrade their facilities to produce Grade A milk, is the first component. To accomplish this objective, the Grade A differential was set at a level that would cover the added cost of producing Grade A milk as opposed to producing Grade B milk. Only Grade A milk can be sold for fluid consumption. Currently, the differential is about $1.10/cwt. Only manufactured products, butter, nonfat dry milk and cheese, are made from Grade B milk. However, technological change, differences in farm size and increased sanitation requirements for Grade B milk now raise questions of whether it still costs more to produce Grade A milk.

♦ The transportation differential, based on the distance from Eau Claire, Wisconsin, is the second component of the Class I differential. Prior to the 1985 Farm Bill, the transportation differential was $0.15/cwt/100 mi. The 1985 Farm Bill increased the transportation differential on a selective-market basis. Controversy arose from the selectiveness of the increase, a simultaneously-mandated reduction in the milk price support level and the installment of a temporary production control program referred to as the DTP. The increased transportation differential averages about $0.23/cwt/100 mi but is not uniform over all federal order markets.

Effective March, 1995, the difference between the Class II and III prices, the Class II differential, was set at $0.30/cwt. Previously, the Class II differential was based on a formula that averaged about $0.10/cwt.

The Class IIIA price is a product formula price that is not directly tied to the Class III price. It is determined primarily by the market price for nonfat dry milk (Outlaw, 1994). In a controversial move in 1993, Class IIIA was established in reaction to California's refusal to raise its price for milk used for butter and nonfat dry milk. The lower Class IIIA price is a factor that encourages processors to utilize nonfat dry milk in making cheese — particularly low-fat cheese.

Structural Changes in the Milk Producer Sector

Over the past four decades, the number of US dairy farms has decreased dramatically, and their average size has increased sharply. While distinct regional differences in structure have been apparent, a trend toward fewer but larger farms has been evident throughout the country (Outlaw, Schwart, Jacobson and Knutson, forthcoming).

There is a great deal of controversy over which definition to use when determining a total number of dairy farms. Three methods used currently and the deficiencies of these methods are:

♦ *Census Definition.* The 1992 Census of Agriculture reported 155,339 farms with at least one milk cow in the United States. However, using a farm with at least one dairy cow as a definition tends to overstate the number of commercial dairy farms in the United States. As an example, a farmer could have a milk cow on his/her farm and be classified as a dairy farmer without ever selling any milk!

♦ *Standard Industrial Classification (SIC) Definition.* Of these 155,339 farms, only 113,412 or 73% of the total were identified as dairy farms according to the SIC. These were farms primarily engaged in the production of cow's milk because 50% or more of their sales was generated by their dairy operations. When using 50% or more of income earned from dairy sales as a definition of a dairy farm, a problem arises because many dairies are diversified into crop and livestock operations. These diversified farms can have a substantial number of cows and still earn the majority of their income from sales of crops or livestock and not from dairy sales. Unfortunately, farms with crop or livestock income (other than dairy) that account for over 50% of their income, cannot be classified as dairy farms by the SIC definition.

◆ *Sales Definition.* In 1992, the number of dairy farms with annual sales of $10,000 or more in farm products was 128,523. The number of dairy farms with sales of $10,000 or more was 17% lower than those with one cow and was approximately 13% higher than the number with at least 50% of sales from dairy. As was expected — the broader the definition, the larger the number of dairy farms.

It is quite apparent that, depending on the definition used, dairy farm numbers vary. However, regardless of the definition used, each indicates that the number of dairy farms in the United States has decreased. Using the census definition of a dairy farm, Table 6.1 indicates that from 1954 to 1992 the number of farms decreased by 95%, from just over 2.9 million to 155,339. The number of dairy farms and the number of dairy cows decreased over the past nine census years, but the average number of cows per farm increased.

While the number of US dairy farms with small herds decreased, the number of farms with large herds increased. In 1964 about 96% of US dairy farms had less than 50 cows (Table 6.2). By 1992, about 60% of the farms had less than 50 cows, but over one-half of the dairy cows were located on farms having over 100 cows (Table 6.3). This meant that, in terms of farm numbers, the US dairy industry was dominated by relatively small herds of less than 50 cows. However, the majority of production took place on farms having more than 100 cows.

The changes in the national dairy herd were significant, but aggregate numbers masked the regional shifts in milk production. From 1980 to 1994, the seven regions in Figure 6.1 accounted for about 90% of US milk production. These designated regions encompassed the major, US milk production areas and were relatively homogeneous with regard to structural trends and production conditions. The upper numbers in Figure 6.1 indicate the percentage of US milk produced in that region in 1980. The lower number represents the percentage of US milk produced in that region in 1994.

TABLE 6.1 Number of US Dairy Farms, Cows on Farms and Cows per Farm, 1954 to 1992, Selected Years

Year	Farms	Cows	Cows Per Farm
1954	2,935,842	20,182,803	7
1959	1,792,393	16,522,026	9
1964	1,133,912	14,622,604	13
1969	568,237	11,174,036	20
1974	403,754	10,654,516	26
1978	312,095	10,221,692	33
1982	277,762	10,849,890	39
1987	202,068	10,849,890	54
1992	155,339	9,491,818	61

Source: US Department of Commerce. Various issues, 1964–92.

TABLE 6.2 Percent of US Dairy Farms With Milk Cows by Size Category, 1964 to 1992, Selected Years

Size of Herd	1964	1969	1974	1978	1982	1987	1992
				Percent			
1 – 19	77.2	64.1	55.5	50.3	41.8	32.5	28.1
20 – 49	18.7	27.4	29.4	30.4	31.9	33.5	31.8
50 – 99	3.3	6.7	11.5	14.4	19.2	23.9	26.9
100 or more	.8	1.8	3.6	4.9	4.1	10.1	13.1

Source: US Department of Commerce. Various issues, 1964–92.

TABLE 6.3   Percent of Milk Cows in US Herds of Different Size, 1964 to 1992

	1964	1969	1974	1978	1982	1987	1992
Size of Herd				Percent			
1 – 19	28.7	17.6	10.1	7.1	5.0	3.4	2.5
20 – 49	43.6	43.2	35.6	31.9	27.2	22.9	17.9
50 – 99	16.4	22.0	27.9	30.1	32.0	31.5	29.0
100 or more	11.3	17.2	26.4	30.9	35.8	42.2	50.5

Source: US Department of Commerce. Various issues, 1964–92.

Production shifted in the direction of those regions having the largest, average herd size, for example, California and Arizona. The shift of US milk production to California is particularly meaningful because it has the lowest producer blend price in the United States (Table 6.4). The underlying reasons for this regional shift in milk production patterns are evident from a combination of USDA cost of production study results and representative farm analyses completed at Texas A&M University (Richardson, Outlaw, Gray, Zimmel, Miller, Smith, Knutson and Schwart, 1995).

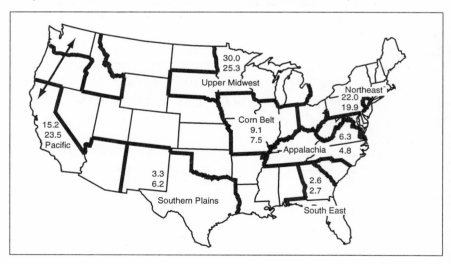

FIGURE 6.1   Percent* of US Milk Produced in Seven Major Regions, 1980 and 1994
*In each region, the upper number is the percent of US milk produced in 1980, and the lower number is the percent of US milk produced in 1994.
Source: USDA, NASS, various years.

TABLE 6.4 Milk Production Values, Costs and Returns per Hundredweight by Regions, 1992

Item	United States	Appalachian	Corn Belt	Northeast	Pacific	Southeast	Southern Plains	Upper Midwest
Output per cow (pounds)[a]	15,419	13,391	14,373	15,498	18,550	14,566	14,867	14,568
Cost & returns per cwt.								
Milk price	$13.15	$14.13	$13.21	$13.78	$11.85	$15.42	$13.70	$13.11
Culled cow value	1.27	1.23	1.53	1.16	.93	.99	1.20	1.56
Other income[b]	.16	.20	.17	.13	.15	.07	.08	.19
Total value	14.58	15.56	14.91	15.07	12.93	16.48	14.98	14.86
Total cash expenses	$10.14	$10.68	$10.37	$10.24	$10.24	$11.65	$11.33	$9.51
Cash receipts less cash expenses	4.44	4.88	4.54	4.83	2.69	4.83	3.65	5.35
Net cash returns[c]	3.15	3.43	2.93	3.29	2.03	3.88	2.52	3.90
Total economic cost	13.94	15.03	16.46	14.97	11.41	13.41	13.31	14.05
Capital replacement	1.29	1.45	1.61	1.54	.66	.95	1.13	1.45
Returns to management & risk	$0.64	$0.53	$-1.55	$0.10	$1.52	$3.07	$1.67	$0.81

[a]Regional average calculated from individual state data.
[b]Includes the dairy enterprise share of receipts from cooperative patronage dividends, assessment refunds, renting or leasing of dairy animals, manure sales and insurance indemnity payments.
[c]Value of production minus cash expenses and capital replacement.
Source: USDA, 1994; USDA, 1992.

Table 6.4 presents the USDA 1992 average cost of production results by region. The data reveals the short- and long-run economic forces affecting the US dairy industry. The returns to management and risk per hundredweight are the highest in the Pacific, Southeast and Southern Plains regions, which are characterized by large farm, drylot dairies. Likewise, these are the regions with the lowest total economic costs per hundredweight. In these three regions, the milk prices range from $11.85/cwt in California to $15.42/cwt in Florida. The low California price reflects the cost- and profit-based policies of its state pricing agency. The high price in Florida reflects federal-order, geographic-based, point-pricing policies that raise the Class I price based on the distance from Eau Claire, Wisconsin. The highest total economic costs are in the Appalachian and Corn Belt regions because they have experienced the greatest decline in the share of US milk production (Figure 6.1).

In 1992, the traditional milk production areas of the Upper Midwest and Northeast had total economic costs in the $14–$15/cwt range. The Northeast received the benefit of a higher milk price because of its distance from the Eau Claire basing point.

Of all regions, the Upper Midwest had the lowest cash expense of $9.51/cwt. This reflects costs to farmers who grow much of their own feed and have plentiful supplies of roughage in the region.[1] It is interesting and important to note that while the Upper Midwest has low-cost roughage, it lacks quality in comparison with the alfalfa hay produced in California. Roughage quality is a major challenge for regions outside the Pacific and Southwest regions.

The low, variable, cash costs in the Upper Midwest suggest that, in comparison with other regions, it is relatively insensitive to short-run declines in milk prices. However, one of the key economic indicators for long-run adjustment in the Upper Midwest is the cost of capital replacement of $1.45/cwt compared with only $0.66/cwt in the Pacific region. The Upper Midwest, like the Northeast Corn Belt and the Appalachian regions, has large amounts of capital tied up in buildings and equipment. Dairy farmers continue dairying up to the point when replacement of physical facilities becomes a necessity, and then many of them drop out of dairying.

However, there is a set of aggressive, next-generation dairy farmers in the Upper Midwest and Northeast. They are developing the means for expanding their operations by utilizing free-stall housing. Representative farm data suggests that these newer and larger dairies are able to compete under the current set of dairy policies (Table 6.5).

The Agricultural and Food Policy Center at Texas A&M University maintains, with the help of producer panels, 22 representative dairy farms located in major production areas across the United States. Farm profitability and net worth growth results are projected annually for these

TABLE 6.5 Selective, Descriptive Income and Net Worth Projections for Representative Dairy Farms Using Bovine Somatotropin Technology, 1992–2000

	Wisconsin		California		New York		Vermont	
Cow numbers	55	190	2,150		110	1000	70	186
Output per cow (pounds)	19,700	21,000	22,700		20,600	20,600	21,500	20,000
Cash receipts ($1,000)	175.8	612.4	6,492.8		319.2	3,129.4	238.6	561.6
Net-cash income ($1,000)	54.3	135.6	1,687.6		-7.3	949.3	39.7	11.7
Percent change in net worth[a]	3.6	33.6	111.7		-78.9	124.0	-21.0	-39.8

[a]Ending real net worth (year 2000) as a percent of beginning net worth (1992).
Source: Richardson et al., 1995.

farms, assuming that current farm policies remain in place (Richardson et al., 1995). Table 6.5 presents selective, descriptive information and projections of financial measures for representative dairy farms using bovine somatotropin technology over the 1992–2000 planning horizon.

The five farms chosen range in size from 55 cows in Wisconsin to 2,150 cows in California. Wisconsin, New York and Vermont each have two representative farms. One is a moderately sized, full-time, dairy farm that is typical of the state. The second is a large farm that, in 1994, was typical of the upper end of farm sizes in the state. For purposes of contrast, the moderately sized 110-cow farm is located in Central New York while the larger farm of 1,000 cows is located in Western New York.

Net cash income reflects whether each farm can pay cash production expenses out of its receipts. Out of this income, the farmer must pay family living expenses, capital replacement and income taxes. As a rule of thumb, a farm must have about $50,000 in net cash income to grow and survive over the long run. Three of the five farms (the New York, 110-cow dairy farm and the two, Vermont dairy farms) do not have this requisite level of income and, therefore, experience declining net worth.

The 55-cow, Wisconsin farm receives an average of $54,300 in net cash income. This income is enough for survival, not enough for growth. These are the types of farms that are discontinuing their operation at a progressively higher rate in Wisconsin. In contrast, the 190-cow, Wisconsin farm does well. It generates $135,600 in net cash income, and its net worth increases by 33.6% from 1992 to 2000.

The 2,150-cow, California farm and the 1,000-cow Western New York farm are typical of modern US dairying. The California farm generates nearly $1.7 million in net cash income from its cows. Its net worth increases by over 100%. The 1,000-cow, New York farm is more profitable on a per-cow basis because of its higher milk price. Its net worth increases by 124% over the 1992–2000 period.

The danger, in an environment of change, is for producers to fall behind in size, technology and efficiency. Generally, the larger farm is on the cutting edge of technology, often having a net cash income that is many times larger than its moderately sized counterparts. For example, the projected net cash income of the 1,000-cow, Western New York farm is 43 times that of the 55-cow, Wisconsin farm; yet the New York farm is only 18 times as large (Richardson, et. al, 1995). These differences are very meaningful in both a trade and supply management context. From a trade perspective and in the absence of price and income supports, the 55-cow, Wisconsin farm must find a way to compete with the 1,000-cow, New York farm and the 2,150-cow, California dairy. In reality, Wisconsin farmers are finding it impossible to compete on the basis of their current, moderately sized farm structure. Unfortunately, and within the past year, Wisconsin's number one position in terms of milk production was surpassed by California. However, Wisconsin will continue to be a major dairy state only if its farmers adjust to producing milk on farms having over 200 cows. These adjustments have occurred within the context of a relatively free market where market prices for milk are well above support levels. One can speculate that, as prices are driven closer to the support price, the trend toward larger farms will accelerate.

Conclusions and Implications for Canada

Until the mid-1980s, the US dairy industry was significantly protected from the forces of change by a high, milk price-support level. During this period, the US dairy industry did change. However, aside from the low prices that characterize the California industry, competitive pressure did not exist to push farmers to a more efficient scale of operation nor did it cause large numbers of inefficient, poorly-managed farms to exit. When these competitive pressures developed, rapid structural change occurred.

The Canadian dairy industry experienced an even higher level of protection under its quota program. Quota values encouraged farmers having a strong equity base to expand and provide a means by which some small farmers could recover equity by exiting the industry (Kaiser, 1986). However, growth is rationed by high quota values to those producers who have access to capital. Likewise, high prices and quota values provide a means by which inefficient producers can stay in business.

Is it possible to forestall the market adjustment process through supply management? In the short run, the answer is yes; in the long run, the answer is no. Without getting into issues of the impacts of production controls, if the US dairy industry continues to adjust its scale of operation in the presence of economies of size while the Canadian industry restrains its size, Canada will continue to regress relative to the United States. This is the case even though other forms of technological progress are pursued in Canada, perhaps at an accelerated pace, such as the employment of embryo transfer technology to produce genetically superior bulls or heifers for export. Each year, the industry of the restraining country falls further behind the competition, either because of limits on the transfer of rights to produce or because of increases in the value of those rights. Each year there are greater incentives for consumers to cross the border for milk products. In the process, these consumers bring back food that otherwise may not have been imported. These are the forces that lead to freer trade and that undermine domestic farm programs.

It is also important to keep in mind that production controls are employed for quite different reasons even though the economic and structural effects are the same. A primary European Union motivation for production controls is the reduction of government costs and the level of antagonism among traders in the world market for dairy products even though smaller farm preservation is undoubtedly also a factor. In the United States, the pressure for controls results from a combination of the desire to lower government costs, raise milk prices and preserve a moderately sized, dairy farm structure. The lack of consensus is the result of diversity and regionalism that is developing within the US dairy industry. Both Republican and Democratic administrations are opposed to production controls because they run counter to freer trade objectives and reduce competitiveness.

This discussion also raises important issues regarding the role of producer-oriented institutions in an economic environment of freer trade. The answer is quite simple. Under conditions of free trade, producer-oriented institutions, such as cooperatives or marketing boards, should do whatever is required to keep farmers competitive in the marketplace. This involves promoting progressiveness in both production and marketing. It begins with support of progressive, basic and applied research at universities. It includes facilitating the transfer of technology to farmers as a means of leveling the playing field. It involves providing milk processors with services (such as forage quality testing, soil testing and cow milker training), product development, minimum quality standards and aggressiveness in exporting as well as in domestic marketing. It is only through such progressive institutions that an industry expects to remain competitive. This is the case regardless of which long-run policy scenario is pursued.

References

Cropp, R. 1995. "The California Milk Price Stabilization Program." *Dairy Markets and Policy: Issues and Options* O–10. Cornell University, Department of Agricultural Economics, Ithaca, NY.

Fallert, R. F., D. P. Blayney, and J. J. Miller. 1990. *Dairy: Background for 1990 Farm Legislation.* Commodity Economics Division, ERS/USDA (March).

GAO/RCED. 1993. *Dairy Industry Potential for and Barriers to Market Development* 94–19. Washington, DC (December).

Gruebele, J. W. 1978. "Effects of Removing the Dairy Price-Support Program." *Illinois Agricultural Economics* 18: 30–38.

Kaiser, H. M. 1986. *Mandatory Supply Management Programs in Canada and Europe.* Cornell Agricultural Economics Staff Paper No. 86–21, Cornell University, Department of Agricultural Economics, Ithaca, NY.

Knutson, R. D., J. B. Penn, and W. T. Boehm. 1995. *Agricultural and Food Policy.* 3rd ed. Englewood Cliffs, NJ: Prentice-Hall, Inc.

Manchester, A. 1983. *The Public Role in the Dairy Economy: Why and How Governments Intervene in the Milk Business.* Boulder, CO: Westview Press.

Outlaw, J. L. 1994. *Analysis of the Consequences of Implementation of Section 102 and Related Issues: Effects on the United States Dairy Industry.* Study prepared for the National Cheese Institute, Washington, DC (1 April).

Outlaw, J. L., R. Schwart, R. Jacobson, and R. D. Knutson. Forthcoming. "The Structure of the US Dairy Sector." *Dairy Markets and Policy* M-4. Cornell University, Department of Agricultural Economics, Ithaca, NY.

Reimund, D. A., T. A. Stucker, and N. L. Brooks. 1987. *Large-Scale Farms in Perspective.* Agriculture Information Bulletin No. 505, ERS/USDA, Washington, DC (February).

Richardson, J. W., J. L. Outlaw, A. W. Gray, P. T. Zimmel, J. W. Miller, E. G. Smith, R. D. Knutson, and R. B. Schwart, Jr. 1995. "Implications of the 1990 Farm Bill and FAPRI January 1995 Baseline on Representative Farms." AFPC Working Paper 95-1, Texas A&M University, Agricultural and Food Policy Center, Department of Agricultural Economics, College Station, TX (February).

USDA. 1992. *Dairy Situation and Outlook* DS–437. ERS/USDA, Washington, DC (October).

_____ . 1994. *Economic Indicators of the Farm Sector: Costs of Production — Major Field Crops & Livestock and Dairy, 1992.* ERS/USDA, Washington, DC (August).

_____ . 1985. *Possible Economic Consequences of Reverting to Permanent Legislation or Eliminating Price and Income Supports.* Agricultural Economic Report No. 526, ERS/USDA, Washington, DC (January).

US Department of Commerce. 1964–1992. *Census of Agriculture.* United States Summary and State Data. Economics and Statistics Administration, Bureau of the Census, Washington, DC.

7

Cost Competitiveness in the Canadian and US Dairy Industries

R. R. Barichello, R. Lambert, T. J. Richards,
R. F. Romain, and B. K. Stennes

Abstract

The international competitiveness of Canadian dairy is now a primary concern with the prospect of gradual elimination of import controls under both the General Agreement on Tariffs and Trade (GATT) and the North American Free Trade Agreement (NAFTA). Little is currently known about the state of either international or interregional competitiveness. Less is known of the factors that contribute to such comparative advantages. This chapter presents the results from three independent studies that deal with the various aspects of the dairy industry in Canada and in the United States. The first study uses comparisons of interspatial and intertemporal productivity differences between Alberta and Wisconsin to show that the Wisconsin dairy sector holds a distinct and growing advantage over Alberta. The second study uses a multilateral cost of production (COP) comparison to show that California is the lowest cost region in North America, followed by Wisconsin, New York, Alberta, Ontario and Québec. The third study investigates COP variations within the Ontario and Québec dairy industries to show that there are no appreciable economies of scale in either dairy. However, the COP for herds of all sizes can be reduced by making them more technically efficient. The first two studies imply that the Canadian dairy sector will undergo a significant rationalization if cost improvements are not made over the period allowed by GATT for tariff reduction. These studies also show that such improvements may accompany larger herd sizes, but this does not appear to be the case for either Ontario or Québec. As a result, the rationalization of the Canadian dairy industry will impact each region differently with production shifting to regions that are able to adapt to the competition.

Introduction

A number of actual and potential changes in trade policy that will affect the Canadian dairy industry provided motivation for the writing of this chapter. The recently completed GATT Uruguay Round Agreement (URA) requires a shift in protection by signatories from nontariff barriers, and in the case of Canada's dairy policy, from import quotas, to tariffs. Even if scheduled URA tariff cuts for milk products will mean little competitive pressure on Canada from imports within the next five years, future GATT rounds may well lead to more significant tariff cuts that will affect farm milk prices in Canada. Another trade policy change that has potential effects on the Canadian dairy industry is NAFTA. Here, the issue is whether NAFTA, rather than URA, has supremacy over tariff cuts for milk and milk products. If so, the tariffs that will be in place following URA tariffication will decrease more quickly and expose the Canadian industry to considerable US price competition and, in general, integrate the North American dairy market.

If this takes place, to what extent will the Canadian industry be competitive with its US counterparts? Will our industry be able to compete with US prices and the potential US milk and milk product flows into Canada? If our industry has difficulties in meeting this competition, to what extent will it contract? Note that it is US competition, not potential New Zealand or European imports, that is the focus of attention. If the main source of competition within the next decade were to arise from URA tariff cuts, then world market competition, that is New Zealand and Europe, would indeed be the focus of our attention. If it is the proposed NAFTA tariff cuts that will govern Canada-US dairy product trade, then we must be concerned with potential US competition.

Knowledge of the factors that affect dairy competitiveness will help guide policymakers and producers in preparing for a future in which dairy products are freely traded. Two related questions face the dairy industry: What are the causes of any differences in costs between Canadian and US dairy? What are some of the causes for differences in costs among producers within each industry?

Not only have answers to these questions eluded a consensus in the past, but methods of arriving at these answers also have remained elusive. Competitiveness comparisons cannot be made between heavily regulated industries on the basis of price. Production cost is often chosen as the metric for competitiveness, but it is difficult to make comparisons between regions that have been isolated by both geography and policy. As economists, we can bring many different methods and perspectives to bear in providing a synthesis of results that will, perhaps, point to a few common conclusions. From this synthesis, a better explanation of the larger picture of Canadian dairy competitiveness will emerge.

Study 1: Productivity Growth Comparison
of Alberta and Wisconsin Dairy (PGA)

Introduction

Ultimately, comparative advantage determines the location of production in a free market. A country has a comparative advantage in producing a good if the relative returns to the good's fixed factors of production are greater than the returns of another country. In a world of policy-distorted trade, competitive advantage, rather than comparative advantage, determines where milk is produced. Competitive advantage is a more political than economic concept; an industry can have a trading advantage because of subsidies, tax breaks, trade protection or other forms of intervention. The prediction of which country's industry will produce and export under free trade requires a measure of advantage that most nearly approximates the traditional notion of comparative, rather than competitive, advantage.

Does Wisconsin or Alberta have a comparative advantage in milk production? Interspatial comparisons of total-factor productivity between Alberta and Wisconsin using farm-level data from 1982 and 1989 directly measure the advantage while comparisons of total-factor productivity growth show the change in advantage over time. Analysis of differences in input use between the two regions indicates how the difference in productivity will affect the industry under free trade and how the difference may be closed.

If North American markets for dairy products and dairy inputs were free from intervention, the access to equivalent technologies and genetic material would lead to similar rates of productivity growth everywhere. In Alberta, however, supply management causes dairy producers to invest significant amounts of financial capital in quota licenses — capital that could otherwise be used to purchase cattle, expand barns, begin an embryo-transfer program or for many other productive uses.[1] Because Alberta producers must divert some capital away from other demands, physical investment and the rate of technological progress have not likely been the same in Alberta and Wisconsin.

Methods and Data

Primal and dual measures of productivity compare performance across space and time. The primal perspective compares the amount of output produced from a standard bundle of inputs, whereas the dual approach compares the relative cost of producing a given amount of output at a constant set of prices. Primal productivity growth is equal to the rate of output growth less the change in an index of total input use. Similarly, the dual rate of productivity growth is found by subtracting the weighted average rate of input price increase and the rate of output growth from the change in total costs. Removal of all measurable cost factors leaves an unexplained residual that is attributed to total-factor productivity growth. Simply replacing each reference to year with region in the index changes the measure from productivity growth to productivity differential.

Farm-level COP survey responses from dairy producers in Alberta and Wisconsin constitute the data set. The Alberta Agriculture/Alberta Milk Producers' Society COP survey data consist of average, annual farm-level observations of 54 producers in 1982 and 58 producers in 1989. Prices and quantities for all inputs are recorded on a dairy enterprise basis.

The Wisconsin data include 1982 and 1989 annual summary reports prepared by the Lakeshore Farm Management Association (various years). The sample consists of 1002 producers in 1982 and 462 producers in 1989. A weighted average of quantities used and produced and prices paid and received for six farm-size groups with less than 30 to more than 125 cattle defines the average Wisconsin farm. As in Alberta, all expenses, other than those directly related to dairy, are allocated on a revenue percentage basis.

Wisconsin and Alberta form a good basis for the comparison of dairy production between the United States and Canada because they are each recognized as relatively low-cost producers and have similar climates, resource endowments, production technology and herd sizes. In Alberta the average herd consisted of 57 head in 1982 and 64 head in 1989, whereas in Wisconsin the sample average size increased from 50 head in 1982 to 65 head in 1989. The comparability of the data, however, is also dependent upon defining farm outputs and inputs in a similar way.

Variable input definitions maintain as much similarity between the data sets as possible. For example, in Alberta the average price is a weighted average of the price paid for all purchased feeds and the imputed market value of all homegrown feeds. For Wisconsin if the total annual feed expenditure is divided by the annual average feed price published in *Wisconsin Dairy Facts* (1991), the equation yields the quantity of feed used. All feed prices are converted to a common currency in each year as are wage rates. In Alberta, the quantity of labor is expressed directly in terms of the number of hours per year, including operator, family nonpaid, family paid and hired labor. An average wage for both Alberta and Wisconsin is imputed by dividing each region's total labor costs by the number of hours per year. The quantity of Wisconsin labor is expressed in terms of man equivalents per year — a man equivalent being defined as 2,080 hours per year. Both Alberta and Wisconsin operator labor values are imputed via a return to equity on net fixed-farm assets. Similar efforts are taken to ensure the comparability of the fixed input prices and quantities reported in the cost surveys.

With respect to physical capital, Ball's (1985) method is used to construct equivalent indices of cost and service flow for each region and year. Further, because they produce for several years, cattle are capital assets and not variable inputs. As such, the total opportunity cost of holding cattle includes an opportunity cost of capital and all maintenance expenses: breeding expenses, vet expenses, cattle transportation and miscellaneous cattle supplies. The number of dairy cattle includes the milking herd and the stock of replacement heifers and dry cows. Adherence to consistent variable definitions between the two regions ensures that any bias resulting from variable measurement error will favor neither region's productivity calculation.

Results and Discussion

With either measurement method, the results show that the rate of change in Wisconsin for total-factor productivity from 1982 to 1989 was double that of the Alberta rate. Furthermore, the results show Wisconsin was more productive in both time periods from the primal perspective, but they also show Alberta was more productive in 1982 using the dual measure. Decomposition of each measure by the change in each input suggests some reasons why growth rates differed.

From a dual perspective, Table 7.1 shows an index of productivity growth for Alberta of 0.9187. This value indicates that Alberta dairy producers produced the same output in 1989 that they produced in 1982 at 92% of the 1982 cost. Therefore, total-factor productivity growth has been responsible for an annual, average reduction in costs of 1.2%. The results for Wisconsin show that Wisconsin dairy producers produced their 1982 milk quantity in 1989 at 83% of the 1982 cost. In annual terms, Wisconsin production costs fell by an average of 2.6% per year over this period.

Disaggregation of the index calculation into changes in cost, output or input prices identifies the source of differences in productivity improvements between Alberta and Wisconsin. Although milk production rose at a faster rate in Alberta than in Wisconsin, higher labor costs outweighed the effects of cost-reducing technological advances. Despite the 50% increase in the price of labor, the rise in Alberta labor share indicates a failure to substitute technology for labor. Embryo transfers, modern breeding programs and optimal, ration-formulation techniques are examples of highly labor-intensive innovations in dairy.

TABLE 7.1   Dual Dairy Productivity Growth Comparison between Alberta and Wisconsin, 1982 and 1989

	Adjusted Cost	Total Cost	Milk Output	Average Price	Cow Price	Labor Price	Feed Price	Capital Price
				1989 Index/1982 Index				
AB[a]	0.9187	1.3718	1.3054	1.3085	1.0066	1.5056	1.1309	1.1870
WI[b]	0.8288	1.1920	1.2930	1.2373	0.9339	0.8028	1.2892	1.0642

[a]Alberta; [b]Wisconsin
Source: Author's calculations.

While the price of cattle inputs fell by 7% in Wisconsin and remained essentially unchanged in Alberta, the share of cost accounted for by cattle fell by 2% of total cost in Wisconsin and over 5% of cost in Alberta. Whereas producers in Alberta substituted labor for cattle, Wisconsin producers moved toward capital-intensive methods. Although the price of capital rose in both regions, the Wisconsin price rose by 12.5% less than the rise in Alberta. In fact, the relative price decrease of capital in Wisconsin caused the cost share of capital to rise by 3%. While this intertemporal analysis clearly shows productivity growth to be stronger in Wisconsin, it makes no comment on the relative level of productivity between Alberta and Wisconsin.

An interspatial, productivity comparison determines which region had a productivity advantage in 1982 and 1989. Wisconsin is defined as the base region in both years, so the interspatial productivity index gives the multiple of Wisconsin costs that were incurred in Alberta to produce the Wisconsin level of output facing Wisconsin input prices. For example, Table 7.2 shows that Alberta dairy producers produced the Wisconsin output for 1.05% below the cost of Wisconsin dairy producers in 1982, given Wisconsin input prices. By 1989, however, Alberta producers faced an 11.39% higher cost to produce the Wisconsin output level at Wisconsin input prices (column 1). Although Alberta producers were more productive in 1982, the advantage was lost by 1989.

TABLE 7.2   Dual Interspatial Productivity Comparison between Alberta and
Wisconsin, 1982 and 1989

	Adjusted Cost	Total Cost	Milk Output	Average Price	Cow Price	Labor Price	Feed Price	Capital Price
			Alberta Index/Wisconsin Index					
1982	0.9895	0.9328	0.8504	1.5566	2.0427	0.7189	1.1356	1.2100
1989	1.1139	1.2338	0.8584	1.6651	2.2018	1.3482	0.9962	1.3497

Source: Author's calculations.

By breaking this difference down by input, the average cost is shown
to be only 7.2% greater in Alberta than in Wisconsin in 1982, but Alberta
farms produced only 85% of the Wisconsin output and faced a
share-weighted, input-price index that was 55.5% higher. Although labor
prices were significantly lower in Alberta (28%), the total opportunity cost
of holding cattle was more than double that of Wisconsin's. In 1989, the
same general input price pattern remained, only Alberta no longer enjoyed
the labor-cost advantage held in 1982. In fact, both labor and capital costs
were 35% higher in Alberta by 1989. A failure to substitute lower-cost inputs
for either labor or capital contributed to the emergent productivity gap.

Evidence from the primal perspective, which measures the proportion
of Wisconsin output that Alberta producers could have made with the
Wisconsin input bundle, supports this hypothesis. For 1982, results show
Alberta producers were less productive, achieving only 98.8% of the
Wisconsin output with the Wisconsin input bundle. In other words,
producers in Alberta made only 85% of the Wisconsin output but used
86% of the Wisconsin input bundle to do so. Almost equal use of cattle
inputs and a 29% greater use of labor were responsible for this gap. In
fact, the productivity gap grew from 1.2% in 1982 to 13.4% in 1989. Alberta
producers continued to use more labor and the same amount of cattle as
Wisconsin producers, but they used 5.5% more feed to produce far less
milk. The lower feed-efficiency level in Alberta suggests that the
productivity gap was due to genetic sources — an inherent inability of
the cattle to produce as much milk as was produced in Wisconsin.

Conclusions

The analysis of this section provides a case-study comparison of productivity in the Alberta and Wisconsin dairy industries. Intertemporal and interspatial measures of productivity show that Wisconsin fluid milk producers possessed a comparative advantage over similar Alberta producers — an advantage that increased from 1982 to 1989. Specifically, the rate of total-factor productivity growth on the sample of Wisconsin dairy farms is shown to be double that of the average Alberta producer from 1982 to 1989. Furthermore, Wisconsin producers are shown to have been able to obtain more milk from a standard bundle of inputs relative to their Alberta counterparts in both 1982 and 1989. At least from a primal perspective, therefore, Wisconsin producers can be said to be more efficient, both in an allocative and technical sense.

Several implications can be drawn from these results. First, it is clear that the relative, competitive position of the Alberta dairy industry, assuming constant transportation costs, eroded significantly over the sample period. Second, this study shows dairy production in Alberta to have been relatively more labor-intensive than production in Wisconsin. Given Alberta producers' failure to substitute capital and cattle for labor when the price of labor rose, the relative decline in competitiveness is not surprising. Third, Alberta dairy production is relatively more cattle-intensive. Intensive use of smaller herds could explain the more rapid increase in average milk yields in Alberta over this period, but the falling relative productivity values suggest that increased yields were not necessarily efficient.

Ultimately, freer trade will result in significant changes to the structure of Alberta dairy. It is likely that Alberta will become more like its competitors in order to match their productivity gains. Therefore, Alberta dairy will become more capital intensive while relying less on labor and cattle specific inputs. Unfortunately, a move to greater capital intensity means the loss of family farms — the typical policy trade-off between maintaining a traditional family farm structure and developing an independent, competitive production sector.

Because this analysis concerns only a bilateral comparison between Alberta and Wisconsin, the generality of the conclusions depend upon each region being representative of its respective national dairy industry. To obtain a more accurate description of the structure of competitiveness in the continent as a whole, Study #2 presents a comparison of dairy competitiveness that includes data from each of the major producing regions in North America.

Study 2: Cost Competitiveness Analysis of
the Canadian Dairy Industry (CCA)

Introduction

Although the notion of competitiveness is widely used in public discussions, the empirical examination of competitiveness, in the case of specific industries, is less commonly undertaken. Several principles should guide such work: 1) cost comparisons should include all segments of the marketing channel; 2) the data must be at a micro-level; 3) cost differences should reflect differences in product quality. Even if these criteria are met, survey cost data from price-regulated industries may depend upon the level of price support, and as so should be treated with a healthy skepticism.

This study investigates only farm-level costs. It does not consider the revenue side nor does it analyze processing costs. Although such comprehensive analysis is beyond the scope of this chapter, an accurate measurement of farm-level costs will determine the competitiveness of the most region-specific input in the marketing chain — raw milk.

Data and Methods

This study examines the cost of producing milk in selected regions in Canada and the United States: Ontario, Québec, Alberta, New York, Wisconsin and California. Provincial dairy cost surveys are used as the principal data source, and production costs are estimated across provinces as consistently as these data permit. Similar procedures are followed using the US data to permit comparisons among these six regions.

Before reviewing the cost data, it is important to recognize the difficulties of the exercise, including both those within and outside the control of the cost analysis. First, measurement of long-run economic costs requires inputs to be valued at their supply price or opportunity cost, the minimum cost to attract those resources to the milk-producing activity.

Second, an annual market rental rate for using the capital asset is applied when possible. Ideally, the capital type and vintage should match the selection and quantities of variable inputs used in calculating variable costs. Finally, economic rents are not costs and should be excluded from land rental values or labor returns.

Even if we follow these basic principles of correctly using opportunity costs, there are six primary reasons why the farm-level data will lead to overstated cost estimates.[2] First, these cost data are often collected for the purpose of determining milk prices, so respondents have an incentive to bias their reported costs upward. Second, small farms often record costs for tax purposes, rather than for management decision support, again causing an upward bias. Third, farms involved in primary production often have a high percentage of value added in their sales, purchasing a relatively small percentage of raw inputs. Fourth, input price varies widely across farms, particularly the supply price of operator labor. Fifth, in this study, new capital (1988 vintage) is combined with average levels of labor and variable costs, again overstating true costs. Finally, the study uses average costs instead of the desired long-run marginal costs.

In this analysis much emphasis will be placed on keeping costs, as far as the data allows, on a consistent basis across the different Canadian and US production areas studied. All costs will be expressed in Canadian dollars assuming an exchange rate of US $0.85 per Canadian dollar. Also all costs will be expressed on a hectoliter (hl) basis with 3.6 kg butterfat/hl.

The dairy enterprise is defined to include milk production (dairy cows) plus replacement animals. Feed produced on-farm will be costed on a production basis, where data exists. Except for hired labor, all other variable costs are taken directly from the survey data.

The hired labor expenditure values reported on the dairy surveys were not used in this analysis. This was largely due to the inconsistency of labor expenditure data between the different surveys. In all cases, except California, a total labor hours per cow rate for paid and unpaid labor was used to cost hired labor.[3]

Although the cash portion of fixed costs is taken directly from the surveys, noncash fixed costs present more of a problem. Noncash fixed costs consist of: a rental cost for owned land, unpaid operator and family labor and the depreciation and opportunity costs associated with the farm's capital stock (excluding production quotas). The cost of land uses annual rental values per acre because the rental value is a direct measure of the annual user cost of land. Unpaid labor is split into operator and family portions. It is assumed that the farms have available 2,400 hours of operator labor and 1,040 hours of family labor (Alberta Agriculture, 1989). Details of the construction of capital cost indices are much more complex and are available through the authors.

Data on each of these variables were obtained from sources gathered by farm management societies or by public agencies. Ontario data were taken from the Ontario Dairy Farm Accounting Project (ODFAP, various years), which included 137 farms from Southern Ontario for 1988. The main source of Québec data is the Groupe de Recherche en Economie et Politique Agricoles Departement d' Economie Rurale (GREPA, 1989) survey. The GREPA survey for 1988 consisted of 164 farms with an average herd size of 35.6, and the average production per cow was 59.61 hl/year. The Alberta survey (Alberta Agriculture, 1989) consisted of 61 dairy producers with an average yield of 67.52 hl/cow, substantially higher than the yields in the ODFAP and GREPA (61.3) surveys.

The main source of data for New York is the Dairy Farm Management Business Summary (Smith, Knoblauch and Putman, 1989), which consisted of 406 farms in 1988. The Wisconsin data were obtained from a survey of 465 farms in eastern Wisconsin (Lakeshore Farm Management Association, 1989) and from a smaller survey consisting of 30 farms prepared by professional appraisers (Schraufnagel and Clark, 1990). Available California survey data for 1988 were averages for five large milk-producing areas in California, and they were used in conjunction with consensus data from the University of California at Davis. Milk yields ranged from 77.82 hl/yr in Del Norte-Humboldt to 87.74 hl/yr in the North Bay. These surveys allow a comparison of COP across regions to be made with some confidence of accuracy.

Results and Discussion

Each analysis of the cost data from the six regions studied — Ontario, Québec, Alberta, New York, Wisconsin and California — uses the same costing procedures insofar as the data would allow. This means that although the caveats noted above still apply, they apply more to the disaggregated cost components, and the aggregate cost differences between regions are very instructive. Costs are still likely to be overstated in all regions; a problem which is most sensitive in cases of small farm sizes — Ontario and Québec averages — and least prevalent in the California data. Table 7.3 shows the results of the cost comparison broken down by component.

TABLE 7.3 Breakdown of the Components of the Average Regional Milk COP*

Region	Herd Size	Feed Cost	Variable Cost	Fixed Cost	Unpaid Labor	Other Revenue	Milk COP
			Canadian Dollars per Hectoliter				
Ontario	43	14.05	23.37	16.59	9.17	5.90	43.36
Québec	36	15.07	24.50	14.57	9.22	4.09	44.21
Alberta	65	13.94	24.85	9.61	5.68	4.26	35.88
New York	69	15.32	23.30	10.08	3.87	5.06	32.18
Wisconsin	50	14.45	21.12	13.52	5.43	8.29	31.77
California	455	15.38	26.13	1.89	0.55	0.00	28.24

*In Canadian Dollars: Cdn. $1 = US $0.85
Source: Authors' estimates based on survey data.

Table 7.3 shows that costs were highest in Ontario and Québec and lowest in California. In fact, these data suggest that the provincial average costs for Ontario and Québec were again about one-half as high as California's. However, average costs in Alberta were just over 10% higher than those in New York and Wisconsin. When cost subcomponents were examined, we noted that variable costs were quite similar across all regions (within the range of $20 to $26 per hectoliter). Variable costs were the highest in California where almost all inputs (including feed and replacements) were purchased, whereas fixed costs were lowest in California by a very wide margin and mostly because of scale economies. Farms in New York and Wisconsin showed less of an advantage in capital costs, but the large farms in New York still enjoyed a savings of almost $10/hl over the Ontario average. Unpaid labor costs showed the same pattern as fixed costs, but these were clearly dependent upon scale because similarly sized farms in all regions had similar, unpaid labor costs per hectoliter.

Despite the extra variable costs that appeared to arise with large-sized operations, the potential savings in unpaid labor and fixed costs were large enough to give cost advantages to larger farms. This means we can expect to see continued growth in the average size of operations as long as there is an adequate supply of managers possessing the necessary skills to run the larger farms. Size appears to place a premium on such managerial skill as yields per cow and milk revenues. Both increase with herd size — neither of which is necessarily a function of scale.

In summary, there are two major reasons behind regional cost differences: herd size and yield per cow. In fact, further analysis of each region shows that yield per cow and herd size explain more than 75% of the difference, with yield per cow usually being more important. In addition to the yield and herd size variables, there are no significant, regional differences in costs when using either a dummy variable by province/state or by country. This result, which is quite striking, is consistent with the variance analysis reported in the paragraph above.

Conclusions

Given our data sources and procedures, it would first appear that the average Canadian dairy farm has higher costs for producing raw milk than its counterparts in the United States. Unless the Canadian milk processing sector is more efficient than that in the United States, the milk industry in Canada would cost more and be less competitive. As a result, reduced trade barriers would lead to less domestic production and a decline in market share for the Canadian industry.

However, closer examination reveals important caveats. If the exchange rate is updated to reflect 1994 conditions, Alberta will have lower costs than New York and Wisconsin. Second, the average data conceals important differences across farms — at current exchange rates the two largest Ontario-size groups (48 cows and larger) have lower costs than New York and Wisconsin. Third, farms with fewer than 45–50 cows have cost levels that are significantly higher than their US rivals. This segment will have to sustain large adjustments in order for it to remain within the industry.

These results identify other needs for future research. The importance of both yield and herd size, in the determination of dairy competitiveness, suggests that a more detailed analysis of the relationship between these two variables and production costs is warranted. Yield and herd size are important not only for their apparent linkage to the current level of competitiveness but also because they represent two key areas of long-term investment for dairy producers. If dairy support has caused producers to lean toward myopic genetic decisions, as some believe, then both of these variables will explain much of the lingering gap between the efficiency of US and Canadian dairy. The final section of this chapter explores the relationship between herd size, yield, efficiency and production cost.

Study 3: Economies of Size and Efficiency
Analysis of Québec and Ontario Dairy (ESEA)

Introduction

Over the last few years, there have been several studies that have analyzed the cost structures of dairy farms in order to determine factors that would help make Canadian farms more competitive. Several studies have indicated that dairy farms in Canada, and in Québec in particular, are smaller and have lower production levels per cow compared to farms in the northeastern United States (Gouin, Lebeau, Hairy and Perraud, 1990; National Dairy Policy Task Force, 1991) — their most likely competitors in the event of freer trade between Canada and the United States. Subsequent to these observations, several studies have concluded that economies of size exist in the industry.

Other studies also note that the average yield per cow increases when the size of the herd increases. This suggests that lower COP results not merely from an increase in farm size but also from an increase in yields per cow or a combination of these two factors. Moreover, most studies indicate that the increased productivity from labor and capital and not from reduced cash costs are primary causes of these economies. In each of these studies, however, there is one common limitation: A relationship between average COP and the size of farms is established without taking into account the fact that the distribution of yields may vary with farm size. Moreover, the concepts of technical and allocative efficiency are not often taken into consideration. Indeed, COP can decrease with improved management of inputs (technical efficiency) and/or a better choice of inputs in the production process according to their relative costs (allocative efficiency).

The objectives of this study are: 1) to analyze COP according to farm size and yield per cow in order to assess whether the economies are caused by larger farm size or higher yield per cow, 2) to determine the level of technical efficiency of dairy farms and analyze the relationship between COP and the level of technical efficiency and 3) to determine the socioeconomic variables that characterize technically efficient farms and farms that produce at the lowest cost.

Data and Methods

First, farms are categorized into several subgroups according to their size and according to several levels of yield per cow. Once the farms have been thus grouped, several cost categories and labor productivity are analyzed for each subgroup. Comparisons are made according to the yield per cow within each farm-size subgroup. This analysis makes it possible to determine how the yield per cow affects average COP and labor productivity. A comparison is also made between the costs and labor productivity for each size subgroup and is further stratified by yield per cow.

Second, a measure of technical efficiency is applied to each of the sample farms. Technical efficiency is generally defined as the maximum level of production that can be obtained from any given combination of factors of production. To determine which farms are the most technically efficient, a production function is estimated using yield per cow as the dependent variable and four major factors of production as independent variables: forage per cow, grains and protein supplements per cow, number of labor hours per cow and the value per cow of equipment required for milk production. Following this initial estimate of the production function, only those farms with yields per cow above the predicted yields are retained for a second estimation This process is repeated several times until the estimated function describes the production relationship for only the most efficient farms. With this equation, it is possible to calculate the potential yield per cow for each farm in the sample. For each farm, the technical efficiency index is defined by the ratio of actual yield per cow to the potential yield per cow. The farms are then divided into three categories of efficiency — less than 75%, between 75 and 90% and over 90% efficient — and their costs of production are compared.

Third, covariance analysis is used to investigate the relationship between the level of technical efficiency, COP and several socioeconomic variables. The socioeconomic variables include: the producer's level of education, the quality of the herd and participation in milk-recording programs.

The data used in this study come from Québec and Ontario.[4] In Québec the data are from 472 Syndicats de gestion members who participated in the AGRITEL data bank in 1990. The sample consists of farms with Holstein herds that received at least 85% of their gross incomes from milk sales. The Ontario data come from the ODFAP data bank. This data bank is used by the Canadian Dairy Commission (CDC) to represent Ontario in their calculations that determine COP of milk in Canada. In 1990, the data bank represented 140 dairy farms — of this sample the 128 Holstein herds are used in this study.

Results and Discussion

To determine the relationships between farm size and yield with COP, the Québec dairy farms are divided into four size subgroups and three yield subgroups while the Ontario farms are divided into three size subgroups and three yield subgroups. Cash costs, COP, labor productivity, interest and depreciation are then compared between each subgroup.

The results of this analysis agree with those reported in other studies. First, in both Ontario and Québec, and for herds of fewer than 50 cows, the average yield per cow does not increase significantly with the size of the farm. Second, in Québec, there is no statistically significant difference in terms of average cash costs before wages and interest for the farm-size groups above 30 cows, whereas these costs are higher for herd size of less than 30 cows than for all other farm-size subgroups. In Ontario, the only significant difference is between small farms (fewer than 35 cows) and large farms (more than 50 cows). Third, all categories of sizes in Québec have similar COP before the return on equity-financed capital and producer and family labor, with the exception of small farms with higher costs than the farms that have between 31 and 40 cows. In Ontario, only small farms indicate significantly higher costs than those of the other sized groupings. Fourth, all sizes of farms in Québec and Ontario record similar interest payments per hectoliter, with the exception of one case in Québec where interest payments for large farms are higher than those for small farms. Fifth, the cost of depreciation decreases with farm size up to 40 cows, rises sharply for farms with 41 to 50 cows and then begins to decrease again with larger-sized farms, suggesting a change in the production technology when the farm operation reaches 40 cows. Finally, in both Québec and Ontario, there is an evident increase in labor productivity as farm size increases.

Two important results emerge when comparing costs by yield category. First, in Québec for farms with fewer than 50 cows, both cash costs and COP drop as the yield per cow increases. Economies of size are only marginal for these farms and nonexistent for larger farms. In Ontario, there do not seem to be economies of yield, and the economies of size are not significant, except for smaller farms (30 cows and fewer) with the highest yields per cow. Second, a sharp increase in labor productivity is noted with both increasing yield per cow and farm size. With this analysis, however, it is not possible to specify the relative effect of each of these factors.

The next set of comparisons results from applying the theory and empirical method of production efficiency to the sample dairy herds. Transcendental production functions for both Ontario and Québec are estimated in log form with ordinary least squares. An index of efficiency for each farm is constructed from the efficient frontier according to the procedure outlined above.

These efficiency indices are then compared to partial productivity and cost values. Labor productivity, as well as the different costs analyzed, generally vary inversely with the level of technical efficiency. These results are much more conclusive in Québec than in Ontario. In a comparison between the most efficient and least efficient farms, COP are 16% lower in Québec and 13% lower in Ontario. Furthermore, when the least efficient farms are compared to the most efficient farms, labor productivity increases by 30% in Québec and 40% in Ontario.

Within each size subgroup, the results are similar to the above-mentioned results: A decrease in costs and an increase in labor productivity are associated with higher levels of technical efficiency but do not indicate significant economies of size. Further, labor productivity generally increases with size for all levels of technical efficiency. This result constitutes the principal finding of this section — cash costs and COP are not affected by the size of the farm when the level of efficiency is taken into account.

Technical efficiency, labor productivity and some cost variables are also compared across three cash-cost levels — high, medium and low. This analysis shows that the level of technical efficiency increases when moving from the high- to the low-cost group, but potential production decreases. This result holds true for both Québec and Ontario. Moreover, average yield per cow is lower for farms with high-cash costs but is similar for the two other groups. These results indicate that farms with higher costs could lower their cash costs and also increase yield per cow by using inputs more efficiently. However, farms have lower COP more as a result of their efficient use of production factors than high yield per cow.

Several factors may account for a farm's technical efficiency level. In this study, technical efficiency is determined by the following socioeconomic variables: herd size, amount of energy from forage to reflect quality of hay, expenses per cow for vet care, participation in a milk-recording program, the year the farm became a member of a management club, level of education, debt per hectoliter, ratio of amount of forage to amount of concentrate and the ratio of labor to capital.

The results of this analysis indicate that in Québec large operations have a slightly higher level of technical efficiency than smaller farms. In Ontario the effect of herd size is not statistically significant. Level of education, participation in a milk-recording program, expenses per cow for veterinary care and artificial insemination, quality of forage and year in which the manager joined a management club are all variables that characterize efficient farms in Québec. Variables that reflect management ability are also statistically significant in Ontario.

An analysis of the characteristics of low COP farms shows that higher levels of technical efficiency help to lower cash costs and COP. In addition, there is a slight correlation between average costs and size of operation. Several other variables also significantly influence production cost. Results from comparing levels of depreciation suggest that a high level of capitalization lowers cash costs in Québec, but the result is not significant in Ontario. In Québec farms with higher forage/concentrate ratios have significantly lower costs. In Ontario a lower ratio of labor to capital reduces cash costs. Finally, the results show that farms with higher yields per cow also have higher average costs. This result reflects increasing average costs as suggested by economic theory. On average, however, farms with higher yields have higher technical efficiency that lowers COP.

Conclusions

This study determines the relationship between COP, herd size, milk yield, technical efficiency and several socioeconomic variables in the dairy operation. Comparisons of cash costs before wages and interest or COP before returns to family labor and equity between farm-size groups reveal no significant economies of size in dairy. However, there do appear to be economies of yield. This result is only apparent in Québec for farms with less than 50 cows. No economies of yield are shown for farms with more than 50 cows in Québec or for any of the size subgroups in Ontario. These results reflect a great variability in COP among operations that could be attributed to different levels of technical efficiency.

An analysis of costs by level of technical efficiency shows that costs decrease with increases in efficiency for all size categories. Taking into account the level of technical efficiency, this study shows that herd size does not greatly affect cash costs or COP. In fact, only farms with less than 30 cows show significantly higher costs than other size categories. Further, for herd sizes of more than 30 cows, an increase in labor productivity according to size is much less apparent when the level of technical efficiency is taken into account.

Dairy farms in Québec and Ontario have considerable potential to improve their technical efficiency through better management because less efficient farms can increase their yields while using the same amount of inputs. Correlations of technical efficiency with various herd and operator characteristics show that such improvements can be achieved through the development of herds with good genetic potential, promotion of milk-recording programs, participation in management clubs and the use of high-quality forage. The resulting increase in efficiency leads to lower farm COP and a greater ability to compete with US milk producers.

Implications of the Three Perspectives

In reviewing the three studies discussed in this chapter, as much can be learned from the areas of agreement between the studies as can be learned from diagnosing the discrepancies. The results presented in this chapter not only foreshadow the nature and extent of future rationalization in Canadian dairy, but they also suggest ways in which the need for adjustment can be reduced over the transition period allowed under GATT. There are four areas that impact dairy competitiveness: herd size, milk yield, labor productivity and managerial quality/efficiency.

Comparisons of both the productivity growth rates and COP between regions in the United States and Canada suggest that the presence of free trade will result in a significant rationalization of the Canadian dairy industry. In fact, productivity growth in Alberta was half of that in Wisconsin during the 1980s. Perhaps more discouraging is the result that Alberta dairy was equally as productive in 1982 but lost ground by 1989. The CCA not only supports this international comparison but also provides an indication of the interregional reallocation that will accompany freer trade in dairy products. Barichello's and Stennes' (1994) analysis shows that California is the low-cost producer in North America followed by Wisconsin, New York, Alberta, Ontario and Québec.[5] The fact that the cost differential between California and Alberta (Cdn. $7.61 at exchange rate US/Cdn. $0.85) is almost the same size as the gap between Alberta and Québec ($7.30 Cdn.) suggests that adjustment will be both southward and westward. The ESEA results are silent on interregional comparisons but do show that herd sizes in Québec are far below those in other areas, lending support to the predicted geographical realignment of production. However, each of these results is only a snapshot taken at one point in time.

As the PGA shows, dairy is a very dynamic industry in which costs and conditions are continually changing. The ESEA suggests ways in which dairy producers in Ontario and Québec can use this dynamism to reduce the impact of freer trade on their industry. Significant cost reductions can best be achieved through improvements in production efficiency where herds of all sizes produce closer to their potential milk output levels. However, ESEA, contrary to PGA and CCA, lends only weak support to the proposition that larger herds in Ontario or Québec will allow the industry to compete with either the United States or Western dairy industries.[6] If it is the case that dairy production in other regions of Canada can take advantage of presently unexploited scale economies, then as the Canadian dairy sector becomes more like the US industry in order to compete, westward adjustment will be accelerated.

Economies of herd size or economies of scale are a contentious issue throughout dairy literature. Within this chapter, CCA clearly shows a high degree of correlation between regions of low-cost milk production and herd size. On a herd-by-herd basis, however, ESEA shows little support for economies of herd size when other sources of high production costs are accounted for — namely, low-labor productivity or technical inefficiencies. The PGA concurs because with cattle defined as a production input, a relative overinvestment in cattle is responsible for lagging productivity growth. The different perspectives show that costs will rise with any input when its level results from an inefficient allocation of inputs relative to the efficient technology for that region. Similar conclusions are reached with respect to labor productivity.

In a partial productivity sense, CCA shows clear support for the ability of a region to lower costs by spreading labor costs over a greater number of cattle and liters of milk. This conclusion is consistent with PGA as much of Alberta's productivity disadvantage is caused by an apparent failure to substitute capital for labor given a relative increase in wages. ESEA shows that these inefficiencies exist in Ontario and Québec as well — herds that make an inefficient use of labor are often those with the highest COP. Perhaps more important than labor productivity in the determination of costs is the partial productivity of the cattle themselves or their milk yield.

Again, CCA provides clear evidence for the ability to reduce costs through higher milk yields. California, New York and Wisconsin enjoy both yield and cost advantages over their Canadian counterparts. However, PGA shows that Alberta's more rapid gains in milk yield, relative to Wisconsin, have done little to restore its competitiveness. If higher yields are generated through an overuse of variable inputs, the productivity of the industry and its competitiveness will be adversely affected. When allowance is made for these inefficiencies in production, ESEA supports the CCA results — farms with high yields also tend to be more technically efficient and have higher gross margins than farms with lower yields when the optimal technology is relatively homogeneous throughout the sample. High yields, therefore, provide evidence of an important complementarity between managerial skill, efficiency, productivity, and low-production costs. Such complementarities found in ESEA help explain the low-cost status of California, in that labor productivity becomes a significant factor in reducing COP only in large herds with high milk yields.

At the bottom line, the best managers in each region will be able to produce milk for the lowest cost possible given the constraints of input prices and technology for that region. CCA makes the complementarity between such skill and herd size clear by demonstrating that COP is lower for the larger herds in part because the larger the herd, the greater the premium for managerial skill. By definition, the rate of total-factor productivity growth reflects the intangible, or unexplained, aspects of production that cause production costs to fall over time. Whether through the adoption of new technologies, superior genetics or investments in human capital, competitiveness now is often created rather than endowed. Many of the socioeconomic variables used in ESEA support this conclusion. Although technical efficiency does not imply allocative efficiency, the two appear to be correlated as better managers tend to achieve both higher yields and lower costs.

Many policy implications follow from these general conclusions. First, any constraints to herd expansion should be avoided in order to realize economies of scale. Better managers will further reduce costs through optimizing their input allocations and adopting new technologies at a faster rate. Second, incentives to improve technical efficiency should be promoted. Management clubs, dairy seminars, cooperative education projects and, most importantly, pricing schemes that promote market sensitive production practices are all avenues to the improvement of production efficiency. With greater production efficiency, increases in milk yield, labor productivity and efficient capitalization will be sure to follow. Many analysts propose exhaustive lists of policy recommendations designed to aid Canadian dairy farmers. However, the central conclusion of these three perspectives is that reformation of one or two key areas, for example, the restoration of incentives for the best managers to enter dairy, will generate the necessary changes throughout the system. In each area, improvements are complementary and should be recognized as such.

Notes

1. In the United States, California is an exception to this comparison as quotas on Class I milk have been used since 1969. However, unlike the case in Alberta, California fluid quotas are not mandatory — only 42% of Class I milk was shipped under quota in 1994.

2. In addition to these caveats regarding cost data at the farm level, questions arise concerning the effects of exchange rates and of public policy on land rents,

especially when international comparisons are being made and differences in "competitiveness" are being assessed. For some discussion of these and related matters see Sharples (1990).

3. Because operator and hired labor were both charged at the same hired-labor wage rate, different mixes of hired and family labor have little effect on total labor costs.

4. The authors would like to thank Mr. Phil Carins of the OMMB and Mr. Real Daigle of the Federation des syndicats de gestion du Québec for their cooperation in compiling the data.

5. The results of this study are roughly in agreement with a similar study conducted by Scott Jeffrey (1992) that shows a ranking of California, Minnesota, Washington, New York, Alberta, Manitoba, Québec, Ontario, British Columbia and Saskatchewan.

6. Using a similar method to Romain and Lambert (1992) and Richards and Jeffrey (1995) shows that the optimal herd size in Alberta is 104 cows, which is 40 cows greater than the current average. Such significant unexploited scale economies suggest that such improvements are available to Alberta producers.

References

Agriculture Canada. 1989. Market Commentary, Farm Inputs and Finance, Policy Branch, Ottawa, Ontario, Canada.

AGRITEL. 1990. Systeme d' information sur l'entreprise agricole. Repertoire des informations disponibles dans AGRITEL. UPA Longueuil.

Alberta Agriculture. Various years. *Economics of Milk Production*. Production Economics Branch, Edmonton, Alberta, Canada.

Ball, V. E. 1985. "Output, Input, and Productivity Measurement in US Agriculture, 1948-79." *American Journal of Agricultural Economics* 67: 475–486.

Barichello, R. 1981. "The Economics of Canadian Dairy Industry Regulation." *Economic Council of Canada Technical Report* E/12. Supply and Services Branch, Ottawa, Ontario, Canada.

Barichello, R., and B. Stennes. 1994. "Cost Competitiveness of the Canadian Dairy Industry: A Farm Level Analysis." Paper presented at the Supply Management in Transition Towards the 21st Century Conference, Macdonald College, McGill University, Ste. Anne de Bellevue, Québec, Canada (28–30 June).

California, State of. 1988. *Milk Production Cost Indices*. Department of Food and Agriculture, Milk Stabilization Branch, Sacramento, CA.

Coughler, P., P. D. Stonehouse, R. Lambert, R. Romain, and G. L. Brinkman. 1992. *Dairy Sector: Review of Literature*. Report to the Canadian Dairy Commission, Ottawa, Ontario, Canada.

Groupe de Recherche en Economie et Politique Agricoles Department d'Economie Rurale. 1988. "Les Coutes de Production des Exploitations Laitieres du Québec." Faculte des Sciences de l'Agriculture et de l'Amentation.

_____. 1989 and 1990. "Québec Dairy Facts." Faculte des Sciences de l'Agriculture et de l'Amentation.

Gouin, D. M., S. Lebeau, D. Hairy, and D. Perraud. 1990. "Analyze structurelle comparee, La production laitiere comparee au Canada, aux Etats-Unis et en Europe." GREPA, Departement d' Economie Rurale.

Jeffrey, S. 1992. "Efficiency in Milk Production: A Canada-US Comparison." Department of Rural Economy, University of Alberta, Edmonton, Alberta, Canada.

Lakeshore Farm Management Association. Various years. "Farm Business Management Association Summary." Val Ders, WI.

National Dairy Policy Task Force. 1991. *Growing Together: A Comparison of the Canadian and US Dairy Industries*. Ottawa, Ontario, Canada: Price-Waterhouse.

Ontario Dairy Farm Accounting Project. 1985-89. *Extension Report 1985-1989*. Agriculture Canada, The Ontario Milk Marketing Board, Ontario Ministry of Agriculture and Food, University of Guelph, Guelph, Ontario, Canada.

_____ . 1985–89. *Annual Summary 1985–1989*. Agriculture Canada, The Ontario Milk Marketing Board, Ontario Ministry of Agriculture and Food, University of Guelph, Guelph, Ontario, Canada.

Ontario Ministry of Agriculture and Food. 1989. *Agricultural Statistics for Ontario 1988*. Prepared by Statistical Services Unit, Economics and Policy Coordination Branch, Ottawa, Ontario, Canada.

Richards, T. J., and S. Jeffrey. 1995. "Efficiency and Herd Size in Alberta Dairy Production." Paper to be presented at CAEFMS meetings, Ottawa, Ontario, Canada (9–12 July).

Romain, R., and R. Lambert. 1992. "Economies of Size, Technical Efficiency, and the Cost of Production in the Dairy Sectors of Québec and Ontario." Research Series No. 21. Groupe de recherche agro-alimentaire, Department d' Economie Rurale, Université Laval, Québec City, Québec, Canada.

Schraufnagel, S., and P. Clark. 1990. "Dairy Farm Financial Parameters, Managing the Farm." University of Wisconsin, Department of Agricultural Economics, Madison, WI.

Sharples, J. A. 1990. "Cost of Production and Productivity in Analyzing Trade and Competitiveness." *American Journal of Agricultural Economics* 72: 1278–1282.

Smith, S. F., W. A. Knoblauch, and L. D. Putnam. 1989. "Dairy Farm Management, Business Summary, New York, 1988." 1988. Cornell University, Department of Agricultural Economics, Ithaca, NY. A.E. Res. 89–12.

Wisconsin Agricultural Statistics Service. 1991. *Wisconsin Dairy Facts*. Wisconsin Department of Agriculture, Trade and Consumer Protection, Madison, WI.

8

Supply Management and Vertical Coordination: The Role of Cooperatives[1]

M. E. Bohman and J. A. Janmaat

Abstract

This chapter studies the impact of supply management on dairy processors, particularly producer cooperatives. The regulations analyzed are output restrictions, minimum producer prices and rules to govern the allocation of milk to processors. The model developed in this chapter shows that supply management regulations increase the incentive for farmers to form cooperatives if they enable producers to capture rents through vertical integration. Milk allocation regulations that allow the first receiver of milk to retain financial ownership, even if the milk is processed by another firm, also increase the incentive for farmers to join cooperatives. In a case where cooperatives cannot transfer rents to farmers, supply management may decrease the incentive for farmers if regulations replace the benefit of having a cooperative with which to bargain with processors.

Introduction

Cooperatives play an important role in the dairy processing industry. They account for approximately 50% of the milk output for all of Canada. The market shares of cooperatives vary across provinces. In Ontario, cooperatives have only an approximate 7% market share, but cooperative members produce 80–85% of the milk in British Columbia and approximately 80% of the milk in Alberta. In both provinces the cooperatives' share of the output market is lower. Exact market shares for dairy cooperatives are not available, but interviews with industry participants suggest an estimate of about 50% for British Columbia.

This chapter develops a model of the milk market that shows why supply management regulations should be consistent with high levels of cooperative market shares.[2] The model also shows that differences in the application of specific regulations could be one reason for variation in cooperative market shares across provinces. Cooperatives have been shown to introduce competition into unregulated, imperfectly competitive markets. The model developed in this chapter assesses whether this also occurs in a regulated market. The regulations studied include those directly tied to the supply management system — production quotas, minimum producer prices, rules for governing the allocation of milk to processors — and pooled pricing, which is operated in conjunction with supply management but is not integral to the system.

The effects of supply management on cooperatives have not been previously documented. Other papers, including many chapters in this volume, estimate the impact of supply management on consumers, producer income, productivity and structural changes in farms. Only a small number of papers, for example, Fulton, Katz and Vercammen (1996), study the effect on the processing sector. Most attention has been focused on the poultry industry. This chapter analyzes the impact of key regulations on dairy processors, particularly producer cooperatives.

A Model of Supply Management Regulations and Processors

The effects of supply management regulations on dairy cooperatives are analyzed using a model of the British Columbia dairy industry. A mathematical version of the model is contained in Janmaat and Bohman (1994). The model includes a single, private dairy processor and a cooperative. Farmers form the cooperative in response to the monopsonistic market with one private firm, and the existence of the cooperative is assumed to make the market competitive. Fulton, Katz and Vercammen's (1996) model of vertical integration is similar to ours, but in their model, supply management can create excess profits at each level of the farm to consumer chain: producer, processor and retailer. This chapter focuses on the relationship between farmers and processors, so the possibility of market power by retailers is not analyzed. Research on retail behavior with supply management is discussed by Schmitz and Schmitz (1994).

The model analyzes the milk market in an individual province. Milk produced by a farmer can be sold in either the fluid milk market or transformed into an industrial milk product. Fluid milk marketing is regulated by provincial marketing boards, and industrial milk marketing is regulated by a national marketing board. A complete model would account for both output markets, but a single-market model is adequate for an illustration of the effects of supply management on cooperatives. The chosen model represents the fluid milk market in British Columbia. A fluid milk market is more simply shown because, in this type of market, industrial milk products can be produced and sold in different provinces. The model is general enough to allow its results to be used in the evaluation of policies in other provinces.

Figure 8.1 illustrates the farm, processing and retail sectors. The demand curve, D_C, is consumer demand where imports are restricted to zero. Without an import ban, the price of imported dairy products would set the domestic price. As the focus of this chapter is not on the welfare effects of supply management, the location of the import price is not identified. The importance of import bans is that they facilitate the implementation of domestic production quotas and prices based on cost of production. Without import bans, quotas or significant tariffs, these programs would require large governmental budget expenditures. The effect of import bans on the market is similar to the effect of quotas and has been well-analyzed in other articles, such as Moschini and Meilke (1991).

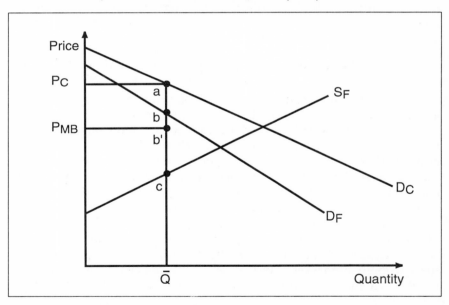

FIGURE 8.1   Supply Management and Dairy Processors

Processor costs are represented by the difference between D_C and the derived demand curve for milk from farmers, D_F. The relationship between D_C and D_F assumes an increasing marginal cost of milk processing. The farm supply curve is given by S_F. The units of milk used at the different levels of the marketing chain are scaled so that one unit of milk from farmers is combined with processing inputs to produce one unit of milk at the retail level.

Restricting output through a production quota equal to $\overline{Q}$ results in a consumer price equal to P_C. At this level of output, the farm-level production costs are given by the supply curve at point c. The processing costs equal ab, and extra profits equal bc. If output restrictions were the only policy, the distribution of these profits between farmers and processors would depend on the relative bargaining position of the two groups.

Cooperatives have been shown to push markets to a competitive equilibrium and to eliminate any market power held by private firms. This effect is known as the "competitive yardstick" and occurs because cooperatives are committed to accepting all production from their members. If profits that are higher than normal exist, members will increase production until a competitive equilibrium is attained. Thus, the internal organization of the cooperative pushes it toward a competitive equilibrium. This result is known as the cooperative incentive problem because each member has an incentive to increase supply at the expense of maximizing the cooperative's total profits. As such, the cooperative cannot attain the profit-maximizing point if it requires internal restrictions on output (Staatz, 1987).

In Figure 8.1, the presence of a cooperative dairy processing firm introduces competition into the market and causes firms to increase the price they pay for milk to the zero profit level or point b on D_F. The benefits to farmers from forming a cooperative are obtained through the market by increasing the price of milk to b.

The output restriction partially alleviates the cooperative's incentive problem, so the cooperative can produce less than the competitive level of output. Production quotas prevent farmers from increasing the amount of milk they deliver to the cooperative when positive profits exist and also prevent them from collectively increasing their production to the point where they earn zero economic profits from their own production as well as the cooperative's activities. While production quotas resolve the cooperative's incentive problem, they also hinder the ability of the cooperative to provide a full pro-competitive effect on the market and all processors can earn positive profits.

Provincial milk marketing boards establish a minimum producer price that determines the division of excess profits between farmers and processors. Assume that the producer price is P_{MB}. The cooperative then has a reduced role in obtaining a better price for farmers.[3] The cooperative no longer has the ability to increase total farm profit by the distance bc, but it can pass on to farmers any excess profits in dairy processing — the expected dividend equals (b - b').

The level of the regulated price has a direct effect on the incentive for farmers to join cooperatives because it determines the amount of processing profits that a cooperative can transfer to farmers. In Figure 8.1 if P_{MB} is less than the intersection of the quota amount and the derived demand for milk, D_F (point b), then excess profits in processing exist. All farmers have an incentive to join the cooperative to obtain a dividend in addition to P_{MB}. In fact, if farmers strictly maximize profits, all farmers have an incentive to join the cooperative to obtain their share of processing profits.

The British Columbia Milk Marketing Board has a production "sleeve," or planned excess production, that ensures the fluid milk market is satisfied. If the "sleeve" is not sufficient, milk is temporarily diverted from industrial to fluid consumption. To the extent that processors can obtain milk for the fluid market on demand, the quantity demanded will equal the intersection of P_{MB} and D_F in Figure 8.1, assuming a competitive market. This implies that no excess profits exist in the fluid milk sector. Dairy processing firms, including cooperatives, may still earn excess profits from sales of industrial milk products.

If the price floor is set at a level where processors cannot make money and as long as a private processor is willing to purchase milk, farmers have no incentive to form a cooperative. However, if a private firm does not exist, farmers will have an incentive if the losses from processing are less than the rents for producing milk.

The regulated price fixes the distribution of profits between farmers and processors and eliminates the role of the cooperative to raise the producer price by making the market competitive. Thus, the regulated price plays the role of cooperatives by countervailing the market power of middlemen, and, it reduces one incentive for farmers to form cooperatives. As pointed out in Fulton, Katz and Vercammen (1996), bargaining over prices occurs within the marketing boards in each province, and cooperatives could lobby for higher producer prices. Anecdotal evidence indicates that cooperatives increase producer bargaining power in the British Columbia Milk Marketing Board.

British Columbia and other provinces also have a set of regulations that allocate milk to specific processors. In terms of efficiency, the optimal policy would allocate the last unit of milk to the firm with the highest margin for processing, based both on processing cost and product value in the marketplace. In British Columbia, a provincial milk pool is established such that all farmers receive the same price and all processors pay the same price for milk used for a specific class of production (British Columbia Milk Marketing Board, 1991). In British Columbia there are five classes of milk[4] with fluid milk having the highest value use. The milk marketing board allocates milk to the processor with the highest class use.

Figure 8.2 illustrates the structure of milk allocation regulations in

British Columbia related to the price pool. In Figure 8.2 solid arrows represent milk flows, and outlined arrows indicate financial flows. Starting from the bottom of Figure 8.2, farmers deliver milk to either the cooperative or private processor. The initial delivery point does not determine which firm processes and sells the milk because the milk board directs milk to the processor who will use it for the highest class use.

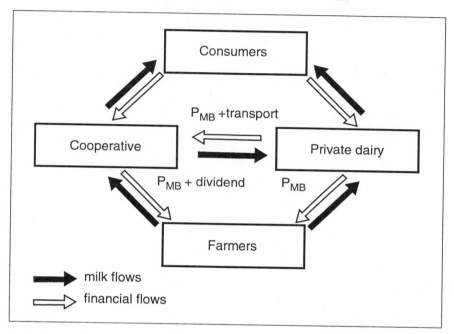

FIGURE 8.2   British Columbia Milk Allocation

The original receiver of the milk transfers it to the processor who transforms the milk. The firm that processes the milk pays the original shipper the regulated price, and the milk board pays the original shipper a fixed amount to cover transportation costs.[5] Firms pay the same price for milk regardless of whether it is shipped directly by farmers or transferred from the original shipper — processors purchase milk at a fixed price from a common pool. The regulations allow milk to be transferred between any two firms, but in practice (as shown in Figure 8.2), milk is only directed from the cooperative to private dairies. The processor sells into wholesale or retail markets in a competitive market.

The price pooling and milk allocation regulations give financial ownership of the milk to the original shipper regardless of who processes and sells the dairy products. This means that the cooperative pays a dividend for all milk shipped by farmers, not just the amount processed by the cooperative. The ownership of all milk shipped by members increases the value of cooperative membership by the amount of any dividend paid by the cooperative. This is an aspect of the regulations that favors cooperatives.

The price pooling and milk allocation system breaks the linkage between the quantity of milk that members ship to the cooperative processor and the amount that the cooperative must transform and sell in the retail market. The cooperative no longer needs to passively process all the milk delivered by its members. This resolves the incentive problem that prevented the cooperative from choosing the efficient level of output. The milk allocation system introduces other incentives for the cooperative to operate efficiently. The ability of the milk marketing board to direct milk from the pool to the highest value user provides an incentive for the cooperative to operate efficiently because it competes for its input requirements under the same conditions as private processors. In order to obtain milk, the cooperative has an incentive to operate at a low cost so that it can find markets for its dairy products.

Net Effect of Supply Management on Cooperatives

The effects of the regulations can be summarized by categorizing the aspects into those that increase incentives for farmers to form cooperatives and those that decrease incentives. Incentives are increased by positive rents or excess profits created by a combination of output restrictions and the level of the floor, or administrative producer price. The ability of the cooperative to transfer rents to farmers is a necessary condition for supply management to increase the incentives for farmers to form cooperatives. Another factor that increases the incentives to form cooperatives is the granting of financial ownership of milk to the original receiver or shipper of the milk. If none of the above conditions are met, then supply management will decrease the incentive for farmers to form cooperatives if it reduces the need for bargaining between farmers and processors through milk marketing boards instead of through markets.

Depending on the regulations and levels of prices, supply management will have either a net positive or negative effect on the incentive to form cooperatives. These factors appear, at least on the surface, to be consistent with variation in dairy cooperative market shares across Canada. Next, the effect on cooperatives for the four largest provinces are discussed.

British Columbia. As mentioned previously, approximately 80% of dairy farmers in British Columbia are members of cooperatives. Supply management regulations that increase incentives to form cooperatives are positive rents in industrial milk processing and milk allocation rules that give financial ownership of milk to the first receiver. Given the existence of a large market share, cooperatives, as the first receiver of milk, are in a slightly privileged position because milk is allocated by class of use and not by specific product. Thus, the first receiver does not have to compete as aggressively for milk within a specific class of use.

Alberta. The Alberta regulations are highly similar to those of British Columbia and suggest the same reasons for a large cooperative presence. The main difference is that Alberta's milk allocation system differentiates between uses of milk within the same class when allocating milk to processors. For example, if a specialty cheese has a higher return than cheddar, then the Alberta system allocates milk to the specialty cheese processor, but the British Columbia system does not move the milk within a class of use. Therefore, in Alberta the benefit to cooperatives of having first call on the milk is somewhat less.

Ontario. Cooperatives have a small market share in Ontario. Two factors from our model indicate that supply management regulations may partially explain this result. 1) Cooperatives do not retain financial ownership of milk used by other processors. 2) Rents from processing are likely to be small.

The Ontario milk allocation system gives the high-value fluid market first claim to all milk, and unrestricted quantities are supplied to plants. This indicates that excess profits do not exist in the fluid milk market. Carton products such as yogurt also have first claim to milk. Manufactured milk products receive the remaining milk through an auction system under the Plant Supply Quota Policy (Ontario Milk Marketing Board, 1993). The plant-level quotas give producers a share of industrial milk left after fluid allocation. The quotas can be transferred on both long- and short-term basis. Plants with higher value use should be able to bid higher than other plants to either buy or lease the quotas. Economic theory predicts that excess profits for processing are bid into quota values, thus eliminating excess profits in dairy processing. This makes farmer investment through a cooperative similar to the investment choice for other industries. In total, the Ontario system may decrease the incentive to join a cooperative because supply management reduces the need for countervailing market power.

Québec. Cooperatives have a strong presence in Québec, both at farmer and processor levels. Québec recently adopted new milk allocation policies that represent a consensus position of stakeholders. It is too early to know the exact implications of the policy, but the major provisions indicate that incentives for farmers to join cooperatives do exist.

Similar to the other provinces, Québec allocates milk to fluid classes on demand. Lower value uses are allocated through a system of processing plant quotas. For industrial milk, the milk marketing board allocates three types of quotas to milk processors. One quota pool is based on historical allocation of milk for the dairy year, prior to the signing of the new agreement. The quota holder captures rents associated with the historic base, and farmers have an incentive to join cooperatives to obtain any excess profits associated with historic base quotas held by cooperatives. The other quota pools are designed to increase the flexibility of the system and the access to milk for new firms. To the extent that these markets are profitable, cooperatives give farmers access to associated rents through vertical integration.

Another factor that increases incentives for Québec farmers to join cooperatives is that, similar to British Columbia, the original shipper of milk retains financial ownership of milk delivered by farmers. Milk is then transferred to the processor that holds the output quota. Different from British Columbia, milk transfers from cooperatives to private firms are observed and vice versa. Transfers from private firms to cooperatives are not explained by our model and indicate a need for further research.

Conclusions

As a whole, supply management regulations appear to be favorable to dairy cooperatives. However, individual aspects of regulations can have either positive or negative effects on the incentives for farmers to form cooperatives. If public policymakers believe that cooperatives provide positive benefits to the dairy sector, then they should be careful not to remove regulations that favor cooperatives while leaving in place other regulations that decrease the incentives to form cooperatives.

Notes

1. Authorship is jointly shared. Research is funded through the Social Sciences and Humanities Research Council (SSHRC) of Canada and the University of British Columbia Humanities and Social Sciences (UBC-HSS) research grants. We would like to thank conference participants and editors for their helpful comments.

2. Sexton and Iskow (1988) discuss economic reasons for the formation of cooperatives.

3. The analysis does not take into account the subsidy paid to farmers as this is not affected by milk allocation policies or the existence of processing cooperatives.

4. The milk classes are: Class 1, fluid; Class 2, cottage cheese, yogurt and sour cream; Class 3, cheese; Class 4, canned evaporated milk, condensed whole milk and condensed skim milk; Class 5, fresh or sterile milk sold in Yukon, Northwest Territories or export; Class 6, nonfluid manufactured products sold in Yukon, Northwest Territories or export; Class 7, manufacture of nonfat milk powder and butter.

5. The milk marketing board makes a fixed payment to the firm that transfers

milk to cover its costs. If this payment is greater than the actual cost, there is a financial benefit in being the original shipper of the milk. Interviews with the British Columbia industry did not produce a consensus on whether the freight reimbursement was generous.

References

British Columbia Milk Marketing Board. 1991. *General Order 133*. Vancouver, British Columbia, Canada.

Fulton, M., M. Katz, and J. Vercammen. 1996. "Vertical and Horizontal Coordination." This volume.

Janmaat, J., and M. Bohman. 1994. "Cooperatives in Regulated Industries: British Columbia Milk." Paper presented at the Canadian Economics Association Annual Meeting, Calgary, Alberta, Canada (June).

Moschini, G., and K. D. Meilke. 1991. "Tariffication with Supply Management: The Case of the US-Canada Chicken Trade." *Canadian Journal of Agricultural Economics* 39: 55–68.

Ontario Milk Marketing Board. 1993. *Summary of Deletions and Amendments to the OMMB's Plant Supply Quota Policy*. Changes effective 1 August.

Schmitz, A., and T. G. Schmitz. 1994. "Supply Management: The Past and Future." *Canadian Journal of Agricultural Economics* 42: 125–148, Guelph, Ontario, Canada.

Sexton, R. J., and J. Iskow. 1988. "Factors Critical to the Success or Failure of Emerging Agricultural Cooperatives." Giannini Foundation Information Series No. 88–3, University of California, Berkeley, CA.

Staatz, J. M. 1987. "The Structural Characteristics of Farmer Cooperatives and Their Behavioral Consequences." *Cooperative Theory: New Approaches*. Agricultural Cooperative Service Report No. 18: 33–60. ASC/USDA, Washington, DC (July).

9

Value–Added Economic Potential

E. van Duren and K. D. Meilke

Abstract

Niche strategies, or the addition of value to traditional raw agricultural products, have often been suggested as an alternative for increasing the profitability of Canada's supply-managed industries. This strategy is appealing because the producer prices of chicken, turkey, eggs and dairy products are higher in Canada than in United States. This largely precludes the use of a low-cost marketing strategy to increase profits. The feasibility of using niche strategies in supply-managed industries will be determined by general economic and demographic conditions, lifefstyle considerations, the trade environment and marketing arrangements. All Canadian food firms face the prospect of selling their products into a slowly growing domestic economy consisting of older, better-educated and smaller households. As a result of multilateral and regional trade agreements, however, these firms have the opportunity to expand sales beyond Canada. Because of higher raw product prices and limited international marketing experience, Canada's supply-managed industries should spend the next several years preparing for increased competition. To the extent that niche strategies can be helpful, they are best pursued through strategic alliances or joint ventures between Canadian firms and multinational organizations.

Introduction

An increase in the value of Canada's raw agricultural products has been suggested as a palatable option for dealing with many of the challenges facing the Canadian agrifood sector. One challenge is determining how the supply-managed industries should respond to the General Agreement on Tariffs and Trade (GATT). However, as a strategic direction for the supply-managed industries, the value added option has evolved with a perplexing twist. Since Canada's supply-managed industries are relatively small by world standards and also have a relatively high cost structure as a result of production quotas, the most appropriate value adding strategy for firms in the sector is a "niche strategy." Presumably, the choice of an appropriate niche, in turn, would generate exports. This notion has considerable popular appeal and support among marketing boards, governments and some analysts. But, is it feasible? We examine this question by first exploring the concepts of value, value added and niche strategies from economic and strategic management perspectives. Then we discuss changes that are likely to occur in the business environment facing the Canadian supply-managed industries into the twenty-first century because of multilateral trade liberalization and other factors. Finally, we assess the Canadian supply-managed industries' capabilities to compete using a niche strategy.

Increasing Value Added: Concepts for Aligning Public Policy and Private Strategy Efforts

At first glance, value added seems a straightforward concept that is commonly understood among economists and businesspeople. Value added is simply the value of a firm's outputs minus the value of the inputs that it purchases from other firms. Thus, it is essentially the sum of factor incomes that accrue to various contributors to the firm's activities and typically include wages, salaries, interest, profits, rent and occasionally "entrepreneurship." The definition is essentially *ex post* in nature, and on its own, it is not particularly conducive to guiding either private strategy or public policy efforts to increase value added. Some of this deficiency results from the lack of a definition for "value."

"Value" has a central role in economics and, since the time of Aristotle, it has been traditional to recognize its two streams of meaning: value in use and value in exchange. "Value in use" deals with a product's[1] capacity to satisfy human wants while "value in exchange" is concerned with a product's worth in terms of its capacity to be exchanged for another product. In classical economics the existence of "value in use" in a product was a prerequisite for it to have "value in exchange." A product had to possess utility for it to be produced and exchanged. In addition, its exchange value was determined by the cost of producing the product including the costs of purchased inputs and factor incomes. The "Marshallian Cross" integrated the two streams of meaning. Market price was determined by the interaction of supply and demand, both of which were derived from a maximization principle. Profit maximization was the relevant behavior for producers creating "supply," and utility maximization was the relevant behavior for consumers creating "demand."[2]

From economic theory we can deduce that consumers want value as "utility," and that producers or firms also want value, as "profit." However, as concepts to guide private business strategy and public policy, these concepts are lacking. This is little wonder since economists have traditionally been interested in developing highly simplified theoretical concepts while those who pursue private strategies and public policies must live in the "real world" of messy, fuzzy, detailed activities and information. Consequently, we also examine the contributions that strategic management researchers have made in defining concepts of value and value added in an effort to develop a definition that can guide public policy and private business strategy efforts to increase value added.

Strategic management researchers define value as what buyers are "willing to pay." Value plays a central role in many, but not all, strategic management paradigms. According to Porter (1985), competitive advantage grows fundamentally from the value that a firm is able to create for its buyers that exceeds the firm's costs of creating it. Consequently, value added or the net marketing margin is the difference between the total value from selling products to buyers and the collective cost of performing the activities that make those products possible. In the provision of products, any firm performs value activities. These are the physically and technologically distinct, but interrelated, functions that create value for buyers. Simultaneously, these activities create costs for the firm. Thus, value added is derived from the value that buyers are "willing to pay" and the costs that firms incur to provide the products that embody that value.

Value to a buyer is an interrelation of buyer perception, product features and performance, stand-alone services and/or services that accompany tangible product features. Like quality, value is notoriously difficult to define because a good portion of it exists in the mind of the buyer, and it is best created by the seller if he/she knows the customer and his/her business. However, product features and performance that often have value for buyers include durability, precision, predictability and ease of use. Similarly, common services that provide value to buyers include timeliness, useful and friendly information and quick and effective response to customer complaints. Buyer perception contributes positively to value if current and previous experiences with the product have been favorable.

Strategic management researchers have made extensive use of the value chain to analyze private strategies (Porter, 1985). Figure 9.1 depicts a simplified value chain. It can be used to assess strategies that focus on creating value as well as strategies that focus on minimizing costs. Activities in the value chain embody purchased inputs, human resources, technology and information. Individually, collectively and through their interrelationships with the value chains of buyers and sellers, they describe how a firm creates value added.

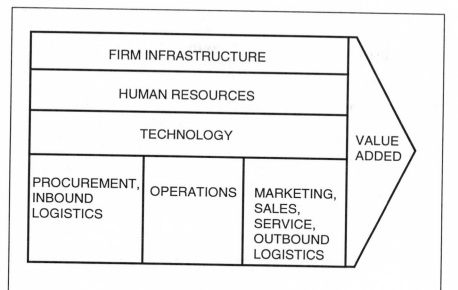

PROCUREMENT, INBOUND LOGISTICS
    purchasing, negotiating with suppliers, inspection of inputs

OPERATIONS
    manufacturing, flexibility-switchovers, scale, seasonality of production

MARKETING, SALES, SERVICE, OUTBOUND LOGISTICS
    pricing, promoting, market selection/creation, delivery, customer service

TECHNOLOGY
    research and development, technology adoption/adaptation

HUMAN RESOURCES
    training, education, compensation, benefits

FIRM INFRASTRUCTURE
    accounting, finance
    legal affairs and political relations
    strategy, planning, product development ...
    management
    leadership

FIGURE 9.1   Value Chain

Value added can be created by a cost-reduction strategy — the cost-minimization objective that occurs in the "black box" of the neoclassical theory of the firm.[3] Figure 9.2 indicates that this "cost leadership" or low-cost strategy is based on broad market coverage. With this strategic approach, all segments of the market are served with the same product that is provided at the lowest possible cost. On the surface it is tempting to conclude that the majority of agricultural products must be produced and marketed using this approach since the options for physical transformation of the raw product are limited. Classic examples of products that are best produced and marketed using a low-cost strategy are tomato paste, standard qualities or grades of grain, fluid milk, boxed meat and chicken parts among others.

Source of Competitive Advantage (Basis for Competing of Adding Value)			
		Cost Leadership (cost for Competing for Value)	Differentiation (creating value for buyers)
Market Coverage	Broad	Cost Leadership	Differentiation
	Selected Segment	Cost Niche	**Differentiation Niche**

FIGURE 9.2   Generic Strategies for Increasing Value Added
   Source: Adapted from Porter (1985)

Value added can also be created through a value creation or differentiation strategy that requires a firm to tailor its products to serve different buyer needs in various segments of the market. Market coverage is broad, as it is with a cost leadership strategy, but the basis for adding value is meeting buyers' different needs in various differentiated segments of the market. Many agrifood products, traditionally assumed to be most suitably produced and marketed with a low-cost strategy, are now being provided through differentiation strategies. Fluid milk is now available at the standard fat level; 2%, 1% and skim fat levels; in chocolate flavors; in special formulations for those with lactose intollerance; and in January, 1995, Adult Foods launched its new Lactantia filtered milk which is purported to taste fresher and has a shelf life that is 50% to 60% longer than other fluid milk. Various tomato purees are now appearing in the market segment between the differentiated, diced and spiced, canned tomato products and standard tomato paste. Chicken continues to be made available in a larger variety of further processed forms.

Two other ways that value added can be created are by focusing a firm's efforts on a given segment or niche of the market by superior differentiation or alternatively by cost reductions beyond those that can be obtained by firms with broad market coverage. Both approaches are "niche strategies," but discussion of a niche strategy typically refers to a differentiation niche strategy. Table 9.1 summarizes factors that typically need to be in place for a firm to be successful using these types of generic strategies for creating value.

A guiding hypothesis for this chapter is that Canada's supply-managed industries will not want to compete in the global marketplace using a cost-reduction strategy unless there are massive changes in both domestic and international agricultural policies. Instead, their best chance to be globally competitive will be using a "value creation" or differentiation strategy that is focused on a given niche in the global market. As Table 9.1 indicates, pursuing such a strategy requires firms to develop several attributes that need to be in place for success. Before turning to our discussion on how trends in the emerging business environment of the twenty-first century will affect firms' capabilities to develop these success factors, we briefly review the record that Canada's supply-managed industries have had vis-à-vis their counterparts in the United States with respect to key value added and other industry-level structure and performance data.

The Supply-Managed Industries' Record
in the Last Decade

At the processing level of the industry, Canada's supply-managed industries' value added and general performance record can be assessed by using various competitiveness indicators (Martin, van Duren and Westgren, 1991). In nominal terms, the Canadian and US poultry industries increased their value added at approximately the same rate from 1980 to 1990 (9.7% versus 9.8% per annum, respectively) while Canada's dairy industry increased its value added at a faster rate (9.2% annually) than its US counterpart (5.8% annually).

As a percentage of sales, the Canadian and US poultry industries also improved their value added at approximately the same rate (7.3% versus 7.4% per annum, respectively). However, Canada's poultry industry continued to have lower value added per dollar of sales than the US industry. The US dairy industry's slower rate of sales growth (4.2% versus

TABLE 9.1   Factors Often Associated with Success with Generic Strategies for
Value Added

Cost Leadership	Differentiation
Procurement Etc.	*Procurement Etc.*
Access to low priced inputs of a minimum quality; relations with suppliers should encourage this	Access to inputs of the appropriate quality at a good price — relations with supplier should encourage this
Operations	*Operations*
Process engineering skills	Economies of scope
Tight cost control	Operations flexibility; switchover expertise
High capacity utilization	
Economies of sale	Tight quality control
Marketing Etc.	*Marketing Etc.*
Low cost distribution system	Creative flair
Strong mass advertising	Strong innovative marketing abilities
Good relations with buyers	Good relations with buyers
Technology	*Technology*
Product design aligned with process engineering	Product development skills
	Reputation for technological leadership
Human Resources	*Human Resources*
Close labor supervision	Compensation based on more subjective measurement and creative incentives
Compensation based on cost and other relevant quantitative information	Amenities to attract highly skilled labor, scientists and creative people
Firm Infrastructure	*Firm Infrastructure*
Tight cost control; frequent reporting	Deeply segmented sales data
Accurate, timely cost and other economic information	Quality and various consumer satisfaction data
Favorable institutional factors, (low taxes)	Favorable institutional factors (public information supportive of product)
General management expertise in reducing costs	General management expertise in serving customer wants
Cost Niche	*Differentiation Niche*
General management expertise and individual value activity in matching target market segment with key cost driver(s) in the industry	General management expertise and individual value activity in matching target market segment with key differentiation driver(s) in the industry

5.2% per annum for Canada) led to value added as a percentage of sales improving somewhat faster in Canada than in the United States (3.8% versus 1.6% per annum, respectively).

Throughout the 1980s, value added per plant remained significantly higher in the US poultry industry than in Canada (Cdn $4.9 million for Canada versus Cdn $12.0 million for the United States), but Canada's industry improved its performance on this indicator more rapidly (8.9% for Canada versus 5.6% per annum for the United States). In the dairy industry, the United States had a higher value added per plant than Canada (Cdn $5.2 million for Canada versus Cdn $6.7 million for the United States). The US industry also added new plants at a much faster rate (1.3% for Canada versus 4.5% per annum for the United States) However, the Canadian dairy industry, even with its slower plant growth, improved its value added per plant more slowly than the US industry over this time (12.2% versus 13.6% per annum, respectively).

On a per production worker basis, Canada's poultry and dairy industries both improved their value added faster than their US counterparts from 1980 to 1990 (6.9% for Canada versus 5.8% per annum for the United States for poultry; 9.0% for Canada versus 7.6% per annum for the United States for dairy). For both industries, Canada continued to perform better than the United States on this measure by 1990 (Cdn $53,600 per worker in Canada versus Cdn $43,200 in the United States for poultry; Cdn $171,500 in Canada versus Cdn $146,500 in the United States for dairy).

These processing-level competitiveness indicators suggest that Canada's poultry industry's value added performance continued to lag its US counterpart during the 1980s with the exception of value added per worker — a measure not particularly important to the low-cost driven firms that dominate the US industry. However, in the dairy industry Canada's value added performance exceeds that of the United States on several indicators.

To obtain an indication of the degree of value added activity occurring throughout all vertical levels of an industry, we computed sales at the processing level of an industry, relative to cash receipts generated at the farm level. Our calculations indicate no firm trend for either the poultry or the dairy industries.

The above discussion suggests that Canada' supply-managed industries may have some advantage in generating value added on a per worker basis but certainly not on a per plant basis. For the dairy industry, Canada has also been able to generate higher value added per dollar of sales than its US counterpart. However, the magnitude of the difference between Canadian and US industries' performance on these measures and their relative growth rates during the 1980s provides insufficient evidence to support the notion that Canada's supply-managed industries should pursue a niche or differentiation strategy for being competitive into the twenty-first century. The data are not available at a sufficiently disaggregated level nor are they representative of the variables that indicate the type of strategies that firms within the industry or segment are pursuing (product development data, Research and Development (R&D), training expenditures, etc.). Therefore, we turn our examination to the type of business environment in which supply-managed firms may find themselves competing as they adjust to a post-Uruguay Round world.

The Emerging Business Environment

In any discussion of the impacts of a trade agreement (especially one with implications as potentially profound as those that the General Agreement on Tariffs and Trade (GATT) agreement may have for the long-run success of Canada's supply-managed industries) it is tempting to ignore the wider context in which these political/legal changes induce their effects.[4] However, to assess Canada's supply-managed industries' value added potential through the use of differentiation strategies in the post-GATT 1994 world, it is imperative that we examine key components of this wider business environment. This section contains this discussion and uses it to set the context for an analysis of the political/legal changes that GATT and its interactions with other trade agreements can be expected to induce.

Economic Trends

The key economic trends to influence the emerging business environment in which Canada's supply-managed industries will compete in the twenty-first century are difficult to ascertain because they are largely the effects of other trends. However, like all industries, the supply-managed industries can expect to have to address the strategic implications of a smaller, "traditional" middle class and the associated changes in spending power and propensities. The decline of this market segment is already making it more difficult to use traditional, branding strategies. Its decline also suggests that advertising with credible information content may become more important.

The results of a 1992 Angus Reid study are also suggestive of the challenge faced by the supply-managed industries. Forty-eight percent of Canadians feel that dairy products are "expensive" while 42% feel they are priced "about right," and only 8% feel they are a "bargain." Twenty-five percent of Canadians feel eggs are "expensive," 56% feel they are priced "about right," and 15% feel they are a "bargain." Thirty-seven percent of Canadians feel that chicken and turkey are "expensive" while 47% feel they are priced "about right" and 12% feel they are a "bargain." Fortunately for dairy and poultry firms, similar patterns exist for fruit and red meat products in terms of consumers' value perceptions. However, consumers feel that bread and vegetables generally provide better value.

Another economic trend that can be examined is consumption patterns of food and beverage products. Canadians consume smaller amounts and a lower proportion of their total caloric intake in the form of products produced by the supply-managed industries than Americans. This suggests that domestic market expansion may be possible. The specific exception is yogurt of which Canadians consume more than Americans but still less than Europeans. Thus, overall there may be room for sales growth, domestically and internationally, but whether that growth is best captured through lower prices, better selection or quality of products is an unknown.

Demographic Trends

Canada's population is expected to grow to 29–32 million by 2011, an annual increase of 0.4% to 0.9% per annum, a rate that makes it impossible to grow in the food industry by "doing business as usual." Therefore, an understanding of other demographic indicators is important to firms in the food business.

Aging. A larger proportion of Canada's and the United States' populations will be older in the twenty-first century. In Canada by the year 2011, between 16% and 18% of the population will be over 65 compared to only 12% in 1991. As people age they tend to eat fewer servings of dairy and meat products although red meat products bear the brunt of the decline. Older consumers shop somewhat more frequently than those in other age groups but purchase less and are more likely to stick to their shopping lists. Older consumers eat lunch and breakfast more frequently than do consumers in other age groups, consume fewer convenience foods, and are less concerned about being able to make a meal quickly. Overall, older people consume fewer calories, are more concerned about nutrition and are more demanding of value than ever before — trends that will strengthen as the baby boomers enter this age group. All this suggests that the food industry should focus on products that embody higher quality, more perceived healthiness and, thus, greater value. Fortunately for North American food firms, Mexico's population has an increasing proportion and number of young people.

Sex-Based Factors. In 1993 for the first time in over four decades, female participation in Canada's labor force declined marginally with the bulk of the reduction occurring in the younger age group (15–24). However, Canada continues to have one of the highest rates of female labor force participation in the developed world. Simultaneously, the gender pay gap continues to narrow, albeit slowly, to the point where adult women earned 68% compared to adult men in the early 1990s. Women head the vast majority of single-income families, and even in two-income families they are much more likely to make the "food decisions" for the household (van Duren and McKay, 1993). Women shop more frequently than do men, spend somewhat less per shopping trip and when purchasing groceries, women find price, package size, product freshness and environmentally friendly packaging slightly more important than do men. Women are also more likely to use all types of information provided on a food product than do men. Not surprisingly, women are much more likely to prepare their own meals (66%) than men (35%). Women also report more "perceived to be correct" food consumption behavior in every category than do men (Angus Reid, 1992). All of the above suggests that firms in the supply-managed industries, like those in all other food industries, need to create products which save consumers time at a price that provides value by facilitating purchasing, meal preparation and nutrition planning.

Household Composition. Household size continues to decline in Canada and more markedly in Québec than elsewhere. As well, two-parent families continue to decrease as a proportion of all households, largely due to their decline in the 18–34 age group. In the future, there will be fewer traditional families in Canada. Consequently, increased flexibility in food purchasing and meal preparation, smaller package sizes and resealable packaging are among the products and services that food firms should aim to make available to consumers in Canada, and internationally, to be competitive in the twenty-first century.

Ethnic Composition. Canada's population continues to become more ethnically diverse as does the United States'. Immigrants tend to consume the foods of their home country, but this tendency diminishes significantly with their children. Throughout North America, Latin American and Asian tastes continue to grow in importance to food producers and marketers; a trend that is somewhat positive for firms in the poultry industry but not promising for those in the dairy industry.

Lifestyle: Psychographic Trends

Attitudes, beliefs, behaviors and lifestyles that characterize the Canadian population, and the larger North American population, will become important in the twenty-first century. They pose a strategic challenge for firms in the supply-managed industries. Existing research on lifestyle and psychographic trends suggests that firms in Canada's dairy and poultry industries must take heed of some key factors if they are going to be successful with a differentiation-niche strategy. At the risk of oversimplification, these factors are as follows.

First, there are several distinct types of food consumers: "kitchen enthusiasts," who at 25% of the population are well-educated and affluent people who regard food and nutrition as an important part of life; "apathetic eaters," who at 17% of the population enjoy eating but tend to stick to traditional fare and pay little attention to physical fitness; and the "fast food socializers," who at 13% of the population eat because it is vital to keep functioning but the quicker and easier the means of doing so, the better (van Duren and McKay, 1993; Angus Reid, 1992). Knowing the tastes and preferences of these various segments of the Canadian food market is ever more important to business success in the food industry as population growth slows and diversity becomes more apparent.

Second, societal trends, such as cocooning, cashing out, pursuit of fantasy adventure and increased personal and social consciousness, provide indications of what products consumers may value in the future. Products need to be easier to prepare at home, sufficiently exotic, environmentally friendly as well as ethically appropriate and "good for you" (Popcorn, 1991).

Information: Technological Factors

Lower cost and more powerful and faster information technologies will facilitate the flexibility that firms in the supply-managed industries, as well as elsewhere in the sector, will need in the coming years in order to satisfy the increasingly segmented food market described above. The development of this capacity will be costly for the supply-managed industries. More than other agrifood industries, they have relied on "inside the trade relations," lobbying of governments and less transparent means for obtaining strategic information. However, developing a market-oriented, strategic-information perspective is imperative to the supply-managed industries' success of a differentiation-niche strategy.

Political-Legal Factors Resulting from GATT

Since the beginning of the Uruguay Round of trade negotiations in 1986, and continuing through the Free Trade Agreement (FTA) and the North American Free Trade Agreement (NAFTA) negotiations, producer organizations in Canada's supply-managed industries have been wary of import liberalization. In the FTA, both Canada and the United States agreed to "put aside" the most difficult negotiating issues in the agrifood sector, leaving these to be resolved by the GATT. For Canada, a high priority was the enshrinement in the FTA of Canada's, then GATT-legal, import quotas for dairy products, poultry and eggs. In return, Canada increased its global import quotas to reflect the level of imports in the late 1980s. However, a successful GATT challenge of Canada's border controls for ice cream and yogurt indicated that the GATT would not sanction import quotas on processed products that were not directly under supply control. This decision was consistent with an earlier GATT panel report that ruled against Japan on a wide range of border measures. Canada blocked the adoption of the panel report on yogurt and ice cream throughout the Uruguay Round of trade negotiations, arguing that it would only consider the panel report after new international trade rules were negotiated. This was a stance taken by several other GATT member countries with respect to adverse panel rulings during the seven years of negotiations.

Canada continued to maintain its defense of import quotas for supply-managed commodities during the NAFTA negotiations. As a result for agrifood, Canada entered into separate bilateral agreements with Mexico and the United States. But, for Canada-United States trade, the terms of the FTA continued to apply. Canadian exports of supply-managed products to Mexico, unlike those of the United States, will face quantitative restrictions. However, Mexico was exempted from Canada's ban on margarine imports. This will be the case at least until the GATT 1994 comes into effect. The above treatment contrasts with that for agrifood trade between the United States and Mexico, which after a phase-in period, will be free of both tariff and nontariff barriers.

Canada's position in the GATT negotiations was bipolar. Canada joined with the United States and the Cairns Group in arguing for liberalized trade in grains, oilseeds and red meats in which Canada has a significant export interest. However, Canada parted company with its major negotiating allies in arguing that Article XI-2(c) should not only be maintained but should be strengthened to allow import quotas on processed products derived from primary products subject to domestic supply control. Canada achieved limited support for its position early in the Uruguay Round of negotiations. Following the tabling of the Dunkel Draft Final Act on Agriculture, which embraced comprehensive tariffication of all border measures in December, 1991, several things became apparent. First, support for Canada's position on Article XI was limited to Japan, Korea and a small handful of other agrifood importing countries. Second, if Canada failed to get support for its position, as seemed likely, then how comprehensive tariffication would be handled with regard to Canada-United States trade, and to a much lesser extent Canada-Mexico trade, was a key question.

This was the case because all tariffs under the FTA are to be eliminated over a negotiated time frame. Hence, if as a result of the GATT negotiations, Canada tariffied its import quotas, would these tariffs have to be eliminated according to the provisions of the FTA? It seemed likely that Canada would argue that GATT tariffs would apply while the United States would argue that they should be subject to the provisions of the

FTA. Canada's position in this argument is bolstered by provisions in the FTA (Article 710) and NAFTA (Annex 703.1) which say that "unless otherwise specifically provided in this Chapter, the Parties retain their rights and obligations with respect to agricultural, food, beverage and certain related goods under the GATT and agreements negotiated under the GATT, including their rights and obligations under Article XI." Canada interprets this provision as saying that any agreement made in the GATT overrides the provisions of the FTA and the NAFTA.

During the last few weeks of the GATT negotiations, when even Japan had reached an accommodation on rice, Canada continued to maintain its lonely position with regard to import controls.[5] Finally, in the last few hours of the negotiations, Canada agreed to comprehensive tariffication. After years of arguing against tariffication, lobbyists for Canada's supply-managed commodities bought into the agreement after they were assured that the negotiated tariff equivalents would be prohibitive and that the GATT minimum access commitments would largely reflect the status quo.

Canada's firm position on import quotas had at least two unfortunate ramifications. First, the status of the newly created tariffs vis-à-vis Canada-United States trade was left unresolved. Second, trade in supply-managed products with the United States became intertwined with a number of other bilateral agrifood trade irritants. These involved exports of Canadian wheat, flour, coarse grains, sugar, tobacco and peanut products into the United States market. Although bilateral negotiations between the United States and Canada have resolved some of these issues, the status of Canadian tariffs on supply-managed products remains unsettled.

Under the provisions of the GATT, trade in supply-managed commodities will be governed by tariff rate quotas. The tariffs applied within the quota are generally modest and will be reduced by 57% over the life of the agreement. Tariffs[6] for over-quota imports are generally prohibitive ranging from 182% for turkey to 351% for butter. These over-quota tariffs are to be reduced by 15% by the year 2000. Some indication of the "true" tariff equivalents can be obtained from the work of the Committee of Experts. This work suggests that the actual tariff equivalent for chicken is in the area of 65% compared to the negotiated tariff equivalent of 280%. Similar figures for eggs and turkey are 60% versus 192% and 75% versus 182%, respectively.

It is impossible to evaluate the potential success of a niche marketing strategy without making some assumptions about what will happen to United States-Canada tariffs, and by implication what will happen to Canadian prices for supply-managed commodities and their products. Hence, for the remainder of this chapter we will assume the following:

1) Canada-US tariffs will remain at GATT-negotiated levels over the life of the agreement although Canada may have to increase its

global import quotas slightly to gain this "concession" from the United States. The exception to this is trade in ice cream and yogurt where, as a result of the GATT panel ruling, Canada may have to offer a major concession.

2)   For poultry and eggs, domestic supply management will be under increasing pressure but not as a direct result of the GATT. Since national supply controls are no longer needed to justify import quotas, several provinces, most notably Ontario and Québec, have given notice that they are going to unilaterally increase their chicken output by substantial amounts. This might be the first step in a process that will lead to more flexible production and pricing arrangements. For eggs, the disposal of table eggs surplus to domestic requirements has been a contentious issue for many years. Again, the threat to supply management is more internal than external.

Finally, the most obvious way to facilitate niche export-oriented marketing — making raw product available at reduced prices for export — has been constrained or eliminated by the GATT, which classifies this type of activity as an export subsidy. Canada, along with all other countries, agreed not to introduce export subsidies on commodities where they did not exist in the base period and to reduce those that did exist. This will involve a cut in explicit dairy product export subsidy expenditures from $126 million in 1986-90 to $81 million in the year 2000. The export-volume commitment involves a reduction from 0.12 mmt to 0.09 mmt. However, the dairy industry is exploring ways to circumvent the export subsidy commitments.

A crucial question is whether there is anything in the GATT that will facilitate Canada's supply-managed industries if they want to pursue a niche marketing strategy, since there is nothing in it that will force them to reduce their cost structure before the end of the century.

On the positive side, the GATT will create new export opportunities, particularly for dairy products, in several nations as a result of the tariffication process. However, the GATT involves no specific product commitments; it allows countries to determine their minimum access quantities based on broad commodity aggregates, for example, dairy products. The individual countries then have complete leeway in allocating these quotas to individual tariff lines. These tariff lines may be advantageous or disadvantageous to Canada.

More importantly, most of the newly created import access will result from the use of tariff rate quotas, that is imports within the quota will face fairly modest tariffs while over-quota imports will face massive tariffs. The import rights for within-quota products will be very valuable. It seems likely that most countries will either allocate or sell these import rights to domestic firms. Hence, in order to export within-quota products, a foreign exporter will need to convince a rent-seeking, import-quota holder to purchase their products. The easiest avenue to obtaining some of the import quota would follow from ownership or association with a multinational firm which has import quota rights. Many countries have "earmarked" all or a large portion of the import quotas for specific countries.

To conclude, there is little in the GATT that will change the business environment within which Canada's supply-managed industries operate.

Assessment

The interaction of changes in the business environment caused by economic, demographic, lifestyle, technological factors, as well as the GATT and other trade agreements, makes it crucial that Canadian agrifood firms begin searching for strategic opportunities beyond Canada's borders and that government and other agrifood institutions assist in this effort. This challenge facing firms in Canada's supply-managed industries is especially complicated. As a result of the GATT, which preserves the degree but not the form of border protection, almost no changes are necessary for compliance with the agreement.

For at least the next six years the supply-managed industries will not face a deluge of low-cost imports: the required increases in minimum access are quite modest and the over-quota tariffs are prohibitive. In addition, the supply-managed industries are considering changes in their marketing arrangements that may allow them, at least temporarily, to circumvent the GATT's disciplines on export subsidies.

For two decades, firms in the supply-managed industries have not faced real international competitive pressure, developing instead an inward looking, self-sufficient strategic perspective. The GATT 1994 does little to change this orientation since it does not effectively open world dairy markets and because the United States is a fierce competitor in the world poultry market. For all these reasons, there will be a strong temptation, by many in the industry, to drag out the status quo and continue to fight change, thereby preserving the huge rents associated with these highly regulated markets for as long as possible.

In our opinion, this is a misguided option. The enormous tariffs that will protect the supply-managed industries over the next six years will be prime targets of low-cost exporters in future multilateral trade negotiations. Pressure will be applied to increase the minimum access quantities, significantly reduce over-quota levies and close loopholes in export subsidy reduction commitments. Hemispheric freer trade will also put increased pressure on Canada to further open its markets to low-cost imports of both raw and processed products. But, of paramount importance is the fact that Canada will want other developed countries to undertake meaningful trade reforms in agriculture, manufacturing, intellectual property and services. As a result, it will be unable to resist them at home. Hence, over time the border protection afforded the supply-managed industries will be reduced. These industries should use the next six years to prepare for the future — to change strategic direction and to negotiate beneficial changes both domestically and in the international fora.

Over the next two decades, firms in these industries will face the same business challenges, resulting from demographic, lifestyle and technological trends, as those in other industries. However, in preparing for the future, they will need to use their unique trade position as a cushion to help them adjust to a more competitive environment, rather than as an umbrella under which to hide.

A differentiation or global differentiation niche strategy based on superior knowledge of one's customer could be successful for dedicated firms in the industry if efforts toward strategic change begin now. Firms already pursuing such a strategic approach are proof of that. Elmira Poultry of Waterloo, Ontario, has successfully navigated many of the obstacles of the poultry supply management system to become a significant domestic producer, as well as an exporter, of high, value added, further-processed chicken and turkey products. Similarly, Roman Cheese of Toronto produces a high-quality lasagna which it exports. These firms have found that the key is to focus on what customers really want and thus are "willing to pay for." Unfortunately, these marketing skills are reduced in managers of supply-managed industries, particularly at the producer level (Howard, Brinkman and McDougal, 1994). However, development of these skills has also been inhibited by marketing boards and government regulations.

The question of whether Canadian consumers are as demanding and sophisticated in their tastes and preferences for dairy and poultry products as consumers in other countries will determine whether firms in Canada's supply-managed industries should focus their new strategic direction on the Canadian market initially or whether foreign markets must also be developed over the next six years. If Canadian consumers do not know what "value" is, relative to consumers in the international marketplace, it would be folly for Canadian firms to focus their efforts on the domestic market.

If we accept that Canadian consumers may not be as sophisticated or value-demanding as their counterparts in the United States or perhaps the European Union (EU), it becomes important to determine whether firms in Canada's supply-managed industries have obtained sufficient access to US and EU markets to develop international niche markets. A determination of this sufficiency is complicated by relatively small minimum access levels in and out of "tariff-rate quota" tariff levels and the proportion of the quota that is locked in by traditional exporters.

For the poultry industry, Canadian firms have never faced quantitative barriers into the United States and it is uncertain what portion of the EU market is available. Firms pursuing a niche marketing strategy into the United States, however, may be constrained by the Canadian system. If these firms want to import US raw product, process it to the specifications demanded by buyers in that niche market and then reexport it to the United States, there may continue to be bureaucratic obstacles to developing the US portion of a niche market that exists in North America. In the past, such problems have existed for firms producing further processed poultry products such as seasoned and tipped chicken wings and chicken that is produced entirely with organic inputs and processes (McKay, 1993).

Another problem facing Canada's supply-managed industries has to do with the regulations put into place at the inception of these programs that carved the domestic market into provincial production shares. These provincial shares proved politically impossible to change and resulted in higher-cost structures and lower-than-optimum processing-plant volumes, even with raw product output controls. These regulations make it more difficult for a firm to follow a "scope" strategy that is facilitated by a large supply of raw product from which market niches can be served. While it seems clear that the provincial market share regulations will be difficult to enforce in the post-Article XI trade environment, it will take several years for national production and processing patterns to be rationalized.

For the dairy industry, the obstacle to accessing the EU and US markets may well be the proportion of quota that will be locked into traditional exporters over the next six years. For example, it appears that Canada's cheese firms, at best, will obtain a proportional increase in their approximate 11% of US import quota from 1995 to 2001. There is also mounting anecdotal evidence that Australia, New Zealand and certain EU countries have locked in significant shares of tariff-rate quota for other products, as well as other countries.

Lack of access to the US and EU markets, as well as the need to develop better knowledge of markets with more demanding consumers and improve marketing abilities, suggests that multinational organizations, strategic alliances or joint ventures between Canadian firms and counterparts in these markets should be seriously considered. Three decades of supply management may well have destroyed our ability to "go it alone" in the twenty-first century.

Notes

1. Throughout this chapter we use the term product to refer to any physical product, its associated services or only a service.

2. Only Marxists, who developed the labor theory of value into a theory of profit or surplus value, retain the separation of "value in use" and "value in exchange" for price determination.

3. For an excellent introduction to the relationship between economic theories of the firm and strategic management research, see Seth and Thomas (1994).

4. Strategic management researchers and teachers typically divide factors in this wider business environment context into groups such as social, technological, economic, political-legal, global, etc.

5. Japan agreed to accept higher minimum access provisions for rice in return for postponing the requirement to tariffy.

6. Canada's tariff equivalents are specified in both ad valorem and specific terms. The higher rate will apply.

References

Angus Reid. 1992. *The Canadian Food Study*. Winnipeg, Manitoba, Canada: Angus Reid.

Howard, W., G. Brinkman, and N. McDougal. 1994. "Identifying Factors Determining Differences in Income Between Farmers in Canada." Unpublished study prepared for the Canadian Farm Business Management Council.

International Agricultural Trade Research Consortium. 1994. "The Uruguay Round Agreement on Agriculture: An Evaluation." Commissioned Paper No. 9 (July).

Martin, L., E. van Duren, R. Westgren. 1991. "Assessing the Competitiveness of Canada's Food Processing Industry." *American Journal of Agricultural Economics* 73 (5): 1456–1464.

McKay, H. 1993. "Strategic Alliances in the Ontario Chicken Industry." Unpublished Master's thesis. University of Guelph, Department of Agricultural Economics and Business, Guelph, Ontario, Canada.

Popcorn, F. 1991. "What is She is Right?" *Report on Business Magazine* 8(4): 66–77 (October).

Porter, M. 1980. *Competitive Strategy*. New York, NY: The Free Press.

Porter, K. 1985. *Competitive Advantage*. New York, NY: The Free Press.

Porter, K. 1990. *The Competitive Advantage of Nations*. New York, NY: The Free Press.

Seth, A., and H. Thomas. 1994. "Theories of the Firm: Implication for Strategy Research." *Journal of Management Studies* 31: 165–191.

van Duren, E., and H. McKay. 1993. *An Analysis of Consumer Trends for Canada's Agrifood Sector*. Agrifood Competitiveness Council. Guelph, Ontario, Canada.

10

Tobacco Supply Management: Examples from the United States and Australia

D. A. Sumner

Abstract

It is sometimes assumed, especially in Canada, that supply management policies require nontariff barriers in order to be effective. However, a supply management scheme has limited US tobacco output, without significant import protection, for more than half a century.

The US tobacco program used production quotas throughout the period during which the United States was the world's largest importer and exporter of tobacco. Ironically, in 1994 at the same time the United States was instituting a new policy of import protection, Australia was in the process of eliminating its own long-standing import barriers under pressure from the ban on nontariff import barriers contained in the Uruguay Round agricultural trade agreement (URA). Australia converted its domestic content rule to a tariff-rate quota for leaf tobacco and reformed its domestic supply management.

This chapter describes the supply management policies in the United States and Australia with respect to the tobacco industries. Particular emphasis is placed on how current policy changes relate to the URA in order to provide insights that may be applied in the context of Canadian supply management schemes.

Introduction

Agricultural supply management policies differ by commodity and location but tend to have similar consequences and a similar basic design (Alston, 1992). It may be instructive, therefore, in the context of a review of Canadian policy, to consider the operation, effects and recent reforms of supply management policies for the tobacco industry in the United States and Australia. These policies are similar in that they have relied on production quotas but are dissimilar with respect to their trade policies.

It is sometimes assumed, especially in Canada, that domestic supply management policies require nontariff barriers in order to be effective. However, such trade policies are clearly not required for low-cost industries or industries protected by natural barriers, such as transport costs. It is true that Canada has tended to use domestic supply management in industries that also employ tight import barriers, which insulate the domestic industry from the international market. But such a linkage between import protection and domestic supply management is not a necessary condition for a farm program to return substantial welfare benefits to farmers or other rural asset owners.

One of the most important supply management schemes in world agriculture has regulated US tobacco supply for more than half a century. Until 1994 this program operated with only relatively low import tariffs (Johnson, 1984). In the case of US tobacco, even the generally low import tariffs overstated the amount of protection that they would provide because tariff revenue was generally rebated through a duty-drawback system. Further, the US tobacco program used production quotas throughout the period in which the United States was the largest importer and exporter of tobacco.

The US tobacco industry began exporting three centuries ago. Its significant world market power was derived initially from a large market share but evolved to rely on a high degree of product differentiation. Throughout recent decades the market position of the US leaf tobacco industry steadily eroded, both in the export and domestic markets in competition with imports. This did not mean that monopoly rents were completely dissipated, but it did signal a change in the balance of pressures on policymakers. After many years of agitation against imports, the industry encouraged the US Congress to restrict imports by instituting a domestic content law that took effect in January of 1994.

Ironically, at the same time the United States was instituting its new policy, Australia was in the process of eliminating its own long-standing domestic content rule under pressure from the ban on nontariff import barriers contained in the URA. For many years Australia has maintained supply management under the protection of a domestic content rule, which limits imported leaf tobacco to a prespecified fraction of a domestic cigarette. Under the URA, Australia has converted the content rule to a tariff-rate quota for leaf tobacco protection and is reforming domestic supply management as well.

Canadian supply management programs continue to regulate domestic supply behind a wall of import protection. However, this is not a requirement of supply management. This chapter describes the supply management policies in the US and Australian tobacco industries and discusses their implications. An emphasis is placed on current policy changes and how they relate to the URA. These insights may be applied in the context of Canadian supply management schemes.

The US Tobacco Program

The US tobacco program, which began its operation in the 1930s, is a classic example of a production quota policy in agriculture. The US program is represented by Figure 10.1. In this standard quota diagram, tobacco exports are shown as an important part of total demand, and imported tobacco is shown to compete primarily with domestic leaf tobacco. However, domestic and imported tobacco are not perfect substitutes even within the same tobacco type.

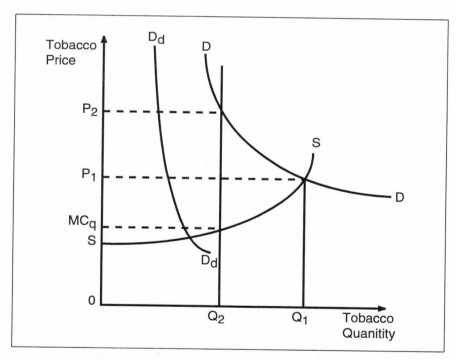

FIGURE 10.1   Effects of the US Tobacco Program on Price, Quantity and Quota Rent

In Figure 10.1, Q_1 and P_1 are competitive output and price, respectively. Q_2 and P_2 are the observed quota-restricted output and price, respectively. The domestic demand is D_d, and total demand is D, which includes exports. The supply curve, S, represents marginal costs and crosses Q_2 to indicate the marginal cost of MC_q at the restricted output quantity. This leaves quota rent per unit as $P_2 - MC_q$.

Welfare consequences of this program are affected by both exports and imports. Exports constitute about 40% of US tobacco leaf production. Further, cigarette exports have expanded to become a significant part of domestic manufactured output. Cigarette imports are negligible, but leaf imports grew from 15% of domestic use in the 1960s to about 35% in 1993.

Background data are provided in Table 10.1 as are estimates of supply and demand parameters from previous research. Export demand elasticities are based on econometric estimates from Sumner and Alston (1987). These estimates are still applicable to the US tobacco situation. Goodwin and Sumner (1992) estimated a supply elasticity of about 4.0 for flue-cured tobacco in North Carolina, which, as a long-run estimate for the whole US industry, is somewhat low. In particular, tobacco supply expansion will not likely be limited much by land use (only a small share of the suitable land is used even in the most tobacco-intensive regions) or by the availability of experienced human capital.

TABLE 10.1   Basic Data and Parameters used to Simulate Tobacco Policy Changes

Variable	Symbol	Most Likely Value
Total Quota Quantity	Q	1.50 billion lbs
Domestic Sales	Q_d	0.90 billion lbs
Exports	Q_e	0.60 billion lbs
Quota Price	R	$0.30/lb
Producer Net Price (P_d-R)	P_p	$1.20/lb
Consumer Price	P_d	$1.50/lb
Export Price	P_e	$1.50/lb
Elasticities		
Supply	ε	5
Domestic Demand	η_d	-1
Export Demand	η_e	-4

Source: Base data and elasticities from United States Department of Agriculture (USDA) sources, Sumner and Alston (1987) and Goodwin and Sumner (1992)

In order to better understand the basic mechanisms of this supply control scheme, these data and parameters may be used to assess the consequences of the program as compared to alternative policies. Some consequences of the output quota program, as compared with an absence of supply management, are provided in Table 10.2. They are based on a simulation of policy removal using the most likely value for each supply and demand parameter. Note in this table that industry output would be up by about 30%; price would be down by about 15%; and total revenue would be up by almost 15%. The elimination of domestic production quotas would substantially reduce imports and increase exports, but total tobacco use would rise only slightly.

TABLE 10.2   Likely Effects of Deregulation on US Tobacco Prices, Quantities and Revenues

Variable	Units	Base Value	Most Likely Change	Most Likely % Change
US Tobacco				
Price	$/lb	1.50	-0.22	-14.8
Marginal Cost	$/lb	1.20	0.08	6.7
Domestic Use	billion lbs	0.90	0.13	14.8
Exports	billion lbs	0.60	0.36	59.2
Output	billion lbs	1.50	0.49	32.5
Revenue	$ billion	2.25	0.30	13.2
Imported Tobacco				
Quantity	billion lbs	0.35	-0.10	-29.6
United States' Use of all Tobacco				
Quantity	billion lbs	1.25	0.03	2.4

Source: Changes in imported quantity were computed using a cross-price elasticity of demand for imports with respect to the US tobacco price of 2.0 (Sumner and Alston, 1987), which is compatible with the other demand elasticities. Changes in the United States' use of all tobaccos were computed by totaling the changes in the uses of domestic and imported tobacco.

One way to assess the welfare impacts of a quota program is to ask what would happen to the economic surpluses under alternative settings of the quota. Table 10.3, based on Alston and Sumner (1988), indicates the effects of alternative quota under the assumed supply and demand conditions compared to an aggregate quota of 1,500 million pounds. Row one shows that with no quota US growers and quota owners lose more than $300 million per year; US consumers, producers and quota owners together lose about $100 million; and the world gains about $73 million. Interestingly, the 1,500 million pound quota is too large to maximize growers' quota-owner surplus. If the quota were further restricted, growers would gain almost $70 million, but this would more than offset the losses by consumers. The United States would lose a total of $70 million. The quota of about 1,500 million pounds is very close to that which maximizes US total welfare. Unfortunately, only $0.4 million is to be gained by slightly enlarging the quota, but foreign buyers would offset the losses by domestic buyers.

Note also in the bottom two rows that price discrimination policies would clearly generate more surplus for the grower-quota owners or for the United States as a whole compared to the simple output-quota policy. In one case the domestic price would be raised to maximize grower returns; in the other case the export price would be raised. The US peanut program uses tight import barriers to practice such price discrimination. These import barriers extract additional rents from domestic buyers but not from foreign buyers (Rucker and Thurman, 1990; Rucker, Thurman and Borges, 1996).

TABLE 10.3   Welfare Effects of Alternative Quota Rules on Different Interest Groups Relative to a Freely Transferable Quota of 1.5 billion lbs

Policy	*Effects on Surplus[a] of:*		
	Producers & *Quota Owners*	*United States* *in Aggregate*	*World* *Total*
Equal Domestic and Export Prices			
Competition (No Quota)	-313.8	-99.3	73.2
Producers and Quota Owners Maximum Surplus	68.2	-70.0	-146.3
US Maximum Surplus	-11.5	0.4	8.4
Price Discrimination			
Producers and Quota Owners Maximum Surplus	206.3	-210.1	-215.4
US Maximum Surplus	-251.8	28.3	2.7

[a]Units: $ million/year
Source: Alston and Sumner (1988)

The evidence has been clear that when tobacco demand or quota quantities shift, the lease rate for quota responds just as economists would expect. The explicit or implicit market for quota leasing reflects any changes in the market for leaf tobacco; more demand for tobacco means a higher quota price, and more quota means a lower quota price. Further, responses that are suppressed in the leaf market are magnified in the quota market. Attempts to limit variability in leaf prices create additional variability in the market for quota (Sumner, 1988). Therefore, as a stabilization policy, it is not clear that the quota program has been successful.

As with all supply management schemes, quota-transfer has been an ongoing issue in the US tobacco program. Periodically, quota has been leased separately from land, but sales of quota were not allowed. At other times, and for some types of tobacco, sales were allowed but leasing was banned. In most cases, transfers are limited to within the original county to which the quota was assigned, but that rule was modified in recent legislation for some tobacco types in some states. Recent work on quota transferability (Rucker, Thurman and Sumner, 1995), which is based on data from county quota markets in North Carolina, has indicated that overall welfare losses from restricting intracounty quota transfers to within countries have been relatively small — only a few percent of total revenue or overall surplus. However, losses associated with potential transfers between quota owners and those who lease-in quota, implied by restrictions in transferability, have been more significant in many markets. In high-quota lease-rate locations, the restriction against allowing quota to enter the county penalizes growers and benefits quota owners. (The opposite is true in relatively low lease-rate locations.) Quota transfer restrictions have been maintained politically because, in many local markets, there are relatively small aggregate welfare losses and offsetting gains and losses. Therefore, there has been no overriding industry interest in removing the restrictions that serve some local political interests.

Overall, the US tobacco program has operated to extract cartel monopoly rents from domestic and foreign consumers to the benefit of domestic quota owners and foreign producers. Domestic consumers have paid about one penny per pack more for their cigarettes because of the supply controls — a fairly trivial cost compared to state and federal excise taxes that generally have been over 60 cents per pack. However, because of the continuing market power of the domestic industry in international markets, the United States generally gains from the quota program while foreign welfare is reduced.

The lesson for Canada is that a larger industry can be supported if domestic quota is large enough to allow the industry to export. Alternatively, the more supply is limited, the larger the quota values and the higher the cost of production, which includes the cost of quota rent. Obviously, supply management can cause industry costs to rise such that an otherwise competitive industry can be priced out of the international market. Given the US policy, let us examine how the recent General Agreement on Tariffs and Trade (GATT) will affect the US tobacco industry.

Implications of the URA for US Tobacco

The net benefits of GATT for the US tobacco industry were clearly understood until the Fall of 1993 (Sumner, 1991; USDA, 1991). As noted above, US tobacco tariffs were generally low, and the United States did not use any nontariff barriers, such as the 22 quotas used for peanuts or dairy products to protect the domestic price-support policy. Further, tobacco export growth has been large in recent years, and international markets for both manufactured tobacco products and leaf tobacco would expand further with the URA because of income growth and improved access.

The implications of the Uruguay Round for tobacco became more complicated in the months leading up to GATT. Unlike the case of peanuts and other crops, the URA as applied to tobacco, did not change as the negotiations neared completion. Rather, US trade policy for tobacco changed. In the Fall of 1993 a few months before the December 15, 1993 completion of negotiations for the Uruguay Round, the United States approved a new nontariff import barrier for tobacco to be implemented January 1, 1994. It stated that, for 1994, cigarettes manufactured in the United States could contain no more than 25% of imported tobacco. Thus, the United States instituted a new agricultural nontariff barrier just two weeks after having successfully completed a seven-year campaign to ban such barriers from international trade in agriculture (Zaini, Beghin and Brown, 1994). Evaluating the URA for tobacco now meant analyzing a new trade barrier.

Not only was this domestic content law (nontariff barrier) inconsistent with the URA, it was found to be a violation of existing GATT rules. The United States did not use its Section 22 waiver or any other specific authority to provide a GATT-legal excuse for the law. Tobacco exporters, including Brazil and Argentina, brought a GATT case that argued that the US law violated Articles III and XI of GATT. The United States made no real legal defense of its new law, it only issued a public statement that no damage had been done to those wishing to export to the United States. Brazil disputed this claim. However, even without a demonstration of damage, the current GATT seems clear regarding the new domestic content laws; it states that these laws violate existing rules and are disallowed in general. The United States lost the GATT tobacco case and in early 1995 is in the process of developing compensation under GATT Article 28 and converting the domestic content rule to a tariff-rate quota.

Oddly, it is not at all clear that the US industry gains from the domestic content law. By placing a restriction on the factor mix, the regulation causes cigarettes to be more costly. Manufacturers would respond by reducing production and by shifting some cigarette manufacturing off shore. Given that a substantial portion of domestic cigarettes are now exported, this second option may be quite important. An analysis in a draft paper by Zaini (1994), based on parameter estimates in Sumner and Alston (1986) among other sources, has indicated that the US leaf producers and quota owners benefited from the content rule, at least in the short run. However, the parameters he used do not reflect the potential for cigarette manufacturers to move off shore or the potential for tobacco importers to retaliate against such a blatant unilateral rejection of the new URA.

Figure 10.2 shows an isoquant diagram illustrating the trade-off between domestic and imported tobaccos. Isoquant A is tangent with the relative price line, at domestic quantity Q_d. The content regulation, however, specifies that input use must lie on or below the diagonal line marked 75%. Therefore quantity Q_d is no longer feasible. Point a on isoquant A complies with the regulation, but a move to this point ignores the fact that the regulation would cause domestic manufacturers to produce less than quantity a. The domestic content rule requires a move to points such as b or c on isoquants B or C. More domestic use would follow if the cigarette output decline was moderate, but if output were to fall all the way to isoquant C, less domestic use would be implied even though the domestic share had increased to 75% of total tobacco use.

In addition to the domestic use of domestic tobacco, the new content regulation may adversely affect the market for US leaf tobacco exports. International leaf importers are likely to view the new US trade barrier as a violation of the United States' commitment to freer trade, and they may respond with more trade barriers of their own.

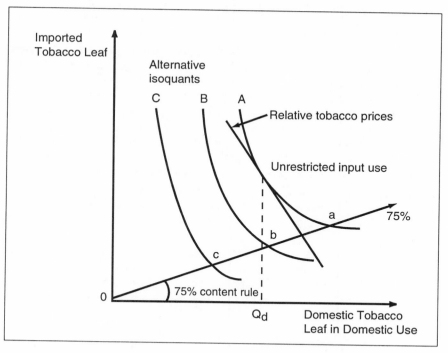

FIGURE 10.2   US Domestic Content Regulation and Domestic Tobacco Use

The domestic content regulation adds a constraint for cigarette manufacturers that may encourage a shift to offshore manufacturing. As the domestic market shrinks, any export market retaliation by foreign countries will become more important. It is unclear how the content policy shifts total demand for US leaf even if domestic demand increases slightly and becomes more inelastic because of the constraint on import supply.

Canadian supply management policies have long relied on nontariff import barriers. The US tobacco industry has attempted to implement such measures but seems to have lost sight of the importance of its export market in the process.

Australian Tobacco Policy and Recent Reforms

As the United States was in the process of instituting its new domestic content law for tobacco, Australia was in the process of dismantling its own domestic content regulation as a part of domestic policy reforms in compliance with the URA. As discussed in Beghin and Sumner (1993) and in Beghin and Hu (1994), the domestic content rule in Australia has limited tobacco imports and allowed the production quota program to maintain domestic tobacco prices well above world market levels.

Australia's domestic production quota program for tobacco is operated independently in Victoria and Queensland. The aggregate quota is set annually at a quantity expected to clear the market. The average quota price is determined by an annual bargaining session, supervised by the government, between growers, representatives and the three domestic cigarette manufacturers. Individual growers are allocated production quota as a share of the national total, and each manufacturer is allocated a share of the market based on historical sales.

Recent descriptions and data from Australia provided by David Goldsworthy (1994 and 1995), an Australian Tobacco Board member and an economist with the Victorian Department of Agriculture, indicate that the domestic quota program in Australia is now being reformed and perhaps eliminated.

In Australia, quota was sold for about $2.50/kg as recently as 1992. However, quota quantity and domestic tobacco use fell by about 30% between 1992 and 1994. Cigarette tax increased and more stringent smoking regulations were introduced in the last few years as they were in many other countries, including Canada and the United States. Unlike the United States, however, Australia has no opportunity to participate in the expanding export market for cigarettes because it is a high-cost, low-quality tobacco producer.

In the State of Victoria, which grows about 40% of national production, the government agreed to buy out a 1.5 million kilogram share of existing production quota at about $2.00/kg. About $3 million was distributed to 70 growers. The rationale for spending state funds was that industry decline was a function of government action. The buy-out was also politically useful, if not necessary, to achieve the regulatory reforms.

The domestic content policy has resulted in imports accounting for 43% of domestic tobacco use. This is a bargained solution, supervised by the government, between a group of domestic users and the domestic producer cartel (Beghin and Sumner, 1993). The GATT caused Australia to replace its domestic content rule with a tariff-rate quota. The first tier, low-tariff quantity was set, under the rule assuring the maintenance of current access, at the average quantity of imported tobacco for the base years, the late 1980s. However, since domestic tobacco use in Australia has been declining, use of an historical base period suggests that imports will now have access to far more than the 43% of total tobacco use upon which the tariff-rate quota itself was based. Unilateral domestic policy changes in smoking regulations and import tariffs are expected to cause even further reductions in the domestic tobacco industry. Thus, a much smaller Australian leaf tobacco industry than the one maintained under current policy will exist in the future.

Canadian supply-managed industries may see themselves as more similar to the Australian tobacco industry than to the US tobacco industry. The Canadian tobacco industry in Ontario has long operated under the shadow of US production quotas and Canadian import barriers. The Victorian example showed that it is possible to buy out a supply management scheme. The political opposition of the industry can be mollified by governmental capital outlays because governments usually find it difficult to simply eliminate the capital assets they have created.

Conclusions

- US tobacco policy pursued supply management without import barriers because quality differences allowed the US tobacco cartel to face downward-sloping demands in world markets, even as overall market share declined.
- Whereas this monopoly power has eroded with the high prices caused by quotas, the rents to the industry remain significant as indicated by (implicit and explicit) quota lease rates of 20–25% of the tobacco market price. Without the quota the industry would expand but profit would fall.
- The United States has had only relatively low tariffs for most types of tobacco, and GATT will expand US exports and export prices by a few percent because of reduced international trade barriers and income growth in importing countries. Until the recent domestic content rule, the US tobacco industry was a clear winner and the domestic policy was essentially unaffected by the Uruguay Round proposed agreement.

- The new domestic content law created a new nontariff barrier starting in 1994, just as GATT had eliminated such nontariff barriers elsewhere. If the law stays in place, the United States will likely be forced to use GATT Article 28 and pay compensation to affected parties.
- Given buyers' reactions and the importance of exports, it is not obvious that the minimum domestic content requirement of 75% benefits the US leaf industry.
- Australia has an output quota policy similar to that of the United States, but it does not export and needs an import barrier to keep domestic prices above potential import prices.
- As in most of the developed world, domestic demand for tobacco in Australia has been falling (in part, because of tax increases and regulations). The domestic quota program is under stress and the government has agreed to buy out (some) grower quota to reduce the losses incurred by growers.
- As a part of the URA, Australia is eliminating its domestic content law and setting a tariff-rate quota at the average of recent imports. However, this quantity will be a large percentage of future domestic demand because the market itself is shrinking. The future outlook is for a much smaller tobacco industry in Australia with low domestic consumption, imports of leaf and no export potential.

Finally, the inconsistency of the United States instituting a nontariff barrier for a major export industry just when other such barriers are being dismantled (in part due to US insistence) is particularly ironic given that the U S tobacco industry has operated a successful cartel and has a future dependent on exports. This new domestic content law will surely cause trade friction and related problems. Meanwhile, Australia has been pressured into eliminating its domestic content law. Its tobacco industry will almost surely decline significantly over the next few years but mainly as a result of declining domestic cigarette sales. Therefore, as Canada faces adjustments in its supply-managed commodity industries, the lessons from the United States and Australia may prove useful.

References

Alston, J. M. 1992. "Economics of Commodity Supply Controls," in Tilman Becker et al., eds., *Improving Agricultural Trade Performance Under the GATT*. Pp. 83–103. Kiel, Germany: Wissenschaftsverlag Vauk.

Alston, J. M., and D. A. Sumner. 1988. "A New Perspective on the Farm Program for US Tobacco." University of California, Department of Agricultural Economics, Davis, CA.

Beghin, J. C., and F. Hu. 1994. "Declining US Tobacco Exports to Australia: A Derived Demand Approach to Competitiveness." International Agricultural Trade Research Consortium Working Paper 94–3, Minneapolis, Minnesota (June).

Beghin, J. C., and D. A. Sumner. 1993. "Content Requirement and Bilateral Monopoly." *Oxford Economic Paper* 44: 306–316.

Goldsworthy, D. 1994. Personal Correspondence (June).

_____. 1995. "Deregulation of the Victorian Tobacco Industry." Paper presented at the Australian Agricultural Economics Society Annual Meeting. Perth, Australia (February).

Goodwin, B. K., and D. A. Sumner. 1992. "Estimation of Market Supply Parameters Under Mandatory Production Quotas." Unpublished paper, North Carolina State University, Raleigh, NC.

Johnson, P. R. 1984. *Economics of the Tobacco Industry.* New York, NY: Praeger Publishers.

Rucker, R. R., and W. N. Thurman. 1990. "The Economic Effects of Supply Controls: The Simple Analytics of the US Peanut Program." *Journal of Law and Economics* 33: 483–516.

Rucker, R. R., W. N. Thurman, and R. B. Borges. 1996. "GATT and the US Peanut Market." This volume.

Rucker, R. R., W. N. Thurman, and D. A. Sumner. 1995. "Restricting the Market for Quota: An Analysis of Tobacco Production Rights With Corroboration from Congressional Testimony." *Journal of Political Economy* 103 (1): 142–175.

Sumner, D. A. 1988. "Stability and the Tobacco Program," in D. A. Sumner, ed., *Agricultural Stability and Farm Programs.* Pp. 113–138. Boulder, CO: Westview Press.

_____. 1991. "Tobacco and the Uruguay Round," in F. Delman, T. Slane, and M. Marion, eds., *Current Issues in Tobacco Economics.* Vol. 4. Pp. 196–198. Princeton, NJ: Tobacco Merchants Association.

Sumner, D. A., and J. M. Alston. 1986. *Effects of the Tobacco Program: An Analysis of Decontrol.* American Enterprise Institute for Public Policy Research, Washington, DC.

_____. 1987. "Substitutability for Farm Commodities: The Demand for US Tobacco in Cigarette Manufacturing." *American Journal of Agricultural Economics* 69 (2): 258–265.

USDA. 1992. *Preliminary Analysis of the Economic Implications of the Dunkel Text for American Agriculture.* Office of Economics, Washington, DC (March).

Zaini, H. 1994. "An Analysis of the US Tobacco Domestic Content Regulations." Seminar paper, North Carolina State University, Raleigh, NC (November).

Zaini, H., J. C. Beghin, and B. Brown. 1994. "Complying or Not With Domestic Content Policies? The Case of the US Cigarette Industry." Working paper, North Carolina State University, Raleigh, NC (August).

11

GATT and the US Peanut Market

R. R. Rucker, W. N. Thurman, and R. B. Borges

Abstract

In the United States, peanut sales for edible use are restricted by a federal quota system, and imports are virtually banned. The Uruguay Round of the General Agreement on Tariffs and Trade (GATT) has obvious importance for producers of such a commodity. We analyze the probable effects of GATT and related agreements on trade in peanut butter between Canada and the United States, on the US peanut trade and on the welfare of US peanut producers. We argue that a GATT side agreement, which halts growth in US imports of Canadian peanut butter, will increase the demand for US-grown peanuts and decrease any treasury costs associated with the US peanut program. We conclude that the primary effect of GATT on US markets will be an increase in raw peanut imports, which will reduce the demand for US-grown peanuts. The net effect of such increased imports on growers will depend upon how US policymakers respond. If there is no response in the aggregate quota or support price, this aspect of GATT will serve mainly to increase costs to the US Treasury. If there are policy responses, as seems likely, then we forecast the annual loss in peanut producer surplus will be in the range of $10–$18 million, which translates into less than one cent per pound of quota peanuts produced. Further, possible expansion in foreign demand for US peanuts as a result of secondary GATT effects may reduce (or even reverse) these losses.

Introduction

In Canada the production of such agricultural commodities as dairy, poultry and tobacco is regulated through the assignment of production quotas to individual producers. In the United States the only two major agricultural commodities whose production are limited by similar production quotas are tobacco and peanuts. In both Canada and the United States there are differences across commodities in the details of the quota-based supply management programs. The details of the US tobacco program are described by Sumner in Chapter 12 of this volume. The details of the US peanut program and the impacts of GATT on US peanut markets are discussed in this chapter.

The GATT, which was signed by more than 100 nations on December 15, 1993, represents the culmination of seven years of negotiations on international trade reforms. Most, if not all, agricultural commodities produced in the United States will be affected in some way by the agreement. In this chapter, we examine the likely effects of GATT on US peanut markets and a side agreement to GATT that limits imports of Canadian peanut butter into the United States.

Although the market value of US peanut production is small compared to many other US agricultural commodities, the impacts of GATT on peanut markets are of considerable interest for at least two reasons. First, because peanut revenues constitute an important component of incomes in many communities in the Southeastern United States, the impacts of GATT on peanut markets are important to farmers and policymakers in these areas. Second, the federal peanut program is unique among US commodity programs with its marketing quota, price supports and import restrictions. An important component of the federal peanut program is an explicit restriction on peanut imports. The restriction has brought attention to trade conditions for peanuts that seem unjustified when the value of the trade in peanuts is considered. During the course of the GATT negotiations, for example, peanut import restrictions were used by other countries as an example of restrictive US trade policies.

The analysis we present below suggests that GATT will have its primary effects on US peanut markets through increased raw peanut imports. This will decrease the demand for US-grown peanuts and will tend to increase the treasury costs of the federal peanut program. A potentially important effect related to GATT will result from a side agreement, which will halt the growth of US imports of Canadian peanut butter. Relative to a scenario in which peanut butter imports are not restricted, such restrictions will increase the demand for US-grown peanuts and reduce any treasury costs associated with the peanut program. A recent analysis by the United States Department of Agriculture (USDA) (March, 1994) suggests that the demand-increasing effects of the reduction in peanut butter imports will outweigh the demand-decreasing effects of the increase in imports, thereby benefiting US peanut growers. This analysis assumes that in the absence of GATT peanut butter imports would have continued their recent rapid growth. In contrast, we argue that peanut butter imports would not have continued to grow, because of political reasons, and the net benefits of GATT to US peanut growers are almost certainly negative, but modest. Our analysis suggests that, at worst, either the treasury costs of the peanut program will increase by about $23 million or annual grower revenues will fall by $10 million–$18 million (less than one cent per pound of quota peanuts produced). Because GATT allows the continued restriction of domestic supply through marketing quota, there will not likely be any offsetting benefits to US consumers.

GATT Mechanisms for Promoting Trade[1]

The GATT identifies four types of impediments to trade: market-access barriers, export subsidies, internal supports and scientifically unjustified sanitary and phytosanitary measures. Commitments are to be made to increase market access, decrease export subsidies and decrease internal support from average levels of prices and quantities during a designated multi-year base period. For peanuts this base period is 1986–88. The commitments made under the Final Act are to be phased in over a six-year period, beginning in 1995.

Of the four areas listed above, the most relevant for the peanut program is market access — imports of raw peanuts have been limited to about one-tenth of 1% of US consumption. Relaxing these restrictions, as required under GATT, will affect US peanut markets and the operation of the peanut program. Because of recent reductions in commodity support levels by the United States under the 1985 and 1990 Farm Bills, the United States is not required to make further reductions in internal support levels to meet its obligations under GATT. Further, the peanut program does not include any export subsidies. Second-order effects on US peanut markets may result from the sanitary and phytosanitary provisions if, for example, these lead to changes in standards relating to aflatoxin content for US imports into European countries. A side agreement to the Final Act that restricts peanut butter imports may also have effects on US peanut markets.

Peanut Markets and the Peanut Program[2]

Peanut production in the United States is concentrated in seven southern states — Alabama, Florida, Georgia, North Carolina, Oklahoma, Texas and Virginia — and is an important source of income in many local economies. There are two peanut markets separated both by end-use of the peanut and by government policy — the edible market and the crush market. Peanuts sold into the edible market are used in such products as salted-in-the-shell peanuts, candy bars and peanut butter. Peanuts sold into the crush market are used to produce peanut oil, cake and meal. Peanut oil is consumed by humans but is not included among the edible uses for policy purposes. Peanut cake and peanut meal are animal feeds.

Government programs regulating peanut prices and production have been in effect since the 1930s. Early incarnations of the peanut program used price supports and set acreage allotments to regulate production. Following large treasury costs in the mid-1970s, the acreage allotment-based peanut program was changed to a poundage quota-based program under the Food and Agricultural Act of 1977. A variety of adjustments to fine-tune the operation of the program have been made in subsequent farm bills. Under the current peanut program, a minimum price is established for peanuts sold into the domestic edible market. Only peanuts with poundage quota can be sold into the edible market. Quota peanuts can also be placed with the Commodity Credit Corporation (CCC). In either case, quota peanuts receive at least the edible support price.

Growers are allowed to grow more than their poundage quota, but these additional peanuts cannot be sold directly into the edible market. Growers who grow additionals have the option of contracting with handlers for sale in the export market.[3] Such contracts must be signed by a legislatively specified date — under current legislation, this date is September 15. Additionals without advance contracts must be placed under loan with the area growers' association. Additionals placed under loan are guaranteed a minimum additional support price that is well below the edible support price. For reasons discussed below, however, growers normally receive a payment for additionals placed in growers' association pools that exceeds the additionals support price and is close to the price received by additionals contracted for export. The high domestic support price is protected by a virtual ban on imports.

Although there are legislated barriers between the domestic edible market and the export and domestic crush markets, these barriers are not leakage-free. Peanut handlers (buyers) can, for example, buy back additionals for use in the domestic edible market. These buy backs require the buyers to make payments to the growers' association equal to the quota price plus a handling charge.

The growers' association takes under loan both quota and additional peanuts. The association can then sell quota peanuts for edible uses or for export or domestic crush. Additional peanuts placed under loan with the association can be bought back for use in the edible market as discussed above, or they can be sold for domestic crush. The association is also allowed to sell both quota and additional peanuts for export, but such sales cannot be made below a USDA-established export support price.[4] Because this export support price typically has been set above the world price, this option rarely is exercised.

The growers' association incurs losses if it buys quota peanuts at the edible support price and then resells them at lower prices for domestic crush or export. Such losses are borne by taxpayers. If, however, the growers' association buys additionals at the relatively low additionals support price and then sells them for the edible support price under the buy back provision, it earns profits. Any such profits are distributed among growers who place additionals in the association pools.[5]

Given the preceding, the effects of the peanut program on peanut markets can be analyzed. We show below that the current restriction on imports of edible-grade peanuts reduces potential gains from trade and that the poundage quota and quota support price maintain an artificially high price in the domestic edible market. This generates a flow of rents to owners of poundage quota but imposes (larger) costs on domestic consumers. We also demonstrate that the buy back provision usually prevents the price of edible peanuts from rising above the quota support price.

To focus on the primary elements of the peanut program we make the following assumptions.[6]

(1) The USDA perfectly enforces all provisions of the peanut program.

(2) There are two uses for peanuts (edible and crush) and two qualities of peanuts (edible-grade and crush-grade). Edible-grade peanuts are of higher quality. They can be used either for edible purposes, or they can be crushed into oil and meal. Crush-grade peanuts are suitable for crush use but cannot be used for edible purposes. US peanut producers grow only edible-grade peanuts while foreign producers grow both edible- and crush-grade peanuts.

(3) The foreign prices of crush- and edible-grade peanuts are exogenous to domestic markets. This assumption is introduced for exposition.

(4) The legislated minimum price for peanuts sold by growers' associations into the export edible market is higher than the world peanut price, so peanuts in association pools are priced too high to be sold in the export edible market.

(5) Imports for edible use are prohibited. Private exports of edible-grade peanuts are not restricted. There are no effective restrictions on the import or export of crush-grade peanuts.

Figure 11.1 shows the domestic demand for edibles, D_e; the domestic demand for crush, D_c; the foreign price of edible-grade peanuts, P^f_e; the price of crush-grade peanuts, P_c; and the domestic supply of peanuts, S^d. In the absence of government restrictions and given our assumption that only edible-grade peanuts are grown in the United States, the domestic edible and crush markets equilibrate separately at prices P^f_e and P_c. The entire domestic crop of Q^s_e is sold for edible uses. Domestic consumption of edibles is Q^d_e; edible exports are X_0; and Q^d_c pounds of crush-grade peanuts are imported for domestic oil and meal uses.

Next, consider an aggregate edible quota in the amount Q^0_q and a support price of P_s, chosen in Figure 11.1 so that the domestic edible market clears. The support price is maintained by both the quota and the ban on imports. In the domestic market, demand is truncated at Q^0_q for prices below P_s. Because the marginal price facing peanut producers is unchanged at P^f_e, however, total domestic production is unchanged. The reduction in domestic edible consumption (from Q^d_e to Q^0_q) is matched one-for-one by an increase in exports from X_0 to X_1. In this scenario, the crush market is unaffected by the quota and support price. The welfare effects with this quota and support price are confined to the domestic edible market where consumer surplus falls by ABEC and producers gain ABDC. The deadweight loss is BDE, the area below the domestic edible demand curve, above the world price and between the quantities demanded at the support price and the world price.

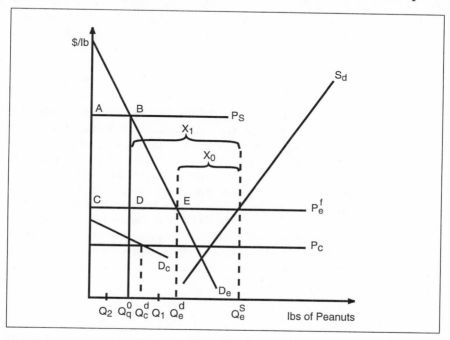

FIGURE 11.1 Effects of US Peanut Program

In the preceding analysis, we assumed that policymakers were able to choose a support price and quota level so that there was neither a shortage nor a surplus of peanuts in the domestic edible market. Because of imperfect knowledge concerning the demand for edibles, however, policymakers may not be able to accomplish this. If, for example, the quota is set equal to Q_1, then an excess supply of peanuts to the domestic edible market results. This excess supply is purchased by the growers' association at the support price and resold into the crush market at the crush price. Treasury costs equal the difference between the quantity demanded at the support price and the quota, multiplied by the difference between the support price and the crush price.

An alternative situation is one in which there is excess demand for edibles, rather than excess supply, at the support price. This situation results if the quota in Figure 11.1 is set at Q_2. Here, the market-clearing price would be above P_s. The buy back provision of the peanut program, however, typically prevents the price in the domestic edible market from rising above P_s. This provision allows for handlers to buy back non-quota peanuts for the domestic edible market at the support price. In Figure 11.1, handlers, who are allowed to buy back as many peanuts as they wish, will request buybacks of Q^0_q - Q_2, thereby effectively shifting the domestic supply of peanuts for the domestic edible market out to Q^0_q. Only if the total available supply of edible-grade quota and additional peanuts under loan is less than Q^0_q will the domestic edible price exceed the support price.

The Effects of GATT on US Peanut Markets

The provision of GATT that most directly affects US peanut markets is the market access provision, which allows substantial increases in imports of raw peanuts over a six-year transition period. Less direct, or secondary, effects may be felt through changes in the foreign price of peanuts, which results from GATT-induced shifts in the demand for US-grown peanuts. A primary concern of the US peanut industry in recent years has been the growth in imports of peanut butter, mainly from Canada. In a separately negotiated side-agreement to GATT, future imports of peanut butter from Canada (as well as other countries) are limited to their 1993 levels. Specific terms of these agreements, as well as their relationships to the Canada–United States Free Trade Agreement (CUSTA) and the North American Free Trade Agreement (NAFTA) are discussed below.

The Effects of GATT on US Imports of Raw Peanuts

Currently, imports of peanuts are virtually banned under Section 22 of the US Agricultural Act of 1933, as amended. The GATT called for a relaxation of this ban to allow imports into the United States equal to 3% of domestic consumption during the 1986–88 base period. This limit is referred to as a minimum access requirement. Allowable imports should rise to 5% of base consumption by the sixth year of implementation of the agreement. For import quantities below the allowable limits, a duty will be charged equal to $60/ton (3¢/lb). Peanuts may be imported above the limits but only if a high tariff is paid. The tariff on shelled peanuts is initially set at 155% of the transactions price; it will decline by 15% to 131.8% by the sixth year of implementation. Peanuts in-the-shell above the minimum access requirement will initially be charged a tariff of 192.7% and will also be reduced by 15% by the sixth year.

An important question is whether it is likely that raw peanuts will be imported into the United States above the minimum access levels.[7] That is, is it likely that the world price of peanuts will be low enough to make it worthwhile to import peanuts into the United States even after paying the 100%-plus tariff? Our answer to the question, in short, is that it is quite unlikely that importing peanuts into the United States on such terms will be profitable. Therefore, the imports to expect under GATT are exactly at minimum access levels. Our conclusion is based on a historical comparison of world peanut prices and the US edible support price.

A monthly series of observations on the Rotterdam (cost, insurance, freight) price for US runner peanuts is displayed in Figure 11.2 for January, 1980, through January, 1994. These cleaned, shelled peanuts are of a quality ready for processing into peanut candies or peanut butter, the former being the primary purpose for peanuts purchased in Rotterdam. The sample mean of the nominal prices is near $1,000/ton. (The effects of deflating the series are discussed below.) The series displays the shape typical for a storable commodity price series (Williams and Wright, 1991) — typical levels are low near $800 with sporadic jumps to quite high levels, and two levels are above $2,000/ton. The two most prominent bursts, one in late 1980 and the other in late 1990, followed short US crops. During both episodes the US import ban was relaxed, and peanuts were imported.[8]

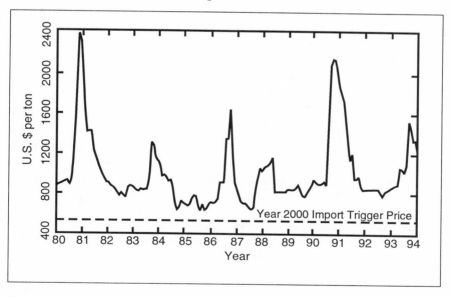

FIGURE 11.2   US Runners on the Rotterdam Market

In addition to the time series of Rotterdam prices, Figure 11.2 displays a horizontal line labeled "Year 2000 Import Trigger Price." It represents the level to which world peanut prices would have to fall in order to trigger imports beyond the minimum access level from world markets into the United States. It is calculated as follows.

The US support price for edible peanuts for the 1994–95 crop year is $678.36/ton on a farmer's stock basis (a price for peanuts in the shell). Using conversion coefficients developed by the US International Trade Commission (ITC) (1993), the shelling of peanuts results in a 25% loss in product weight while the cleaning and culling result in an additional 12% loss. Therefore, $678.36/ton in-the-shell is equivalent to a shelled price of $678.36/[(.75)(.88)] = $1,027.82. After adding the $200 estimate for shelling costs that is used by the ITC, the US shelled-and-cleaned equivalent price is $1,227.82. This price should be compared to the price of imported peanuts after paying the tariff for peanuts above the minimum access level.

The price of imported peanuts will be related to the Rotterdam price of peanuts as:

$$P_{imports} = (P_{Rotterdam} + \Delta \text{ transport cost}) (1 + r)$$

where r is the tariff rate — 1.55 in 1995, falling to 1.318 in 2000. Δ transport cost is the difference in cost between shipping peanuts from, say, Argentina to Rotterdam and from Argentina to the United States.

As a reasonable approximation, assume that there is no change in transport cost if peanuts are redirected from Rotterdam to the United States. We can find the Rotterdam price that will trigger imports, $P_{trigger}$, by equating the expression for $P_{imports}$ with the US shelled-and-cleaned equivalent price:

$$P_{trigger}(1 + r) = \$1,227.82$$

or

$$P_{trigger} = \$1,227.82/(1 + r).$$

The trigger price calculated with the year 2000 tariff rate of 1.318 is $529.69 and is displayed in Figure 11.2. For earlier years, the trigger price is lower.

Note that in Figure 11.2 the Rotterdam price has not fallen to the trigger level over the last decade and a half. Further, if one accounted for the decline in the purchasing power of the dollar over the period, one would inflate earlier year Rotterdam prices. Thus, our conclusion is that imports of raw peanuts under GATT should equal only the minimum access levels.

Our conclusion follows from the assumption that the sum of quota production and additionals placement into the pools is sufficient to meet US domestic demand at the quota price. This condition ensures that the US domestic price will not rise above the support price. What would happen if the US domestic price did rise above that level? What will happen when world market conditions are as they were in 1980 and 1990? If the US crop is short and there exists a large enough price differential at the border to induce imports, then it is possible that imports will enter the country. This requires both an unusually short domestic crop and unusually low world prices. Because the domestic crop is assumed to be small, however, the role of imported peanuts would be to augment US consumption to the level demanded at the support price rather than to displace US peanuts in the domestic market. The effect on US growers would be to lower their received price from a level higher than the support level to a level closer or equal to the support level.

The Effects of GATT on US Imports of Peanut Butter

Imports of raw peanuts have virtually been banned under the authority of Section 22 of the 1933 Agricultural Adjustment Act. Imports of processed peanut products have been treated differently. Historically, there have been no restrictions on imports of peanut butter in particular. The difference between US and foreign peanut prices would appear to create considerable incentive to purchase peanuts, produce peanut butter outside the United States and then import the peanut butter into the country. Until recently, however, peanut butter imports have been insignificant. Table 11.1 shows that imports of peanut butter, primarily from Canada and Argentina, comprised less than 1% of estimated domestic consumption until 1990. After that time Canadian imports rose rapidly.

Total imports comprised 5.5% of domestic consumption in crop year 1993. The bulk of those imports came from Canada. Though imports slowed in 1994, a decrease in domestic demand continued to increase the market share controlled by imports. Throughout this period and because of CUSTA, the tariff on imported Canadian peanut butter has decreased by annual 10% decrements from $60/ton in 1989.

The rapid increase in Canadian peanut butter imports has concerned US peanut producers because of its effect on the derived demand for US peanuts (Nail, 1994; Hollis, 1994). Purchases of US peanuts by peanut butter manufacturers declined in crop year 1993, and the decline was attributed, in part, to the surge in imports. Producer concern led to an ITC investigation on whether or not a Section 22 action to limit Canadian peanut butter imports was warranted.[9]

Evidence that policymakers have responded to peanut growers' concern is found not only in the ITC investigation but also in statements from the executive branch. The Secretary of Agriculture responded to one congressperson's concerns over the deleterious effects that NAFTA may have on US peanut growers by saying that he would "work vigorously ... to limit the volume of Canadian exports of peanut butter and paste, which would include your suggestion of a cap at 1 percent of US domestic consumption" (Orden, 1994).

US grower concern over imports led to a side agreement in which the most important provision limits Canadian peanut butter imports into the United States to 31.9 million pounds per year, an amount roughly equal to that imported in the 1993 calendar year. Imports above that level are subject to a tariff rate of 155%.[10] This agreement coincides with GATT and is scheduled

TABLE 11.1 Peanut Butter: US Imports from Canada and Argentina and Domestic Consumption

Year	Peanut Butter Imports			Peanut Butter Consumption			Food Use of Peanuts	
	Canada	Argentina	Total	Estimated Domestic Consumption	Imports as a %	Price	Total Food Use	In Peanut Butter
1982	0.1		0.1	685	0.0%		1,849	700
1983	0.0		0.0	670	0.0%	$1.48	1,856	696
1984	0.1	0.1	0.2	794	0.0%	$1.53	1,911	723
1985	0.3	0.5	0.8	700	0.1%	$1.55	2,023	726
1986	0.2	3.9	4.5	689	0.7%	$1.73	2,073	713
1987	0.6	2.5	3.5	721	0.5%	$1.80	2,071	747
1988	1.7	3.4	5.3	832	0.6%	$1.79	2,254	860
1989	3.7	2.2	5.9	864	0.7%	$1.84	2,312	897
1990	7.9	6.0	14.3	721	2.0%	$2.07	2,020	742
1991	11.6	6.7	19.6	863	2.3%	$2.03	2,207	886
1992	28.1	6.9	35.7	793	4.5%	$1.87	2,122	798
1993	30.2	6.8	39.6	725	5.5%	$1.82	2,035	727

Notes: Units are millions of pounds.
 Import data are from Commerce Department, August–July crop years.
 Price data are crop year averages of retail CPI component.
 1993 quantities are only through November, 1993.
 Total food use of peanuts: farmers' stock (in-shell) basis.
 Peanuts in peanut butter: shelled basis.
 Estimated peanut butter consumption = (peanuts used in peanut butter)(.88/.9) + imports — exports. The factor .88
 accounts for cleaning losses. The factor .9 accounts for non-peanut inputs into peanut butter.

to take effect on July 1, 1995. This leaves Canadian (and other) peanut butter imports uncontrolled for the 1994 crop year, a fact of obvious concern to US growers.

With the discussion above as background, consider the task of analyzing the effects of GATT on peanut trade. Should we attribute to GATT the freezing of Canadian peanut butter imports at the 1993 level under the side agreement just discussed? This is the approach taken by the USDA (March, 1994) whose analysts assume that recent trends in peanut butter imports would have continued without GATT. They predict that by the year 2000 imports will have increased roughly fivefold over the 1993 level, the level to which the side-agreement restricts imports. If viewed as an effect of GATT, this restriction is an important counter, in terms of US peanut producer welfare, to the deleterious effect of increased raw peanut imports. In fact, the USDA argues that the restrictions on peanut butter imports are worth more to domestic producers than the losses from expansion of raw peanut imports. They also argue that GATT is a boon to domestic producers.

We find this to be a strange interpretation of the effects of GATT. Peanut producers clearly were concerned about peanut butter imports before the Uruguay Round was near completion, and some action to restrict Canadian peanut butter imports seemingly would have been inevitable, with or without GATT. Therefore, our best estimate of the effect of GATT on peanut butter imports is zero.

In the GATT impact estimates presented below, we humor the reader who disagrees with our political prediction and consider two scenarios for the effects of GATT on peanut butter imports. One scenario corresponds to our belief that GATT will have no effect on peanut butter imports — political pressures will lead to restrictions on imports even without the side agreement to GATT. The other scenario corresponds to a situation where, in the absence of GATT, peanut butter imports would increase at a rate that we consider highly unlikely given current political conditions.

The Estimated Effects of GATT on Domestic Demand, Producer Surplus, and Treasury Costs

The estimated effects of GATT on the domestic demand for US-grown peanuts, producer surplus and treasury costs under four different scenarios are presented in Tables 11.2 and 11.3. The calculations underlying these tables can be understood by referring to Figure 11.3. In that figure the initial domestic demand for US-grown peanuts is D_e^0, and the initial national quota and quota support price are Q_0 and P_s, respectively. In this situation, the quantity demanded at the support price is exactly equal to the quantity supplied, implying that initially there will be neither treasury costs nor buybacks.

TABLE 11.2   Estimated Effects of GATT

Scenario Assumptions:

(I)   Without GATT, no increase in peanut butter imports, and
(II)   (A)   Treasury costs are allowed to increase (aggregate quota remains constant).
        (B)   Treasury costs are not allowed to increase (aggregate quota is reduced).

	Change in Domestic Demand (million lbs) Due to:			Net Change in Demand (million lbs)	Change in Producer Surplus ($ million)		Change in Treasury Costs ($ million)	
Year	Increased Imports of Raw Peanuts	Reduced Imports of Peanut Butter	Increased US Income		(A)	(B)	(A)	(B)
1995	-72.6	—	+11.6	-61.0	0	-10.1	+12.6	0
1996	-82.5	—	+11.6	-70.9	0	-11.7	+14.7	0
1997	-92.4	—	+11.6	-80.8	0	-13.3	+16.7	0
1998	-102.3	—	+11.6	-90.7	0	-15.0	+18.8	0
1999	-112.2	—	+11.6	-100.6	0	-16.6	+20.8	0
2000	-122.1	—	+11.6	-110.5	0	-18.2	+22.9	0

Additional assumptions that apply to Tables 11.2 and 11.3:

Initial US consumption of peanuts for food use = 2,100 million lbs.

Initial level of buybacks = 0.

Raw peanut imports increase as prescribed in GATT, and there is a one-for-one trade-off between foreign and US-grown peanuts.

GATT increases US income by 1%; income elasticity of US demand for peanuts = 0.55 (Rucker and Thurman, 1990).

Real support price remains constant.

The treasury cost per pound of quota peanuts placed in the pool and sold for crush (the difference between the quota and crush prices) is 20.7¢ (1992 dollars) for the period 1995–2000. This amount (20.7¢) is the average of the differences between these prices for the years 1987–92. Source for quota price and crush price (a weighted average of peanut oil and meal prices) is *Oil Crops Situation and Outlook* (USDA, Various issues).

The difference between the quota price and the foreign price is 16.5¢/lb (1992 dollars) for the period 1995–2000. (This is the average difference in these prices for 1987–92.) The measure used for foreign price is the contract additionals price obtained from surveys of county extension agents.

TABLE 11.3 Estimated Effects of GATT

Scenario Assumptions:

(I) Without GATT, total peanut butter imports increase at an annual rate of 25% (from an assumed base of 42 million lbs in 1993). With GATT, peanut butter imports remain constant at 42 million lbs. Also, each pound of imported peanut butter is assumed to displace .9/.88 = 1.023 lbs of domestic peanuts. (See notes under Table 11.1.)

(II) (A) Treasury costs are allowed to increase. (Aggregate quota remains constant.)
(B) Treasury costs are not allowed to increase. (Aggregate quota is reduced.)

	Change in Domestic Demand (million lbs) Due to:			Net Change in Demand (million lbs)	Change in Producer Surplus ($ million)		Change in Treasury Costs ($ million)	
Year	Increased Imports of Raw Peanuts	Reduced Imports of Peanut Butter	Increased US Income		(A)	(B)	(A)	(B)
1995	-72.6	+24.2	+11.6	-36.8	0	-6.1	+7.6	0
1996	-82.5	+40.9	+11.6	-30.0	0	-5.0	+6.2	0
1997	-92.4	+61.9	+11.6	-18.9	0	-3.1	+3.9	0
1998	-102.3	+88.1	+11.6	-2.6	0	-0.4	+0.5	0
1999	-112.2	+120.9	+11.6	+20.3	0	+3.3	0.0	0
2000	-122.1	+161.9	+11.6	+51.4	0	+8.5	0.0	0

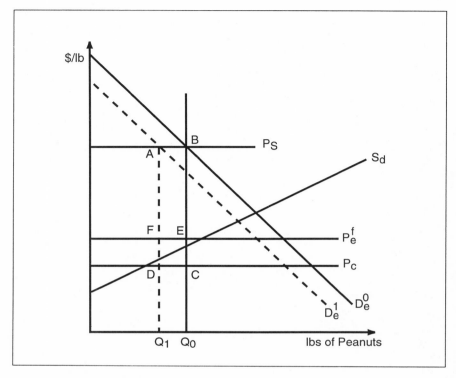

FIGURE 11.3   Graphical Depiction of the Effects of GATT

As the preceding discussion suggests, the primary impact of GATT on US peanut markets will be its effects on the domestic demand for US-grown peanuts. These effects originate from three distinct sources: decreased demand due to increased imports of foreign-grown raw peanuts; increased demand due to (possible) reductions in peanut butter imports; and increased demand resulting from GATT-induced increases in US income levels. In Tables 11.2 and 11.3 the net effects of these factors are shown under the assumptions listed. In both tables it is assumed that raw peanut imports increase as prescribed in the minimum access provisions of GATT and that the effect of GATT is to increase aggregate US income by 1%.[11] The factors, which are allowed to vary, are the effects of GATT on peanut butter imports and whether treasury costs are allowed to increase.

In Table 11.2, columns A, it is assumed that GATT has no effect on peanut butter imports. (As indicated above, this is based on the belief that political factors would limit these imports without GATT.) It is also assumed that the aggregate quota and price support do not change, so that the primary effect of reductions in demand is to increase treasury costs. As seen in the column of Table 11.2 labeled "net change in demand," the net effect of the increased imports of raw peanuts and the increased US income is a decrease in the domestic demand for US-grown peanuts. In Figure 11.3 this corresponds to a shift in demand from D_e^0 to D. With no change in either the quota level or the support price, producer surplus remains constant, and treasury costs increase by the area ABCD. (Q_0 - Q_1 quota peanuts are purchased by the association pools for the quota support price and then sold at the crush price, the dollar value of which appears in the last column labeled A in Table 11.2.)

In Table 11.2, columns B, an alternative scenario is presented in which the assumption is maintained that GATT will not affect peanut butter imports, but treasury costs are no longer allowed to increase. Here, as demand falls, the quota level is reduced, and producer surplus falls. In Figure 11.3 as demand decreases from D_e^0 to D_e^1, the quota is reduced from Q_0 to Q_1. The quantity of peanuts, Q_0 - Q_1, is now sold at the foreign price, P_e^f, rather than the quota price, P_s, and producer surplus falls by ABEF or the dollar values indicated in the final column labeled B in Table 11.2.

Table 11.3, columns A, show the effects of GATT under the assumptions that the quota level and support price remain constant and that peanut butter imports increase by 25% per year from their (assumed) 1993 level in the absence of GATT. This assumed level of increase in peanut butter imports, which we view as highly improbable given recent political discussions concerning this issue, is a level at which demand-increasing factors roughly offset demand-decreasing factors late in the 1995–2000 period. In this scenario treasury costs are seen to increase (by declining amounts) until 1999 after which the net change in demand becomes positive, and treasury costs are unaffected relative to the no-GATT situation.[12]

Table 11.3, columns B, show the effects of GATT under assumptions similar to those made in columns A of the table, except that the quota level is adjusted to maintain the initial level of treasury costs (zero in Figure 11.3). In this case, producer surplus is seen to decrease (by declining amounts) until 1999 when the net change in demand becomes positive. For 1999 and 2000, producer surplus actually increases.[13]

In Table 11.2 it is assumed that the recent growth in peanut butter imports will not continue even in the absence of GATT. This can be viewed as approximating a worst case scenario for the effects of GATT from the perspective of peanut growers. Another political unknown is the viability of a peanut program that incurs large treasury costs. Historically, political factors have limited the size of the allowable treasury costs resulting from this program (Rucker and Thurman, 1990). Whether the program would remain intact with the $12–23 million increases in annual treasury costs shown in the final column A of Table 11.2 is unclear. If political pressures would keep treasury costs from increasing by such amounts, then quota levels likely would be reduced causing producer surplus to decline. Given our view that it is highly unlikely that absent GATT annual increases in peanut butter imports will be as large as 25%, reasonable upper limits on the annual reductions in producer surplus are those shown in columns B of Table 11.2 — $10 million–$18 million. Such reductions translate into less than 1¢/lb of quota peanuts produced.

Other Effects of GATT

The primary effects of GATT and related treaty agreements are increased imports of raw peanuts and decreased imports of peanut butter, which act to shift the demand for US-grown peanuts in opposite directions. Secondary GATT impacts may arise that act to shift the foreign demand for US-grown peanuts. For example, Thailand, Korea and Switzerland all have agreed to reduce or eliminate tariffs on imported peanuts. The requirements that countries must have a scientific basis for establishing health-based measures also may increase the European demand for US peanuts insofar as these requirements lead to less stringent aflatoxin standards for imported peanuts.

Finally, GATT is expected to increase foreign incomes. An increase in foreign demand from any of these sources will increase the foreign price of US peanuts and increase domestic producer surplus. Although the magnitude of any increase in foreign prices is purely speculative, it is straightforward to obtain a lower-bound estimate on the increase in domestic producer surplus associated with any given increase in foreign price. This estimate is simply the pre-GATT level of exports, multiplied by the increase in the foreign price. The average of US edible peanut exports over the last ten years is about 570 million pounds. An increase in the foreign price of, say, 3¢/lb (about a 15% increase) would increase domestic producer surplus by at least $17 million.[14]

Relations Among Peanut Provisions of GATT, CUSTA, and NAFTA

In addition to being affected by GATT, US peanut markets have also been affected by CUSTA and NAFTA. A full understanding of the effects of GATT requires knowledge of how GATT peanut provisions affect the relevant provisions of CUSTA and NAFTA.

Peanut imports have been restricted under Section 22. GATT, NAFTA and CUSTA relax, either directly or indirectly, the nature of the protection provided for US peanut markets. The CUSTA, which went into effect in 1989, affected US peanut markets through its provisions regarding peanut butter imports. Prior to 1989, the duty (tariff) on peanut butter from any source was $60/ton (6.6¢/kg or 3¢/lb). For Canada the initial duty under CUSTA was set at $53.52/ton (5.9¢/kg or 2.7¢/lb) and was scheduled to decrease by $5.99/ton (.66 ¢/kg or .3¢/lb) per year until the duty fell to zero. The only restriction on imports from Canada is that some form of processing must take place in Canada. Thus, raw peanuts can be imported into Canada, processed into peanut butter, and exported to the United States at the lower Canadian duty rate. Peanut butter made in Argentina, transferred into Canada and then to the United States, however, is taxed at the higher rate of $60/ton.

Under NAFTA, which was implemented on January 1, 1994, Mexico was granted an initial duty-free import quota of 3,723 tons (7.45 million pounds) of peanuts. This quota is scheduled to increase by about 3% annually for fifteen years. In 2007 the quota should be 5,466 tons (10.9 million pounds). In 2008 there is no limit to imports without tariff. Before then imports above the annual quota will be subject to tariffs equal to the maximum of 120% ad valorem or $710.40/ton for shelled peanuts and the maximum of 181.4% ad valorem or $469/ton for peanuts in-the-shell. These rates will be reduced by 15% over the first six years and then eliminated completely by linear annual decrements over the next nine years. All peanut imports from Mexico must be Mexican-grown, and all peanut butter imported from Mexico must be made from Mexican peanuts.[15]

The side-agreement to GATT restricting peanut butter imports alters the related provisions of CUSTA. The declining tariff rate schedule under CUSTA will apply only to the minimum access quota set equal to the level of 1993 imports. In addition, an overquota ad valorem tariff rate of 155% has been established. The terms of NAFTA concerning peanut imports from Mexico are not affected by GATT — the Mexican import quotas and the tariff schedule described above will apply. However, Mexican peanut imports, while not directly limited by GATT, do count against GATT quota and so constrain imports from other countries.

Conclusions

A final evaluation of how US peanut growers fared in the political negotiations leading to GATT and related trade agreements will require the passage of time to reveal the actual impacts of several uncertain factors.[16] At present, however, it appears that on the negative side (from US growers' perspective), increases in raw peanut imports under GATT's market access requirements will have detrimental effects. Such effects, however, are likely to be small when combined with modest assumptions concerning GATT's impacts on US incomes. Our worst case estimates indicate losses of less than one cent per pound of quota peanuts grown.

On the positive side (from US growers' perspectives), the basic structure of the US peanut program — a structure that growers strongly support — remains intact. Although raw peanut imports will increase, tariff rates have been set at levels high enough that, at least for the near term, overquota imports are unlikely. In addition, growers have succeeded in curtailing recent increases in Canadian peanut butter imports, thereby preserving domestic grower rents. Finally, if GATT's ultimate effect on the foreign demand for US peanuts is considerable, then US growers may eventually view GATT as a boon.

Notes

1. USDA is the primary source for most of the material in this section. (January, 1994; March, 1994).

2. Much of the material in this section is adapted from Rucker and Thurman (1990), which contains a more detailed analysis of the effects of the peanut program.

3. Growers are also allowed to sell their additional peanuts into the domestic crush market, but because the price received in this market is lower than other alternatives, this option is not economically relevant.

4. This support price applies only to sales of peanuts by the growers' association and not to contract additional sales by growers. Peanuts sold by the growers' association to processors for domestic crush cannot be exported by those processors.

5. See Rucker and Thurman (1990) for details on past and current rules for distributing these profits.

6. Detailed discussions of the justifications for these assumptions can be found in Rucker and Thurman (1990).

7. Given world prices and assuming foreign peanuts are of sufficient quality, it is clear that raw peanuts will be imported up to the minimum access levels at least.

8. Under normal circumstances, the domestic support price serves as not only a floor to the US price but as a ceiling as well. This results from the buy back provision of the peanut program described in section II above. The circumstances under which the US price can rise above the support price are those in which the US crop is so short that the sum of quota peanuts and additionals bought back is insufficient to meet domestic edible demand at the support price. This is what happened in 1980 and 1990.

9. The ITC investigation began in January, 1994. The initial hearings were postponed pursuant to requests from the National Peanut Growers Group. In late June, 1994, the investigation was suspended indefinitely at the request of the President. At this time, the side agreement to GATT concerning peanut butter had already been negotiated.

10. This agreement also limits imports of peanut butter from other sources as follows: Argentina, 8.03 million pounds; less developed countries, 1.65 million pounds (increasing to 3.52 million pounds after six years); and other countries, .55 million pounds.

11. This impact of GATT on aggregate US income is consistent with the assumptions made in the analysis by the USDA (March, 1994, Appendix I).

12. Note that treasury costs will not decrease as a result of GATT in 1999 and 2000. This is because the buy back provisions of the peanut program take effect when the net effect of GATT is an increase in the domestic demand for peanuts (from the initial situation shown in Figure 11.3).

13. Again, after 1999 the buy back provisions of the peanut program would take effect. As demonstrated in Rucker and Thurman (1990), the increases in

producer surplus in 1999 and 2000 would actually be dissipated through competition among growers for the increased revenues associated with selling their additionals for the quota support price. The present analysis assumes that the quota level is not increased when demand shifts to the right of D_e^0 in Figure 11.3.

14. In Figure 11.3, it can be seen that an increase in the foreign price increases domestic producer surplus by a trapezoid-shaped area bounded by the level of quota, the old and new foreign prices and the domestic supply. The estimate in the text of the increased producer surplus from an increased foreign price would be exact if the domestic supply were vertical. The actual increase in producer surplus increases with the elasticity of the domestic supply.

15. The extent of the future growth in imports of Mexican peanuts is highly uncertain. Recent statements by US industry leaders indicate plans by US processors to set up operations in Mexico to take advantage of lower peanut prices (*Orlando Sentinel*, 1995).

16. The following comments presume that growers and quota owners are not separate groups.

References

Hollis, P. 1994. "Global Market is Key to Farm Income." *Southeast Farm Press* 21(4): 16.

Nail, C. 1994. "Solution Sought to Limit Imports." *Peanut Farmer* 30(3): 12.

Orden, D. 1994. "Agricultural Interest Group Efforts to Affect the Congressional NAFTA Debate," in G. W. Williams, and T. Grennes, eds., *NAFTA and Agriculture: Will the Experiment Work?* Papers presented December 12–13, 1993, to the International Trade Research Consortium Conference, San Diego, CA.

Orlando Sentinel. 1995. "Rising peanut prices driving up imports." Orlando, FL (25 February).

Rucker, R. R., and W. N. Thurman. 1990. "The Economic Effects of Supply Controls: The Simple Analytics of the US Peanut Program." *Journal of Law and Economics* 33: 483–515.

USDA. Various issues, 1987–92. *Oil Crops Situation and Outlook.* ERS/USDA, Washington, DC.

_____. 1994. *Effects of the Uruguay Round Agreement on US Agricultural Commodities.* ERS/USDA, Washington, DC (March).

_____. 1994. Foreign Agricultural Service. *Agricultural Provisions of the Uruguay Round.* FAS/USDA, Washington, DC (January).

US International Trade Commission. 1993. *The Economic Effects of Significant US Imports Restraints.* Publication No. 2699, Washington, DC (November).

Williams, J. C., and B. D. Wright. 1991. *Storage and Commodity Markets.* Cambridge, MA: Cambridge University Press.

12

The US Sugar Industry: The Free Trade Debate

A. Schmitz and D. Christian

Abstract

The US sugar program, which began in 1934, has been largely unaffected by the General Agreement on Tariffs and Trade (GATT), The North American Free Trade Agreement (NAFTA) and the Canada–United States Free Trade Agreement (CUSTA). The sugar program has been the subject of numerous debates. In resolving many of the issues surrounding the US sugar program, it is important to understand why, historically, the sugar program has received strong political support. This support can be attributed to many factors, including the wide geographical dispersion of various components of both the sugar and corn sweetener industries. A strong tie between these two industries also contributes to strong political support.

Introduction

The US sugar program has been in operation since 1934 with the exception of a few short intervals. Its effects on producers and the general public have been the center of many debates (Schmitz and Christian, 1993), but these are not discussed here. This chapter outlines the basic elements of the US sugar program and discusses the effects of GATT, NAFTA and CUSTA on the program. Sugar trade between the United States and Canada is becoming a contentious issue. Canada alleges that the US sugar program creates unwelcome trade distortions. Some major factors surrounding this sugar trade dispute are discussed.

Like Canadian supply management, the US sugar program has received strong political support. We show in this chapter some of the reasons why this is the case by developing a regional model of the US sugar program's effects. There is broad-based support for the sugar program, partly because many of the regions in the United States are net corn sweetener and/or sugar exporters. Because of the resulting net economic gains, legislators from these regions generally vote for the program.

The US Sugar Program, GATT, and NAFTA

The United States imports sugar to supplement its domestic production. The US Sugar Act, which was implemented in 1934, supported and regulated this industry. The program expired in 1974, but regulations and support were reinstated for 1977–79 and again in the 1981 Farm Bill. The Sugar Program, still in effect as the 1995 Farm Bill is debated, includes a combination of nonrecourse loans, tariff-rate quotas, specific import quotas and, when necessary, sugar marketing allotments. A loan rate of 18¢/lb for raw cane sugar was provided in the 1985 and 1990 farm bills. The sugar price support program was extended through the 1997 crop year.

The essence of the US sugar program has been outlined by Johnson and Ortego:

> Price support for raw cane sugar has been constant at 18¢ per pound since 1985. The refined beet sugar loan rate, which was 23.6¢ in 1993, has increased slightly based on an increasing ratio of the returns to cane growers. Price support is achieved through the Commodity Credit Corporation (CCC) nonrecourse loans available to sugar processors. Current legislation specifies that the loan program be operated at no cost to the federal government. To ensure no forfeitures of sugar to the CCC, sugar imports are restricted through a tariff-rate quota (TRQ) to comply with the Uruguay Round of GATT. The quota is set to ensure that domestic prices remain above the loan price plus interest. The TRQ replaced an absolute quota which regulated the entry of sugar into the United States from 1982 to 1990.
>
> The TRQ is a two-tiered duty. The lower tier of imports is subject to a nominal or zero duty for quantities that balance domestic supply with projected sugar use at the supported price. This amount is allocated on a country by-country basis. The allocation of import quotas among countries is based on historical import, but quotas have occasionally been changed to achieve foreign policy objectives. Sugar imported above the low duty quantities is subject to a second-tier duty of 16¢ a pound, raw basis. The high tariff duty will be reduced under the Uruguay Round agreement to 14.45¢ a pound by the end of the century. This lowered duty will still price US sugar import above the 18¢ a pound raw sugar loan rate.
>
> The 1990 farm bill established provisions for marketing allotments to processors of domestic sugar whenever imports were estimated to fall below 1.25 million short tons. On June 30,1993, the United States Department of Agriculture (USDA) announced sugar marketing allotments

for fiscal 1993. The allotments limited domestic processor sugar marketings based on their sales history. Marketing allotments have not been needed for fiscal year 1994, but sugar imports are so close to the 1.25 million ton minimum that marketing allotments remain a probability in future years.

The NAFTA gives Mexico duty-free access to the US sugar market beyond the present quota if Mexico has 'net surplus production'. A side agreement defines net surplus production such that Mexico production must exceed consumption of both sugar and HFCS for Mexico to be considered a net surplus producer. The chances of Mexico's becoming a net surplus producer using this definition is unlikely. (1995)

There are similarities between the US sugar program and Canadian supply management, four of which are mentioned below. 1) Both use import quotas to protect prices and regulate quantities to be used in processing. The latter has only recently become a phenomenon in the United States with the introduction of marketing allotments, but these are in no way as restrictive as production controls in Canada's supply management. 2) In both cases, exporters of sugar to the United States and exporters of supply management commodities to Canada receive import quota rents. 3) Sugar producers in the United States and producers of supply-managed commodities in Canada have been largely unaffected, at least in the short run, by NAFTA, CUSTA and GATT. 4) Lastly, the costs and benefits resulting from these programs have been the subject of heated debates.

Canada–US Sugar Disputes

Canada is a large net importer of sugar. Roughly 90% of its annual needs are imported largely as raw cane sugar while the remainder is supplied by domestically produced sugar beets. Canada produces roughly 1.5 million tonnes of refined sugar in Vancouver, BC Sugar; Toronto, Redpath Sugars; and Montreal and St. John, Lantic Sugar. Refined beet sugar is manufactured by two BC Sugar subsidiaries in Taber, Alberta and Winnipeg, Manitoba.

The United States is a major supplier of sugar to Canada. In recent years there has been a significant increase in US shipments of refined sugar to Canada (Figure 12.1). US shipments of both sugar and sugar-containing products to Canada have also risen sharply (Figure 12.2). As the numbers show, the overall Canada — US trade balance in sugar and sugar-containing products is in the United States' favor.

According to Marsden of the Canadian Sugar Institute:

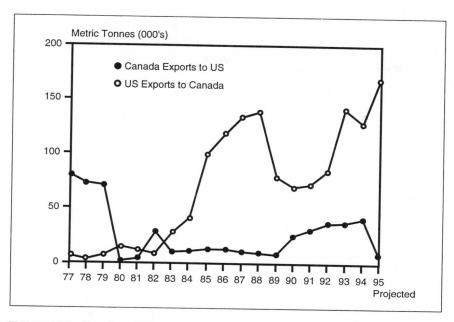

FIGURE 12.1   Canada — US Trade, Refined Sugar
   Source: Canadian Sugar Institute (February, 1995)

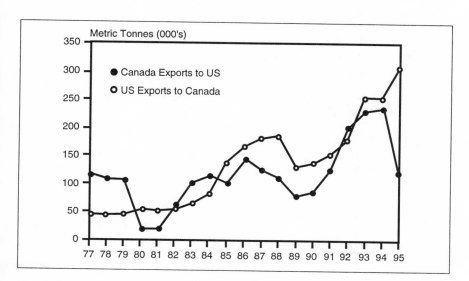

FIGURE 12.2   Canada — US Trade, Sugar + Sugar-Containing Products
   Source: Canadian Sugar Institute (February, 1995)

For Canada the fundamental problem with the US Sugar Program is that the guarantee of high returns to the US domestic industry has led to overproduction. The result has been the tightening of US import restrictions and the imposition of marketing allocations to control domestic sales. Since the early 1980's, the Canadian sugar industry has had to adjust to the trade imbalance perpetuated by a variety of US measures to restrict Canadian exports of refined sugar and sugar-containing products and to the growth in imports of US surplus refined sugar production.

The impact of trade distortions resulting from the operation of the US sugar program is having devastating consequences in the Canadian market — a market that largely serves its domestic consumers without the burden of high internal prices. For many years the Canadian sugar industry has managed its trade within the boundaries of the US sugar program. The industry can no longer manage in the face of growing uncertainty over access to the US market for refined sugar and sugar-containing products, or the increasing impact of unfairly traded imports from the US and other countries that stimulate excess production under high price regimes. (1995)

In addition the Institute states:

[1] When the FTA was introduced in 1989, the US retained a variety of restrictive quotas and duties on refined sugar and sugar-containing food products. Canada continued to maintain its open market policy. The only benefit of the FTA was that the US could not apply a quota to refined sugar. This will change in October 1995, when Canada's refined beet sugar exports will become subject to a restrictive global quota of just 22,000 tonnes global quota.

[2] The NAFTA was implemented on January 1, 1994, however the agricultural provisions of the FTA continue to govern Canada – US bilateral trade on sugar and sugar-containing products. With implementation of the NAFTA, the US unilaterally reduced Canada's access for certain sugar-containing products. This change was made without prior notice or consultation with the Canadian government.

[3] Late last fall the US government indicated its intention to impose new restrictive definitions and product classification changes for certain sugar-containing products as part of its GATT tariff schedule. Examples of the types of products affected include crystal drink mixes, iced tea and sweetened cocoa mixes.

The US imposed these changes on January 1, 1995 even though the market access provisions for other agricultural commodities will not be implemented before July 1 or October 1. The new US quota on refined sugar will be implemented in October 1. These changes for sugar and sugar containing products, are unique in that they will further reduce existing market access contrary to the GATT and NAFTA principles of expanding market access and reducing government created trade distortions. (1995)

On February 10, 1995, Canadian trade officials requested talks with the United States to discuss trade restrictions that have affected Canadian exports of refined sugar and sugar-containing products. Their major concern was that on January 1, the United States restricted imports of sugar-containing foods to a global quota of 64,000 tonnes, and on October 1, they are planning to implement a global quota of 22,000 tonnes on refined sugar.

Canada's position on sugar obviously favors freer Canada – US trade, but there is a problem when trade is viewed in a broad perspective. As was documented by Schmitz, de Gorter and Schmitz (1996), supply-managed commodities in Canada are highly protected from US exports into Canada. GATT did not significantly reduce these protection levels. This poses an interesting dilemma that will be difficult to resolve. In addition, there are other broader questions to address, such as: Is the US sugar program the only trade irritant in the Canada – US sugar trade?

Regional Effects and Policy Determination[1]

Introduction

US agricultural commodity programs redistribute wealth from one group of individuals (taxpayers or consumers) to another (producers). These programs generally impose a net cost on society since deadweight losses incurred in the redistribution result in producers' benefits being less than the losses incurred by consumers/taxpayers. It is increasingly common to attribute the enactment of these net cost policies to the disproportionate influence of special interest groups on the policymaking process. These interest groups are presumed to expend monetary and other resources in an effort to persuade policymakers to act on their behalf. Their success is determined by their efficiency in applying pressure relative to the efficiency of opposing interest groups, for example, Zusman (1976) and Zusman and Amiad (1977). There is also a body of literature, for example, Becker (1983, 1985) and Gardner (1987), that considers the social cost, or deadweight loss, of the redistribution.

These studies have for the most part neglected an important set of factors affecting policy determination, namely the differing regional effects of national policies and corresponding regional representation in Congress. Senators and representatives do not represent the nation at large but rather an individual state or district for which the economic impact of a given policy may differ from the aggregate nationwide effect. Moreover, if individual legislators weight equally the welfare of their own constituencies, then proposed legislation providing net benefits to a majority of legislative districts will be enacted, even in the absence of special interest groups and irrespective of the policy's aggregate effect.

In other words, it is possible for the varying regional effects of national policies to combine with regional representation in Congress to sustain a policy that redistributes wealth in such a way that a majority of legislative districts enjoy net benefits.

This perspective emphasizes the importance of the following: 1) Policies have varying impacts across regions within the United States; 2) US legislators are elected on a regional basis and are likely to support policies that are beneficial to their constituencies, whether or not these policies confer a net gain or loss on the country as a whole; and 3) Policy adoption requires a simple majority in each house of Congress. Previous studies have generally neglected these points, particularly the varying regional impact of national policies and regional representation in Congress. One implication of our perspective is that the benefits of political reform, such as limiting the amount of campaign contributions that legislators can accept, may be overstated. It is generally implied that complete or perfect reform, eliminating the influence of special interests, would subsequently cause the rejection of welfare-reducing policies by legislators truly representing the best interests of their constituents.

The US sugar program benefits sweetener industries, including the corn sweetener industry, by imposing quotas on sugar and sugar-containing products. Since the United States is a net importer of sugar, the costs to consumers exceed producer benefits, thereby imposing a net cost on society. Open to question, however, is the size of the net cost which clearly depends on the world price. Given the world sugar price of 14¢/lb in April, 1995, the net costs of the sugar program are much less than they were when sugar was 4¢/lb in 1985 (Schmitz and Christian, 1993). Importantly, the beneficiaries of the sugar program — producers of sugarcane, sugar beets, corn used in the production of high fructose corn syrup (HFCS) and other sweeteners and domestic refiners of sugar and corn sweeteners — are geographically dispersed throughout the country, yet concentrated enough in some areas that their benefits outweigh the costs borne by local consumers. Such a configuration makes the sugar program a potential example of a net cost policy that could be enacted in the absence of the influence of special interest groups.

We do not contend that this has always been the case for sugar. In the 1970s, for example, because of the absence of the corn sweetener industry, special interest groups played a major role in sugar policy legislation. However, as our results show, the importance of this influence has been greatly diminished because of broad-based economic support from industries that have grown because of the sugar program and whose futures critically depend on it.

Aggregate Versus Regional Effects of the US Sugar Program

The sugar program is arguably the most criticized of all US farm programs. Criticism has focused on the high-net, domestic social cost of the program, which has generally maintained domestic sugar prices at a higher level than world prices. Using a variety of assumptions about the quota price premium, the elasticity of the world excess supply curve and the substitutability of sugar and HFCS, Leu, Schmitz and Knutson (1987) estimated that the US sugar program cost consumers between $372 million and $4 billion in 1983. These costs far outweighed the gains to producers with the net societal cost estimated between $203 million and $3.1 billion depending upon fluctuations in the world sugar price. According to the US General Accounting Office, however, net costs have fallen in more recent years to an average of only $276 million for 1989–91.

As with Leu, Schmitz and Knutson (1987), most analyses of the economic impact of the US sugar program have been confined to estimating the aggregate effects on US producers, processors and consumers. These studies use a model similar to the one shown in Figure 12.3 in which the United States is depicted as a net sugar importer. Under a free trade regime at a world sweetener price of P^*, US consumers demand a certain amount of sweeteners, domestic producers provide that amount and the balance is imported. Implementation of a policy that supports a domestic price of $P^T > P^*$ causes a loss to consumers equal to areas $a+b+c+d$ but increases producer surplus by area a. Since foreign sugar exporters receive the quota rents (area c), a net welfare loss of $b+c+d$ *is* incurred by the United States.[2]

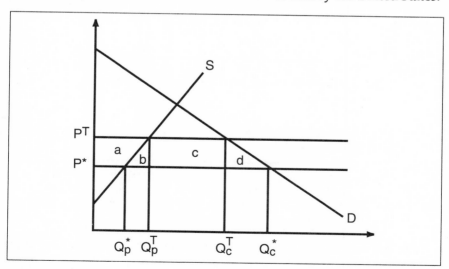

FIGURE 12.3   Effect of Sugar Policy on Aggregate US Economy

High sugar prices supported by import quotas have also spurred the growth of the corn sweetener industry, an effect partially incorporated in the dynamic model of Leu, Schmitz and Knutson (1987). Consequently, corn producers and refiners in regions which are net exporters of corn sweeteners, will benefit more than consumers are hurt by a policy supporting high sugar prices.

This model and the corresponding empirical results on net losses are consistent with special interest theory arguments since a policy is supported which imposes net social costs.[3] However, we argue that regional analyses integrated with the policymaking process provide additional arguments for legislative support for the sugar program.

The economic impact of the sugar program on individual regions within the United States can differ significantly. In particular, in a region that produces more sugar and/or corn sweetener than it consumes, a policy that raises internal US sugar prices will benefit sugar producers and processors more than it costs sugar consumers. The appropriate model for analyzing the effect of the sugar program on the economies of net-exporting regions is presented in Figure 12.4. Here, a quota increases producer surplus for sugar by areas $e + f$; this is larger than the loss in consumer surplus of area e.

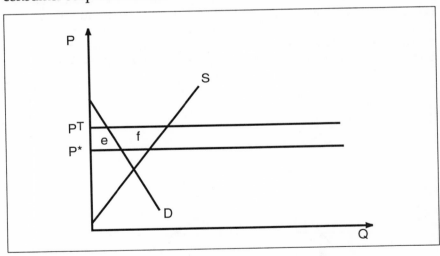

FIGURE 12.4   Effect of Sugar Policy on Sweetener Exporting Regions

The fact that sugar policy — or any policy — imposes a net loss on the United States as a whole does not necessarily imply that each region suffers a net loss. In the text that follows, we show that many sugar- and corn-producing regions in the United States enjoy net benefits from sugar import quotas. These regions also generate strong political support for the program.

The Policymaking Process

The successive stages in the adoption of sugar and sweetener legislation are put into motion by a coalition of industry representatives. This coalition is comprised of sugar beet and sugarcane growers, refiners of domestic beet and cane sugar, corn sweetener refiners and, since the recent 1990 legislation, refiners of imported cane sugar. Each subgroup of industry representatives develops proposed legislation expressing their preferred support levels, the method to achieve these levels and other policy variables and settings. The proposed legislation that emerges from the coalition represents the outcome of bargaining between its various subgroups, each of which is affected differently by a given sugar policy. The success of this bargaining process is crucial. Because each of these subgroups is relatively small, the united support of the entire industry is needed to obtain passage of supporting legislation.

Proposed sugar and sweetener legislation is usually introduced simultaneously in the appropriate subcommittees of the House of Representatives Committee on Agriculture and the Senate Committee on Agriculture, Nutrition and Forestry. The Senate subcommittee and the Subcommittee on Agricultural Production and Stabilization of Prices are dominated by senators from leading agricultural states while the House subcommittee and the Rice, Cotton and Sugar Subcommittee (RCSS) are dominated by representatives from states that produce those commodities. Because they revise the proposed legislation and control the agenda, it may be argued that subcommittee members are more influential than other legislators in the policymaking process. In addition, grouping several commodities in the House subcommittee ensures that legislation supporting one commodity receives at least tacit approval from representatives of areas where the other commodities are produced. For example, adoption of sugar policy by the House RCSS requires support not only from legislators representing sugar and sweetener constituencies but also from those representing rice and cotton production areas.

Subcommittee hearings are held to allow interested parties to express their opinions regarding the proposed policy. The bill is then redrafted, approved by the subcommittee and forwarded to the respective House or Senate agriculture committee. The agriculture committees may make further, minor amendments to proposed commodity-specific legislation before including it in the comprehensive farm bill, which is then forwarded to the House or Senate floor.

Since farm legislation is drafted by the subcommittees with an eye toward what the full House and Senate will accept, most floor debate can be interpreted as political posturing because the final vote is little more than a formality. Occasionally, however, a particular commodity program is singled out for a floor vote on an amendment to weaken or eliminate its support level. The sugar program was the object of eight floor votes between 1981 and 1990. In the following section, we examine these votes to illustrate that a policy, regardless of its effect, may be adopted as a result of differential regional effects and congressional voting rules.

Regional Representation by Legislators

With the growth of the corn sweetener industry, there is a wide geographical dispersion of sugar and sweetener producers and refiners (Figure 12.5). The sugar program, therefore, may be an example of a policy for which legislators, who represent regions deriving net benefits, comprise a majority of Congress. If congressional representatives and senators equally weight the welfare of sweetener consumers, producers and processors in their constituency,[4] then the sugar program will receive support from legislators representing net-exporting regions where the loss to consumers from sugar quotas is less than the gain enjoyed by producers and processors. Since policy is determined by a simple majority in each house of Congress, the sugar program could be sustained even without the interference of special interest groups if legislators from net-exporting regions outnumber legislators from net-importing regions.

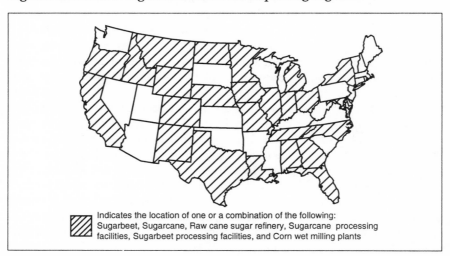

Indicates the location of one or a combination of the following: Sugarbeet, Sugarcane, Raw cane sugar refinery, Sugarcane processing facilities, Sugarbeet processing facilities, and Corn wet milling plants

FIGURE 12.5   US Sugar Crops Production Regions and Processing Facilities

In this section, we 1) identify which states are net exporters of sweeteners and therefore net beneficiaries of the sugar program in order to determine if their legislative representatives comprise a majority of Congress and 2) contrast the voting records of legislators who represent net-exporting states with those who represent net-importing states in an effort to determine if their votes are correlated with the sugar program's economic impact on their states.

Net Sweetener-Exporting States

Sweeteners are produced from one of the three commodities: sugarcane, sugar beets and corn. Sugarcane is produced in four states: Hawaii, Florida, Louisiana and Texas. All but Texas are estimated to be net exporters of sugar.[5] Each of these three states annually produces at least twice as much sugar as their populations consume.

Sugar beets are currently produced in 13 states: California, Oregon, Idaho, Montana, Wyoming, New Mexico, Colorado, Nebraska, Texas, North Dakota, Minnesota, Michigan and Ohio. Six of these states are estimated, based on beet production, to be net exporters of sugar: Idaho, Montana, Wyoming, Nebraska, North Dakota and Minnesota.

High domestic sugar prices supported by quotas on sugar imports have fostered the development and growth of the corn sweetener industry. Corn sweeteners' share of total sweetener deliveries increased from less than 20% in 1980 to more than 50% in 1989. Thus, corn producers and corn sweetener refiners are also beneficiaries of the sugar program. At least six states are estimated to be net exporters of sweeteners based upon their corn production:[6] Illinois, Indiana, Iowa, Kansas, South Dakota and Wisconsin. Presumably Minnesota, Nebraska and North Dakota — and perhaps Idaho, Montana and Wyoming — are also net exporters of corn-based sweeteners. However, estimates were not calculated for these states since it is established, based upon their sugar beet production, that they are net exporters of sweeteners.

Several states produce or refine more than one type of sweetener. For example, California derives dual benefits from the current sugar program based upon its own sugar beet production and its refining of Hawaiian cane sugar. Colorado and Michigan, as producers of both beets and corn, are also net exporters of sweeteners.

In total, 18 states are estimated to be net exporters of sweeteners. Senators from these states comprise 36% of the Senate, and the 168 representatives from these states for the 1982–92 period comprised nearly 39% of the House. Thus, although these states provide a strong foundation of support for the sugar program, they do not comprise the legislative majority needed to enact sweetener support legislation the basis of regional economic benefits alone.

Evidence from Voting Records

Eight congressional votes on amendments to sugar legislation recorded between September, 1981, and July, 1990, are summarized in Tables 12.1 and 12.2. Five votes occurred in the Senate, and three votes occurred in the House of Representatives.[7] We correlate these voting records with estimates of the sugar program's regional effects.

Legislators representing the 18 states estimated to be net exporters of sweeteners were expected to have supported the sugar program. Support by legislators representing sugar-producing states was particularly strong (Table 12.1 and 12.2). In all five Senate ballots, only two votes by senators from sugar-exporting states were cast against legislation favorable to sugar. In the House, each sugar program vote received support from more than 75% of sugar-state representatives.[8]

Support of sugar legislation by corn sweetener-exporting states was less strong, particularly in the early votes. However, support from these states increased between 1981 and 1990 as evidenced, for example, by the votes of the Illinois, Wisconsin and Ohio House delegations. This voting behavior is consistent with the dramatic increase in the corn sweetener consumption over this period: Corn sweeteners' share of total sweetener deliveries increased from less than 20% in 1980 to over 50% in 1989.

Finally, additional support for sugar legislation in the House is evident from areas where sweetener production and/or refining is important to local economies. The delegations from Ohio, Tennessee, Texas, Missouri and North Carolina, states in which HFCS refining and corn production are significant, showed mixed support. Interestingly, after opposing earlier programs, House delegations representing Georgia and Maryland supported the 1990 sugar legislation, which included a minimum import provision protecting sugar cane refiners located in those states. Similarly, the New York delegation's previously strong opposition to the sugar program weakened considerably in 1990.

To test the hypothesized relationships regarding legislators' support or opposition to the program, cross-tabulations of the eight roll call votes by legislators' hypothesized positions on the sugar program are presented in Tables 12.3 and 12.4. The chi-square statistics for the House votes range from 21.8 to 62.8 with six degrees of freedom. All are significant at the 1% level, indicating a rejection of the null hypothesis that legislators' voting behavior is independent of the presence of sugar program beneficiaries in the state. In the Senate, the chi-square statistics range from 13.0 to 21.4 with four degrees of freedom. Four of the statistics are significant at the 1% level, and the fifth is significant at the 2.5% level. This implies a strong association between voting behavior and the economic impact of the sugar program on regional economies. The above should be interpreted in light of the economic importance of the industry.

TABLE 12.1   Senate Votes on Proposed Sugar Legislation, By State

	Vote 1	Vote 2	Vote 3	Vote 4	Vote 5
	(votes for policy favorable to sweetener industries to votes against; vacancies, absences and no votes in parentheses)				
Probable Supporters					
Net Exporters of Sugar*					
California	1 - 0 (1)	1 - 0 (1)	2 - 0	2 - 0	2 - 0
Colorado	1 - 0 (1)	1 - 0 (1)	2 - 0	1 - 1	1 - 1
Florida	2 - 0	2 - 0	2 - 0	2 - 0	2 - 0
Hawaii	2 - 0	2 - 0	2 - 0	2 - 0	2 - 0
Idaho	2 - 0	2 - 0	2 - 0	2 - 0	2 - 0
Louisiana	2 - 0	2 - 0	2 - 0	2 - 0	2 - 0
Michigan	2 - 0	2 - 0	2 - 0	2 - 0	2 - 0
Minnesota	2 - 0	2 - 0	2 - 0	2 - 0	2 - 0
Montana	2 - 0	2 - 0	1 - 0 (1)	2 - 0	2 - 0
North Dakota	2 - 0	2 - 0	2 - 0	2 - 0	2 - 0
Wyoming	2 - 0	2 - 0	1 - 0 (1)	2 - 0	2 - 0
Net Exporters of Corn Sweeteners*					
Illinois	1 - 1	0 - 2	1 - 1	2 - 0	2 - 0
Indiana	0 - 2	0 - 2	0 - 2	0 - 2	0 - 2
Iowa	2 - 0	2 - 0	2 - 0	2 - 0	1 - 1
Kansas	2 - 0	2 - 0	2 - 0	2 - 0	0 - 2
Nebraska	2 - 0	2 - 0	2 - 0	1 - 0 (1)	2 - 0
South Dakota	2 - 0	2 - 0	2 - 0	2 - 0	1 - 0 (1)
Wisconsin	0 - 2	0 - 2	0 - 2	0 - 2	0 - 2
Probable Opponents					
Alabama	1 - 1	2 - 0	0 - 1 (1)	2 - 0	2 - 0
Alaska	2 - 0	1 - 0 (1)	2 - 0	2 - 0	1 - 1
Arizona	2 - 0	2 - 0	1 - 1	1 - 0 (1)	1 - 1
Arkansas	1 - 0 (1)	2 - 0	0 - 2	2 - 0	2 - 0
Connecticut	1 - 1	1 - 1	0 - 0 (2)	0 - 1 (1)	1 - 1
Delaware	0 - 2	0 - 2	1 - 1	0 - 2	0 - 2
Georgia	1 - 1	1 - 1	0 - 2	1 - 1	1 - 1
Kentucky	2 - 0	2 - 0	2 - 0	2 - 0	1 - 1
Maine	2 - 0	2 - 0	0 - 2	0 - 2	0 - 2
Maryland	0 - 2	1 - 1	2 - 0	0 - 2	1 - 1
Massachusetts	0 - 2	0 - 1 (1)	1 - 1	0 - 2	0 - 2
Mississippi	1 - 0 (1)	2 - 0	2 - 0	2 - 0	2 - 0
Missouri	0 - 2	0 - 2	1 - 1	0 - 2	0 - 2
Nevada	2 - 0	2 - 0	2 - 0	2 - 0	0 - 2
New Hampshire	0 - 2	0 - 2	0 - 1 (1)	0 - 2	0 - 2
New Jersey	0 - 1 (1)	0 - 1 (1)	0 - 2	0 - 2	0 - 2
New Mexico	1 - 1	0 - 2	1 - 0 (1)	1 - 0 (1)	0 - 2
New York	0 - 2	0 - 2	0 - 1 (1)	0 - 2	1 - 1
North Carolina	2 - 0	2 - 0	2 - 0	1 - 0 (1)	2 - 0
Ohio	0 - 2	1 - 1	2 - 0	0 - 1 (1)	0 - 2
Oklahoma	1 - 1	1 - 1	1 - 1	1 - 1	1 - 1
Oregon	0 - 2	0 - 2	2 - 0	1 - 1	0 - 2
Pennsylvania	0 - 2	0 - 2	0 - 2	0 - 0 (2)	0 - 2
Rhode Island	0 - 2	0 - 2	0 - 2	0 - 2	0 - 2
South Carolina	2 - 0	2 - 0	2 - 0	2 - 0	2 - 0
Tennessee	2 - 0	2 - 0	1 - 0 (1)	2 - 0	2 - 0
Texas	2 - 0	2 - 0	2 - 0	2 - 0	2 - 0
Utah	2 - 0	2 - 0	0 - 2	2 - 0	1 - 0 (1)
Vermont	1 - 0 (1)	1 - 0 (1)	0 - 2	0 - 2	1 - 1
Virginia	1 - 1	1 - 1	1 - 1	2 - 0	0 - 2
Washington	1 - 1	2 - 0	2 - 0	0 - 2	1 - 1
West Virginia	2 - 0	2 - 0	1 - 1	2 - 0	2 - 0

Vote 1: To table amendment to eliminate sugar program: passed 61–33; September 17, 1981
Vote 2: To table amendment to reduce sugar loan rate; passed 64–30; September 17, 1981.
Vote 3: To table amendment to reduce sugar loan rate; passed 60–31; September 23, 1982.
Vote 4: To table amendment to allow reduction in sugar loan rate; passed 60–32; November 22, 1985.
Vote 5: To table amendment to reduce sugar loan rate; passed 54–44; July 24, 1990.
*Calculations available from authors upon request.
Sources: Calculations available from authors upon request

TABLE 12.2 House of Representatives Votes on Proposed Sugar Legislation, By State

	Vote 1	Vote 2	Vote 3
(votes for policy favorable to sweetener industries to votes against; vacancies, absences and no votes in parentheses)			

Probable Supporters

Net Exporters of Sugar[a]

	Vote 1	Vote 2	Vote 3
California	26 - 10 (7)	34 - 7 (4)	33 - 12
Colorado	4 - 1	4 - 1 (1)	5 - 1
Florida	10 - 3 (2)	13 - 4 (2)	12 - 5 (2)
Hawaii	2 - 0	2 - 0	1 - 0 (1)
Idaho	1 - 0 (1)	2 - 0	2 - 0
Louisiana	8 - 0	8 - 0	8 - 0
Michigan	11 - 6 (2)	16 - 1 (1)	13 - 4 (1)
Minnesota	6 - 2	8 - 0	7 - 1
Montana	2 - 0	2 - 0	2 - 0
North Dakota	1 - 0	1 - 0	1 - 0
Wyoming	1 - 0	0 - 1	1 - 0

Net Exporters of Corn Sweeteners[a]

Illinois	0 - 23 (1)	10 - 6 (6)	13 - 9
Indiana	3 - 8	3 - 6 (1)	4 - 6
Iowa	5 - 1	6 - 0	5 - 1
Kansas	5 - 0	4 - 1	3 - 2
Nebraska	3 - 0	3 - 0	3 - 0
South Dakota	2 - 0	1 - 0	1 - 0
Wisconsin	2 - 7	5 - 3 (1)	6 - 3

Mixed[b]

Missouri	8 - 2	9 - 0	7 - 2
North Carolina	8 - 1 (2)	11 - 0	11 - 0
Ohio	4 - 19	10 - 10 (1)	9 - 12
Tennessee	5 - 3	5 - 4	4 - 4 (1)
Texas	17 - 6 (1)	19 - 7 (1)	21 - 6

Probable Opponents

Alabama	5 - 2	4 - 2 (1)	7 - 0
Alaska	1 - 0	1 - 0	1 - 0
Arkansas	3 - 1	3 - 0 (1)	3 - 1
Arizona	3 - 1	2 - 1 (2)	1 - 4
Connecticut	0 - 4 (2)	0 - 6	0 - 5 (1)
Delaware	0 - 1	0 - 1	0 - 1
Georgia	1 - 7 (2)	8 - 1 (1)	8 - 2
Kentucky	6 - 1	6 - 1	6 - 1
Maine	0 - 2	0 - 2	0 - 2
Maryland	2 - 6	1 - 7	7 - 0 (1)
Massachusetts	0 - 10 (2)	1 - 9 (1)	1 - 10
Mississippi	5 - 0	5 - 0	5 - 0
Nevada	0 - 0 (1)	1 - 1	2 - 0
New Hampshire	0 - 2	0 - 2	0 - 2
New Jersey	0 - 12 (3)	2 - 12	2 - 11
New Mexico	2 - 0	3 - 0	3 - 0
New York	1 - 36 (2)	7 - 25 (2)	15 - 18 (1)
Oklahoma	5 - 1	6 - 0	3 - 2 (1)
Oregon	2 - 2	4 - 0 (1)	5 - 0
Pennsylvania	4 - 20 (1)	9 - 13 (1)	9 - 13 (1)
Rhode Island	0 - 2	0 - 2	0 - 2
South Carolina	3 - 1 (2)	6 - 0	5 - 1
Utah	1 - 1	1 - 1 (1)	0 - 2 (1)
Vermont	1 - 0	1 - 0	1 - 0
Virginia	4 - 6	5 - 4 (1)	7 - 2 (1)
Washington	4 - 2 (1)	7 - 1	4 - 3 (1)
West Virginia	3 - 1	4 - 0	4 - 0

Vote 1: Amendment to eliminate sugar program: passed 213–190; October 15, 1981
Vote 2: Amendment to reduce sugar loan rate; defeated 263–142; September 26, 1985.
Vote 3: Amendment to reduce sugar loan rate; defeated 271–150; July 24, 1990.
[a]Calculations available from authors upon request.
[b]States in which sweetener production and/or refining important to local economies.
Sources: Calculations available from authors upon request

TABLE 12.3   Cross-Tabulations of Senate Votes on Proposed Sugar Legislation, By State *(See Table 12.1 for description of proposed legislation.)*

	Senate Vote 1			
	Proponents		*Opponents*	*Total*
	Cane or Beets	*Corn*		
Supported	20	9	32	61
Opposed	0	5	28	33
No Vote	2	0	4	6
	22	14	64	100

chi square (4 dof) = 15.250   (1%)

	Senate Vote 2			
	Proponents		*Opponents*	*Total*
	Cane or Beets	*Corn*		
Supported	20	8	36	64
Opposed	0	6	24	30
No Vote	2	0	4	6
	22	14	64	100

chi square (4 dof) = 12.961   (2.5%)

	Senate Vote 3			
	Proponents		*Opponents*	*Total*
	Cane or Beets	*Corn*		
Supported	20	9	31	60
Opposed	0	5	26	31
No Vote	2	0	7	9
	22	14	64	100

chi square (4 dof) = 15.332   (1%)

	Senate Vote 4			
	Proponents		*Opponents*	*Total*
	Cane or Beets	*Corn*		
Supported	21	9	30	60
Opposed	1	4	27	32
No Vote	0	1	7	8
	22	14	64	100

chi square (4 dof) = 16.262   (1 %)

	Senate Vote 5			
	Proponents		*Opponents*	*Total*
	Cane or Beets	*Corn*		
Supported	21	6	27	54
Opposed	1	7	36	44
No Vote	0	1	1	2
	22	14	64	100

chi square (4 dof) = 21.410   (1%)

Sources: Calculations available from authors upon request

As Table 12.5 shows, the entire US sweetener industry is large by any standard (Landell Mills, 1994).

TABLE 12.4   Cross-Tabulations of House Votes on Proposed Sugar Legislation, By State *(See Table 12.2 for description of proposed legislation.)*

	Proponents		House Vote 1 Mixed	Opponents	Total
	Cane or Beets	*Corn*			
Supported	72	20	42	56	190
Opposed	22	39	31	121	213
No Vote	12	1	3	16	32
	106	60	76	193	435
			chi square (6 dof) = 62.752 (1%)		

	Proponents		House Vote 2 Mixed	Opponents	Total
	Cane or Beets	*Corn*			
Supported	90	32	54	87	263
Opposed	14	16	21	91	142
No Vote	8	8	2	12	30
	112	56	77	190	435
			chi square (6 dof) = 50.414 (1%)		

	Proponents		House Vote 3 Mixed	Opponents	Total
	Cane or Beets	*Corn*			
Supported	85	35	52	99	271
Opposed	23	21	24	82	150
No Vote	4	0	1	9	14
	112	56	77	190	435
			chi square (6 dof) = 21.776 (1%)		

Sources: Calculations available from authors upon request

Implications and Conclusions

In this chapter, we have added to the literature as to why strong political support for the US sugar program has existed in the past. Even if one accepts as given that the US sugar program results in net societal costs (although such results are open to question), such a policy still may be adopted by Congress — even in the absence of pressure from special interest groups — if legislators weight equally the welfare of their constituent interest groups and the policy redistributes wealth in such a way that a majority of legislative districts enjoy net benefits.[9] We analyzed the regional effects of US sugar policy and correlated these effects with congressional voting records on eight amendments intended to weaken the sugar program. We found that legislators representing regions that form the sugar program do not comprise a majority of Congress but that they do constitute a sizable minority of over one-third in each house. Furthermore, voting behavior appeared to be consistent with the sugar program's net economic impact: Legislators representing net sweetener-exporting regions, the beneficiaries of US sugar policy, tended to be strong supporters of the sugar program; Support for the sugar program from corn-growing and refining regions increased between 1980 and 1990 consistent with the growth of HFCS market share; and Legislators, who represented states that refined imported cane sugar, switched from opposing the sugar program to supporting it following the introduction of a minimum import provision in the 1990 sugar legislation.

TABLE 12.5   US Sweetener Industry — Economic Impact and Jobs, by Sector

	Direct Economic Value (Million $)	Direct & Indirect Economic Impact (Million $)	Direct Jobs (Full-time Equivalent)	Direct & Indirect Jobs (Full-time Equivalent)
Beet Growing	1,080.8	2,701.9	26,692	66,729
Beet Processing	1,349.4	3,373.6	8,585	21,463
Beet Total[a]	2,430.2	6,075.6	35,510	88,775
Cane Growing & Harvesting	848.3	2,120.8	22,488	56,221
Cane Milling	733.5	1,833.7	6,268	15,669
Cane Refining	716.3	1,790.8	4,231	10,577
Cane Total[a]	2,298.1	5,745.3	33,229	83,072
Sugar Total	4,728.3	11,820.8	68,739	171,847
Corn Growing for Sweetener	1,626.5	4,066.3	90,537	226,343
Corn Processing for Sweetener	4,132.1	10,330.3	8,549	21,371
Corn Sweetener Total	5, 758.7	14,396.6	99,086	247,715
Grand Total	10,487	26,217.5	167,825	419,562

[a]Including other types of employment (for example, Company HQ, Regional Sales Offices)
Source: Landell Mills, 1994

Based purely on positive regional effects, a strong foundation of support for the sugar program exists although there was not a majority of votes needed to sustain it in Congress. Some legislators, who presumably would oppose the program on the basis of its deleterious impact on their constituencies, must be persuaded to vote in favor of the sugar program. For example, Lopez (1989) argues that legislators are more likely to support the sugar program because taxpayers' costs are zero. Nevertheless, the broad base of support for sugar legislation, generated by the relatively large number of beneficiary regions, suggests that sweetener interest groups can either obtain a given level of support with less lobbying than other commodity interest groups or, for a given level of lobbying, can increase the level of support. Several implications of our approach appear to be in contrast to the findings of some recent studies. For example, Gardner (1987) found that a commodity's support level decreased as the dispersion of its producers increased. He concluded that the increased cost of organizing an effective political lobby of geographically dispersed producers was greater than the potential benefits of having producers spread across many congressional districts. Our analysis suggests that it is how producers are dispersed that is important, not simply the degree of dispersion. More specifically, dispersion may be advantageous to producers if there are a large number of localities in which their benefits outweigh the costs borne by local consumers.

In a study of House votes on proposed amendments to farm legislation in 1985–86, Abler (1989) found that sugar proponents joined in a vote-trading coalition with rice, cotton and peanut proponents. He concluded that sugar was one of the commodities that received the greatest amount of support from proponents of other commodities. Stratmann (1992), studying some of the same 1985 House votes, reached the similar conclusion that there was mutual logrolling between peanut and sugar proponents. These findings seemingly contradict our implication that sugar is less dependent on logrolling than most commodities. Indeed, Abler (1989) states that one "would certainly expect a group with limited representation in Congress to make more use of vote trading and/or campaign contributions," implying that vote trading by sugar proponents was an indication of relatively weak congressional representation. However, both Abler (1989) and Stratmann (1992) assume independence between the issue being voted upon and the outcome of the vote. They analyze the vote outcomes as if they were unrelated to the level of support being voted upon. In fact, given the relatively high level of support enjoyed by the sweetener industries, it may be that the vote trading and logrolling observed by past researchers were a manifestation of sugar proponents parlaying their strong base of support into rejection of even a marginal reduction (5.5%) in the sugar support price.

This approach helps explain the form of the present program. For example, when sugar import quotas were reimplemented in 1981, there was no provision for limiting domestic production of sugar as in previous programs. This policy reflected the sweetener environment of the time: The quantity of imported sugar was substantial. The geographical distribution of domestic production was in transition, and perhaps most importantly, the corn sweetener industry was in its infancy. There was little or no resistance within the sugar industry to limiting imports and strong opposition to limiting domestic production given the rapidly changing production pattern. By 1990, however, the situation had changed dramatically. Imports of sugar had fallen precipitously, from over 5 million short tons raw value (s.t.r.v.) in 1979 to approximately 1.25 million s.t.r.v. in 1989. In addition, corn-derived sweeteners, primarily HFCS, had captured over half the domestic-caloric sweetener market and threatened to capture substantially more upon the imminent development of a crystalline HFCS. Reflecting this new reality, proposed 1990 legislation provided for a minimum level of imports (1.25 million s.t.r.v.), quotas on domestic production of cane and beet sugar if necessary and a 200,000 ton limit on marketing of crystalline fructose. These provisions preserved strong congressional support for the sugar program. The minimum-import provision extended the number of congressional supporters to include those representing refiners of imported cane.

In conclusion, national policy does not usually affect all regions of a country equally. Consequently, opposition or support for a proposed policy is likely to vary by region, and its adoption or rejection will depend on how these regions are represented in the policymaking body. In the United States, the result may be adoption or continuation of a policy that may impose a net cost on society at large, simply because a majority of legislators represent constituencies that enjoy a net benefit. This implies that limiting the access of special interest groups may not be as effective as other studies suggest because some policies, such as the sugar program, have a strong base of support, even in the absence of special interest groups. The geographical dispersion of sugar production and processing, corn production and HFCS processing will likely provide broad, regional backing for sugar support well into the future. It is important to keep in mind that the entire sweetener industry (not just sugar) is a significant part of the US agricultural sector.

Notes

1. We do not infer that this model necessarily will apply to the new sugar program under the 1995 Farm Bill.

2. A net domestic welfare loss would persist even if tariffs rather than quotas were employed as policy instruments although in this scenario the government would receive area c in the form of tariff revenues.

3. See the papers by Schmitz (1995) and Schmitz and Vercammen (1995).

4. Since consumers typically outnumber producers and processors, equally weighting their welfare levels would appear to be irrational for a popularly

elected legislator. However, in local economies, it is possible for consumers to feel that the indirect benefits of the sugar program outweigh the direct costs if sugar production and/or processing is important to the local economy, for example, through multiplier effects. Furthermore, local consumers may feel that the bulk of the economic burden is being borne by consumers from other parts of the country. In other words, they may not perceive the sugar program as transferring wealth from themselves to producers and processors but rather as transferring wealth from other parts of the country to their region. For example, although Florida produces approximately twice as much sugar as it consumes, there are only a handful of producers and refiners who are direct beneficiaries of the sugar program. Yet over the 1980–90 period, Florida's legislators were among the sugar program's strongest supporters.

5. Data on sugar production and per capita consumption of sweeteners were obtained from the USDA (various issues). (Except where otherwise indicated, figures are averages of the period 1985–89.) Sugar beet production was converted to a refined sugar basis using the average recovery rate. Consumption levels are the average per capita consumption of refined sugar and corn sweeteners multiplied by the estimated population in 1988 as reported by the US Bureau of the Census.

6. State-level data on the use of corn in sweetener production are not available. In approximating the level of corn used for sweetener production in each state, it was assumed that 1) an equal proportion of each state's corn production, as reported in USDA (various issues), was refined into sweeteners and 2) one unit of corn sweeteners requires $1^{2}/_{3}$ units of corn (by weight).

7. All votes were in favor of the sugar and sweetener industry with the exception of the 1981 House vote to eliminate the sugar program, which was later struck in the House-Senate Conference Committee.

8. To be precise, the analysis of House votes should define regions as Congressional districts rather than states since the net economic impact of the sugar program may vary across districts within a state. On the other hand, there are a number of reasons why representatives of net-importing districts within net-exporting states may support the sugar program. For example, the definition of a "region" is somewhat subjective; constituents and legislators alike might identify their states, rather than temporary and imprecisely known legislative districts, as their "region." Also, election to the House of Representatives is often a stepping stone to a future state office, further cause for a representative to define the entire state as his/her region.

In any case, an analysis of the 1990 House vote along these lines is presented in Schmitz and Christian (1993). The significance of the tests we present here are not substantially different.

9. Note that this idea is not necessarily inconsistent with Rausser's and Freebairn's (1974) concept of a revealed policy preference function in which policy actions reveal the government's weighting of the welfare levels of various interest groups comprising the political economy. In this formulation, unequal welfare weights may arise in two ways. Rausser and Freebairn discuss the pressure applied by special interest groups in an effort to influence the position of policymakers as a "type of bargaining game between political representatives

and interested pressure groups" (1974), the outcome of which is revealed by the (unequal) weights in the government's policy preference function. Another source of uneven welfare weights, not mentioned by Rausser and Freebairn, is the unequal representation of regions (and/or interest groups) indirectly brought about by regional representation in Congress. This differs from the former in that the government's unequal welfare weights arise not from the unequal welfare weights of individual policymakers but from the way in which the votes of legislators, who weight equally the welfare of their individual constituent interest groups, are aggregated.

References

Abler, D. G. 1989. "Vote Trading on Farm Legislation in the US House." *American Journal of Agricultural Economics* 71: 583–591.

Becker, G. S. 1983. "A Theory of Competition among Pressure Groups for Political Influence." *Quarterly Journal of Economics* 98: 371–400.

_____. 1985. "Public Policies, Pressure Groups, and Dead Weight Costs." *Journal of Public Economics* 28: 329–347.

Canadian Sugar Institute. 1995. *Canadian Sugar Institute Backgrounder.* Toronto, Ontario, Canada.

Gardner, B. L. 1987. "Causes of US Farm Commodity Programs." *Journal of Political Economy* 95: 290–310.

Johnson, R. G., and A. Ortego. 1995. "Sugar and Honey Policy," in R. D. Knutson, ed., *1995 Farm Bill: Policy Options and Consequences.* Pp. 74. Texas Agricultural Extension Service, Texas A&M University, College Station, TX.

Landell Mills Commodities Studies. 1994. *The Importance of the Sugar and Corn Sweetener Industry to the US Economy.* Study prepared for American Sugar Alliance, Washington, DC.

Leu, G. J., A. Schmitz, and R. D. Knutson. 1987. "Gains and Losses of Sugar Program Policy Options." *American Journal of Agricultural Economics* 69: 591–602.

Lopez, R. 1989. "Political Economy of US Sugar Policies." *American Journal of Agricultural Economics* 71: 20–31.

Marsden, S. 1995. "Arguments for Global Reform." Pp. 1–3. Paper presented at International Policy Council on Agriculture, Food and Trade: Sweeteners Forum, Washington, DC (27 March).

Rausser, G. C., and J. W. Freebairn. 1974. "Estimation of Policy Preference Functions: An Application to US Beef Import Quotas." *Review of Economics and Statistics* 56: 437–449.

Schmitz, A. 1995. "The Free Trade Myth and the Reality of European Subsidies." Paper presented at the 12th International Sweetener Symposium, US Sweetener Policy in the Farm Bill: Charting a Course for the Future, Washington, DC.

Schmitz, A., and D. Christian. 1993. "US Sugar Policy," in K. Maskus and S. Marks, eds., *The Economics and Politics of World Sugar Policies.* Ann Arbor, MI: University of Michigan Press.

Schmitz, A., and J. Vercammen. 1995. "Efficiency of Farm Programs and Their Trade-distorting Effects," in G. C. Rausser, ed., *GATT: Negotiations and the*

Political Economy of Reform. Berlin, Germany: Springer-Verlag.

Schmitz, A., H. de Gorter, and T. Schmitz. 1996. "Consequences of Tariffication." This volume.

Stratmann, T. 1992. "The Effects of Logrolling on Congressional Voting." *American Economic Review* 82: 1162–1176.

US General Accounting Office. 1993. *Sugar Program under Changing Conditions*. GAO/RCED 93-84, Washington, DC.

USDA. Various issues. *Agricultural Statistics*. Washington, DC.

_____ . Various issues. *Sugar and Sweetener Outlook and Situation Report*. USDA/ ERS, Washington, DC.

Zusman, P. 1976. "The Incorporation and Measurement of Social Power in Economic Models." *International Economic Review* 17: 447–462.

Zusman, P., and A. Amiad. 1977. "A Quantitative Investigation of a Political Economy — The Israeli Dairy Program." *American Journal of Agricultural Economics* 59: 88–98.

SECTION THREE

Regulation and Supply Management

13

Supply Management Canadian Style[1]

H. G. Coffin, R. Saint-Louis, and K. A. Rosaasen

Abstract

The regulatory systems used to govern supplies and influence prices in Canada's dairy and poultry sectors are examined in terms of their organizational and jurisdictional features. The conditions prevailing in these sectors prior to supply management are explored as factors contributing to the implementation of such a system. This chapter also outlines the implications for provincial cost/price relationships and opportunistic behavior under the price leveling and regulatory aspects of national marketing plans.

Introduction

Extreme price variability and economic hardship in the dairy and poultry sectors during the 1960s, coupled with concerns about vertical integration and the possible loss of the poultry sector to import competition, prompted governments and producers in Canada to seek radical solutions such as national supply management. A solution recommended at the time by the Task Force on Agriculture, which consisted of four academics and an independent businessman,[2] was viewed as a kind of industrial strategy. The implementation of market-sharing agreements began with industrial milk in 1970 and has progressed through eggs in 1973, turkey in 1974, chicken in 1978 and broiler hatching eggs in 1986.

The system of supply management in these sectors is both unique and controversial, involving power-sharing arrangements between provincial and federal levels of government and with agricultural producers. The use of quotas to control production and imports[3] in order to stabilize producer prices has been criticized as costly to consumers and as an inefficient means of transferring income to farmers (Arcus, 1981; Barichello, 1981; Josling, 1981; Veeman, 1982; Green, 1983). On the other hand, Coffin, Romain and Douglas (1989) observed that producer and wholesale price differences for chicken in Canada and in the United States have been smaller in real terms since the implementation of supply management than was the case when imports were not restricted by quota. Proulx, Gouin and Saint-Louis (1992) also found little empirical evidence to support some of the usual criticisms of supply management. With regard to performance indicators such as milk production per cow and per farm and eggs produced per hen, Canadian producers were found to have been gaining ground on their counterparts in comparable areas of the United States.

Both the dairy and poultry sectors have been subject to comprehensive policy review (National Poultry Task Force, 1991; Task Force on National Dairy Policy, 1990). While the debate on the merits, demerits and future of supply management continues, Canadian dairy and poultry producers have lobbied effectively to maintain protection through the establishment of relatively high tariffs with which to replace import quota under the new General Agreement on Tariffs and Trade (GATT) agreement.[4] Although tariffication may precipitate some adjustments, the main pressures for change will come from within these sectors because of growing tension over such issues as market share, pricing policies, determination of market requirements and surplus removal.

While these regulatory systems came into existence partly because of market failure during the 1960s and 1970s, the constitutional division of powers between the federal and provincial governments was also a contributing factor. In the latter case it was seen as a solution to a long-standing jurisdictional conflict between the two levels of government and among the provinces themselves over matters of marketing legislation. To better understand the nature and ramifications of these systems, we turn first to a review of their organizational and legal framework.

Organizational and Legal Framework

Supply management has a rather specific connotation in Canadian agriculture as the policy framework under which the dairy and poultry sectors operate. It refers to the systematic use of production or marketing and import quota (or tariffs) to manage national supplies to satisfy projected demand at a target price that corresponds to the estimated cost of production (COP) at the farm level. In this context supply management is distinct from occasional interventions by government, such as the buy-out of dairy herds or the set-aside or diversion of crop land in order to reduce surpluses — actions to which the term is sometimes applied in the literature. In the Canadian case the emphasis is placed on stabilization of farm prices and incomes by preventing surplus from occurring in the first place.

Canadian supply management is also distinct in that its powers are essentially exercised by producers and producer agencies. The desired national output is allocated among provinces on the basis of historical production and other criteria. Provincial shares are further divided among producers by provincial boards — where methods of quota transfer and limits on maximum quota holding per enterprise are determined by the provincial agencies. (See Rosaasen, Lokken and Richards, 1996).

Green (1983) found it almost unthinkable in view of the potential for conflict of interest that such extraordinary powers were granted to producer marketing boards. In his economic and legal analysis of the history of agricultural marketing boards, which he calls "long, tangled and complicated," Green argues that two ingredients on which their development and power depend are the "constitutional evolution and the behaviour of the political process in a pluralistic society." He also suggests that a "significant part of Canadian constitutional interpretation emanates from marketing board cases."

The constitutional framework for supply management began with the British North America (BNA) Act under which Canada was founded in 1867. The BNA Act provided for the federal union of four of the current provinces (Québec, Ontario, Nova Scotia and New Brunswick). It also defined federal and provincial jurisdiction, giving provincial governments authority over production and marketing within their own boundaries and the federal government jurisdiction over interprovincial and international trade (Figure 13.1). This division of power was used to strike down early unilateral attempts at regulatory legislation by both the federal and provincial levels of government in the 1930s.

Gilson and Saint Louis (1986) argue that four sections of the Canadian Constitution are relevant to jurisdiction in agriculture: Sections 91, 92, 95 and 121. Section 91.2 grants the federal government authority to regulate trade and commerce among the provinces for agricultural commodities but not to regulate intraprovincial trade. Section 92 lists those areas in which provinces can exclusively make laws. Section 121 of the BNA Act is perhaps most important from the standpoint of agricultural marketing legislation in Canada in that it prohibits the establishment of trade barriers at provincial boundaries.

According to Lemieux, (1993),[5] another area of past constitutional dispute was the power of a federal agency to collect levies from producers to cover its cost of operation. These levies were considered a form of taxation that fell within provincial jurisdiction. However, in 1978, Chief Justice Bora Laskin of the Supreme Court of Canada ruled that such levies were not taxes.[6] Instead, he argued that they should be viewed as continuous costs of operating a marketing plan. This meant that both the federal and the provincial governments had authority over such matters.

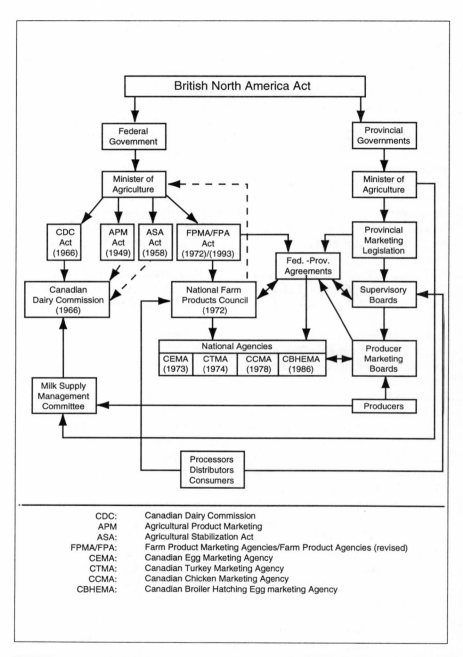

FIGURE 13.1   Legal and Organizational Framework for Supply Management for Poultry and Industrial Milk.

Court decisions, such as the aforementioned, have had wider implications for the Canadian marketplace. Gilson and Saint-Louis (1986) concluded, from the examination of judgements made by various courts on agricultural marketing issues that the Canadian agricultural economy is closer to being a customs union than a more integrated common market. The constitutional guarantee of an agricultural common market in Canada is incomplete in the sense that, although interprovincial customs and/or other explicit border taxes are prohibited, no explicit reference to other nontariff barriers is made in the Constitution itself. This is reflected in the refusal of the Supreme Court of Canada to bar the exercise of powers by various producer marketing boards under agricultural supply management rules. This suggests that, since the 1960s, constitutional issues have been at the heart of the perennial compromising process of the supply-managed sectors.

The division of powers in the Canadian Constitution resulted in two models of organization in support of a national marketing plan: The Canadian Dairy Commission (CDC) on the one hand and the National Farm Products Council (NFPC), which supervises producer agencies on the other. In both cases either one level of government delegates its powers to the other, or both levels of government must reach a federal-provincial agreement as a contract between or among sovereign powers. Both models are illustrated in Figure 13.1.

In the case of the CDC, which was established in 1966 to deal with milk production for the manufacture of dairy products, there exists a cross-delegation of powers using three pieces of federal legislation: the Canadian Dairy Commission Act (CDCA) of 1966; the Agricultural Products Marketing Act (APMA) of 1949; and the Agricultural Stabilization Act (ASA) of 1958. The CDCA did not give the CDC the power to fix milk prices at the provincial level. Instead, it set a national target price for industrial milk and supported it by offering to purchase butter and skim milk powder from processors under the ASA when wholesale prices fell below the equivalent target price for milk. In a similar vein, the federal-provincial agreement invoked the APMA to commit the provincial authorities to collect the producer levies and pass them on to the CDC.[7] Finally, the CDCA empowers the CDC to fix the level of output, which it does on the advice of the Canadian Milk Supply Management Committee (CMSMC). This committee consists mostly of producer and provincial government representatives. The authority for pricing and production of fluid milk remains with the provinces.

The NFPC model applies to the poultry agencies and is based on the enabling legislation of the Farm Products Marketing Agencies Act (FPMAA) of 1972 and corresponding provincial legislation. (In 1993 this act was revised and renamed the Farm Products Agencies Act (FPAA)).

When marketing plans are approved under this model, federal-provincial agreements are struck to govern the creation and operation of these national agencies. These agencies, consisting largely of provincially elected producer representatives, determine the level of output subject to approval by the NFPC, which reports to the Federal Minister of Agriculture. The NFPC also allocates production among the provinces. Price determination remains at the provincial level[8] and products may move freely among provinces.

Beyond its monitoring and supervisory role, the NFPC is empowered by the FPAA to conduct the procedures under which new agencies may be established and to advise the Federal Minister of Agriculture on all matters pertaining to the establishment and operation of those agencies. The establishment of a national supply management plan is not a simple matter. Producers wishing to establish such an agency must develop a marketing plan and earn acceptance of that plan so that the NFPC is convinced of the need to call public hearings about the plan. If the NFPC then wishes to recommend that the plan be implemented, it must be satisfied that a majority of producers favor implementation. This may require a national or regional plebiscite among all affected producers.[9] If it passes this test, up to 32 signatories[10] must negotiate a federal-provincial agreement before the supply management agency can be established.

The NFPC and its provincial counterparts, the supervisory boards, have important roles in protecting the public interest. For example, approval of national production levels falls within the NFPC mandate that requires agencies to "promote strong, efficient and competitive production and marketing" of agricultural products under national marketing plans, having "due regard to the interests of producers and consumers" . . . of these products (National Farm Products Marketing Agencies Act (NFPMAA), 1972). It is clear that the NFPC must be satisfied that planned production adequately protects the interests of consumers as well as producers. Despite these provisions in the FPAA, however, Green (1989) found that the accountability of the provincial boards that drive the system to be "limited and weak." This may be a result of how the frontiers of empowerment are established.

National marketing plans are set up under conditions beyond those which legislative empowerment by the federal government to the provincial governments can accommodate. The vehicle of this empowerment is the federal-provincial agreement. In return, the provincial governments bestow the same principles upon the provincial supervisory boards and the laws governing their authority. The results are that federal and provincial agencies remain at arm's length from each other and from provincial/producer boards.

In practice, the provincial producer marketing boards consider themselves regulated more by their respective provincial policing agencies than by the federal system, including the relevant national policing agencies. But the fact that provincial producer boards maintain linkages with three other levels of authority in this system has probably allowed them to occupy some grey zones as the system has evolved over time. In their own opinion, expressed before the 1982 Inquiry Panel, they went as far as admitting that this dispersion of policing authority gave them greater chances to behave opportunistically than might otherwise be the case (Saint-Louis, Ivison, Brechin, Vielfaure and Williams, 1982).

In turn, the national marketing agencies consider themselves regulated more closely by the federal government and the relevant federal agencies than by the provincial governments and/or agencies. The fact that national agencies also maintain linkages with three other levels of authority in this system probably also provides them with some margin for opportunistic behavior.

So each level of government in the agricultural supply-managed sectors of Canada polices its own portion of that spider's web, depending upon the nature and the extent of the issues on the table at a given moment. Thus, each level of government will likely continue trying to occupy as much field as possible as the system evolves over time and as the issues become more complex and more chaotic in the aftermath of GATT.

It is perhaps not surprising that such institutional realities may seem rather strange in relation to the United States. In the US system, national law is always superior to state regulations. Therefore, the US regulatory authorities probably leave less grey zones between levels of organized marketing systems in agriculture than do their Canadian counterparts.

Returning to the alternative models of supply management in Canada, an important structural difference should be noted. In the case of the commission model, for example, the CDC, the commissioners are appointed by the federal cabinet and have no particular geographical constituency. On the other hand, in the national agency model the directors of the poultry agencies are producers elected from the ranks of the producer marketing boards in each province. As such, they have a distinct geographical constituency, and they have a territorial interest to defend.

Skogstad (1993) discussed the implications of this kind of structural difference. One consequence may be that the national agencies model is more democratic in terms of producer input than the commission model, but it is also more rigid and less capable of change, particularly with respect to regional shares of production. Moreover, the necessity for a federal-provincial agreement leads some provinces to view these industries as instruments of regional economic development. If they yield autonomy by signing on to the national marketing plan, they expect a piece of the action in return.

To recap then, supply management systems require either delegation or sharing of federal and provincial powers through agreement with many signatories. They are relatively difficult to establish because producers must have a strong case and a strong consensus among themselves that supply management represents the best solution. Once established, however, even though such systems include a measure of governance by representatives of the processing, distribution and consumer groups of stakeholders, they tend to be run primarily by producers through agencies, which may have structural impediments to change and may find avenues for opportunistic behavior.

Causes of Supply Management

Given the complexity of the legal and administrative framework just discussed, one might question why Canadian policymakers have chosen to create supply management systems. The authors of this chapter contend that whatever the current economic judgement is and whatever indications of rent-seeking emerge, at any level, this course of action was chosen and has been pursued as a good faith attempt to address the failures of the marketing system.[11]

Supply management programs for dairy and poultry had their origins in the 1960s. Although there had been some attempts to organize, especially in the dairy industry, as early as the 1930s, most legislation was challenged and struck down on jurisdictional grounds.[12] By the 1960s, however, producer marketing boards with varying degrees of power were fairly well-established at the provincial level, particularly for fluid milk with as many as 35 regional boards reported (Hiscock and Walker, 1969).[13]

The 1960s were troubled times in Canadian agriculture. Surpluses began to accumulate, and prices and incomes were low and unstable. Conditions were especially difficult in the egg sector. One former Minister of Agriculture, Honorable H. A. Olson, described the situation as chaotic, recalling egg prices as low as 8¢/doz during the 1960s.[14]

A portrait of 26 years of monthly average egg prices to producers in Canada, centered around the implementation of supply management in 1973, is shown in Figures 13.2 and 13.3. Expressed in 1993 dollars, the level of prices in the 1960s does not seem so low in relation to more recent values. However, the extreme variability of prices and the seeming failure of interprovincial arbitrage are clear features of egg prices in the pre-supply management period, 1960–72. (Figure 13.2)

The coefficient of variation, calculated on the basis of two-year moving averages of monthly prices to Canadian producers, frequently yielded values in excess of 20% in the 1960s when most other prices were relatively stable. As indicated by the shaded background and right-hand scale in Figures 13.2 and 13.3, the coefficient of variation for monthly egg prices has fallen from those values in the pre-supply management period to less than 5% in the first 13 years following the introduction of supply management and has remained low since then.

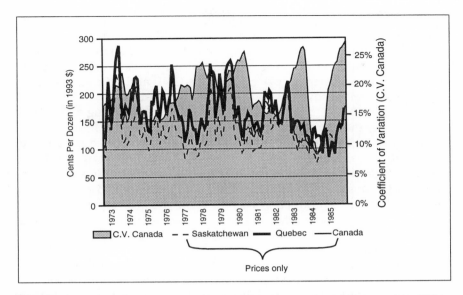

FIGURE 13.2   Real Monthly Weighted Average Egg Prices, Paid to Producers and Coefficient of Variation, Canada (1960–72) (Before Supply Management).

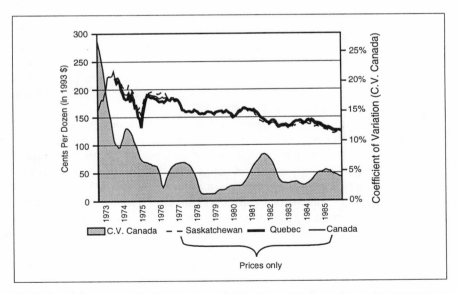

FIGURE 13.3   Real Monthly Weighted Average Egg Prices, Paid to Producers and Coefficient of Variation, Canada (1973–85) (After Supply Management).

With regard to interprovincial arbitrage, price spreads shown between Saskatchewan and Québec in Figures 13.2 and 13.3 are much larger and less consistent in the pre-supply management period than they have been since. This reflects, of course, the initial imposition under supply management of a formula price based on COP and adjusted among surplus and deficit provinces by transportation costs.[15]

The fledgling broiler chicken industry was also encountering difficulties. As the third farmer in Saskatchewan to enter the business in 1959, Flaten[16] recalls prices as low as 16½ ¢/lb (36 ¢/kg) in 1960 with feed conversion and growth rates at little more than half of what they are today. Hill (1963) reports that producer prices for broilers reached an all time low in 1961 in the United States and Canada and that "both industry and government recognized that overproduction was the basic cause of the trouble." He refers to "many proposals for both government and privately-operated marketing boards . . ." and the difficulty of planning production to meet projected demand.[17]

By 1969, chronic overproduction and low prices prompted the provincial chicken marketing boards to form a national association known as the Canadian Broiler Council. Faced with mounting stocks of frozen chicken, they attempted voluntary supply restraint and agreed that production should be reduced by 10% in all provinces. By the time the strategy reached the level of individual producers, most decided it would be a good time to increase production if everyone else was cutting output. The result was predictably a complete failure of the strategy of voluntary supply restraint.[18] This experience of broiler producers is typical of the information problem faced by all agricultural producers. In describing conditions that led to national supply management of industrial milk, Mestern (1972) observed that, "The root cause of cyclical overproduction in the industrial (milk) sector was the lack of a guide to the size of the market available to individual producers."

Throughout this period of depressed prices, deficiency payments and government purchases helped support farm prices and incomes. Under the ASA, the federal government made direct payments or intervened through purchases to support prices for poultry, egg and dairy producers. Payments were made to dairy producers every year from 1958 until a new dairy subsidy was formalized under the CDC in 1966. By 1963–64, the annual payments to dairy producers under the ASA reached $117 million, a substantial sum for the time (Veeman, 1975).[19]

The size and continuous nature of these stabilization payments may have contributed to a decision by the federal government to create the CDC in 1966. Other factors that contributed to this action were the loss of export markets when the United Kingdom joined the European Community and the need for a stable economic environment in which to plan and execute structural changes required within the dairy industry at the time. Trant (1994) cites technological advances, such as the adoption of the bulk tank, as important factors that precipitated structural change.

Initially, the CDC enticed producers to register in supply management by offering a "subsidy eligibility quota" based on their previous year's production with subsidy holdback penalties for production in excess of that quota. The failure of this approach to supply management induced the Dairy Farmers of Canada to develop a national plan which Mestern (1972) describes as a "comprehensive and equitable market sharing program." This proposal became the basis of a federal-provincial agreement that established the market share quota for industrial milk with compulsory registration and more severe penalties for overquota production. This agreement, which was struck in 1970, was the first successful national supply management program in Canada.[20]

The adoption of supply management by the poultry industry did not progress as quickly and as smoothly as it had in the dairy industry. This may have been partly because of the nature of the egg and poultry industry itself. Kirk[21] describes the industry as "more piratical" when it comes to issues such as market sharing or matters which require a cooperative approach. This cutthroat attitude may be related to poultry production requiring "less-extended investments" in management than does dairy production, in terms of breeding, selection, development and maintenance of the herd.

The egg sector was the first within the poultry industry to move toward supply management. A national conference was conducted in 1968 following a resolution by the Canadian Federation of Agriculture in support of supply management. The egg producers requested a national agency similar to the CDC model. However, one of the circumstances of the time was that agriculture, and especially controversial legislation, did not rank very high among the priorities of the federal government. The ruling Liberal Party was preparing for a change of leadership (from Pearson to Trudeau) celebrating the centennial anniversary of the country and trying to decide how to deal with the increasingly militant actions of the independence movement in Québec. Thus, changes transpired slowly in the area of agricultural policy.

The frustration of farmers over the generally disastrous conditions in agriculture and the slowness of the federal government to act resulted in a large-scale march in Ottawa with vigorous demonstrations. The response of the Hon. J. J. Greene, Minister of Agriculture, was to appoint a five-man task force to make a "comprehensive assessment of Canadian agriculture in terms of its contribution to the achievement of national goals" (Campbell, Comtois, Gilson, MacFarlane and Thain, 1969). The task force was also mandated to give particular attention to the income and welfare of farmers and to "study and make recommendations concerning agricultural policies required to achieve long-range national and agricultural goals, taking account of the interests of farmers and consumers" (Campbell et al., 1969).

Within its comprehensive review of Canadian agriculture, the task force devoted one chapter of its final report to marketing boards and supply management. The principal recommendation on those subjects was that the federal government should introduce legislation "to permit the creation of national commodity marketing boards" (Campbell et al., 1969). Among the other recommendations was one that suggested that any quotas imposed on inputs or sales should be nationally negotiable, rather than provincial in nature, so as to permit the relocation of production to areas of lowest cost. This was not done, and it is one of the contentious issues of today that the dairy sector is attempting to address (Balcaen, Doyle, Matte, Morin and Sherwood, 1992).

By the time the task force submitted its report at the end of December, 1969, the Minister of Agriculture was the Honorable H. A. Olson, now a member of the Senate. It was he who eventually steered the enabling legislation through the Canadian Parliament in 1972 but only after he was convinced that the provincial governments were prepared to delegate powers to a national agency through federal-provincial agreements. Both he and the Honorable E. F. Whelan , who succeeded him as Minister of Agriculture, independently credit the Honorable William Stewart, then Minister of Agriculture for Ontario, with influencing his provincial counterparts to commit their support to national supply management.[22]

One of the concerns of provincial governments that drove them toward supply management was a fear of losing a large part of the poultry and egg industry, first to corporate control through vertical integration, then possibly to the United States through imports.[23] According to Kirk,[24] supply management was viewed as an industrial strategy to maintain production in Canada and especially in the traditional family farm structure.[25]

There was also a need to find a mechanism to resolve nasty provincial disputes, such as the Chicken and Egg Wars of 1970–71. This conflict arose during a long period of depressed prices when the Québec Egg Producers Marketing Board attempted to ease the situation for its own members by introducing a provincial surplus disposal scheme. This program attracted shipments of eggs from Ontario and Manitoba, adding to the cost burden on Québec producers. Attempts by Québec producers to interfere with the inflow of eggs from other provinces were met by retaliation from Ontario producers who blocked shipments of chicken from Québec. Lawsuits ensued leading to a Supreme Court ruling that provinces could restrict export of products from their own territory but could not restrict imports. It became apparent that a national mechanism had to be found. Federal legislation to create national agencies was a potential solution agreed to by provinces.

When the NFPMAA was passed in 1972 on the second attempt and after much debate, the egg producers moved quickly to seek approval of their plan. Public hearings were held, and a federal-provincial agreement was negotiated by the signatories, creating the Canadian Egg Marketing Agency (CEMA) in January, 1973. Despite the speed with which the process moved in the final stages, however, nearly five years had passed from the time the concept was first agreed to by egg producers. During that time, producers continued to expand production, and importers continued to increase imports in order to establish a larger base volume in anticipation of an eventual quota system. Hence, by the time the CEMA was created, there were already too many eggs, too much production capacity and no effective means of supply control.

The initial years of operation for the CEMA were turbulent ones, marked by eggs spoiling in storage, financial difficulties for the Agency, several changes in management and intense public scrutiny spearheaded by the Food Prices Review Board.[26] It was a climate in which a price increase of even 1¢/doz was considered worthy of a front-page newspaper story.

The initial difficulties encountered by the CEMA did not impede the establishment of the Canadian Turkey Marketing Agency (CTMA) which followed closely in 1974 but may have delayed final acceptance of the marketing plan for chicken which did not evolve until 1978 — at which time the Canadian Chicken Marketing Agency (CCMA) was established. The latest agency to be established under a supply management marketing plan was the Canadian Broiler Hatching Egg Marketing Agency (CBHEMA), which was launched in November, 1986. As had been the case with chicken and eggs, two of the main motivations for pursuing supply management in the hatching egg business were stabilizing prices and improving income through import control and countervailing power with hatcheries.

The issue of countervailing or bargaining power has been an important theme for farmers throughout the development of producer marketing boards and supply management agencies in Canada. In general, the food processing industries to whom farmers sell their products exhibit a more highly concentrated structure than do corresponding US industries. As reported by Lanoie (1985), for example, in Canada the CR4 for poultry processors on a national basis as of 1978 was 39% compared to approximately 15% in the United States. Similarly for dairy processors, the Canadian CR4 was 35% compared to the US measure of 26%. Given the economic geography and regionalization of industries in Canada, the concentration of these sectors on a regional basis is much greater, at both the processing and retail levels (Coffin et al., 1989)

Institutionalized Compromise

Many of the consequences of supply management have been illuminated in the comprehensive review articles by Schmitz (1983) and Schmitz and Schmitz (1994).[27] This section deals primarily with the manner in which the Canadian tradition of federal-provincial compromise is realized in supply management systems, an issue which has heretofore received little attention.

In 1991 Ruppel, Boader and Peterson (1991) investigated the complex question of the total costs of negotiating, ratifying and implementing international agreements among federated countries that were characterized by differences in the particular interests of the federation members and the broader interest of the federation itself. They concluded that when federated nations are signatory to international economic agreements, problems often arise because of inherent conflicts between federal law and legitimate, provincial or state interests.

In this chapter, particular attention is given to federal and provincial constitutional issues as they relate to the problems of implementation of provincial producer marketing boards and federal regulatory agencies within supply management in Canada. The main purpose is to explain the unique set of problems faced by this web of federal-provincial government and its parastatal institutions in a somewhat peculiar constitutional context. It is a context where perhaps too few strong institutional mechanisms will remain available to allow coalitions of provincial boards and/or federal regulatory agencies to offset opportunistic behavior by any member of a marketing plan.

From a behavioral study of these marketing systems and the moving ground of the constituent interests of both the stakeholders and the regulatory agencies, the authors of this chapter suggest that one of the key factors in maintaining cohesiveness between the provincial boards and the federal plans is the implicit, if not explicit, price-leveling mechanism that fuses them together. To visualize the process, consider the two sets of data reported in Figure 13.4. The first set of data, depicted by histograms, illustrates the real accounting COP of eggs at the farm level averaged from representative provincial samples of egg farms, province by province from the West Coast to the East Coast. The second set of data, depicted by a full line, illustrates the weighted average reference price calculated by CEMA from a standardized simulation model under singularized sets of input price levels prevailing in each province at specific points in time. These two sets of data were at the heart of a 1982 public inquiry carried out by a five-member panel at the request of the National Farm Products Marketing Council (NFPMC). The inquiry reviewed the COP formula used by CEMA that established producer prices for Grade A large eggs.[28]

As in all other agricultural supply-managed sectors in Canada, a domestic egg market share had been established for each provincial board. Those crucial shares were based on the historical performance of each province, plus or minus very minor correcting factors. Moreover, very little erosion of those provincial market shares was foreseen under the Federal-Provincial Agreement (FPA) which was signed by the federal Minister of Agriculture, NFPMC, CEMA, the provincial ministers of agriculture, the provincial supervisory boards and finally each provincial egg-marketing board (NFPMC, 1976). Thus, surplus and deficit provinces were expected to remain in their respective categories, barring a major change in the local market.

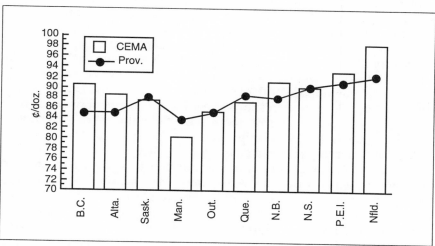

FIGURE 13.4 Egg Production Costs: A comparison of Provincial COP for Grade A Large Eggs and CEMA Farm Gate Price.
Source: Saint-Louis et al., 1982 and Touche, Ross & Partners, 1981.

In the case at hand, the issue had been identified as a pricing problem stemming from the rigidity of the technical formula-pricing model itself. More precisely, the use of a uniquely-weighted set of simulated costs ensured that Saskatchewan, Manitoba and Québec had reference prices that were set above their respective true average costs. Even with the addition of relevant and fair transportation costs to other provinces, prices in those other provinces would be below their respective COPs.

As the Panel members expected, submissions made by the producer boards attempted to push the price line upward across all provinces, including those in which reference prices (as calculated by the model) were already above their respective costs of production. This negotiation strategy can be labeled a price-leveling mechanism.

From the standpoint of economic welfare, this type of price adjustment across provinces could not be a first- nor a second-best solution. Consequently, the Panel of Inquiry offered a compromise solution. It stated that should the COP price for a deficit province result in a price differential insufficient to cover the cost of moving eggs across the relevant provinces, the CEMA would set a levy on eggs produced in the deficit province in order to provide payment for those eggs moving into that province. In this way, the integrity of the pricing mechanism was maintained, and interprovincial movement was provided for.

This set of real circumstances illustrates both the subtlety and the pragmatism of such mediation exercises. This good faith bargaining is continuous in Canadian agricultural supply-managed sectors, between regulators and regulated, and it is hardly a misnomer. In this case, the compromise is that higher COP provinces have to accept less than their COP in order to maintain their share of production.

One could perhaps argue that Canadian laws allowing such publically supervised price-leveling mechanisms to openly operate are probably more flexible than those of the US federal system. In the United States, the principle of preemption seems to bar all state regulations concerning interstate pricing mechanisms because the federal level of government already occupies the field (Ruppel et al., 1991).

Institutional compromise in this system has also taken other forms. The chicken sector, for example, has accommodated first Alberta and subsequently British Columbia, remaining outside the agreement for much of its existence. The compromising ability of all systems is being tested increasingly by emerging economic conditions, such as trade liberalization, differential market growth rates and COP differentials across provinces.

Summary and Conclusions

The national supply management systems, which govern output and pricing in the Canadian dairy and poultry sectors, are unique and controversial. Their legal and organizational features are unique, especially the role played by producers themselves in operating these systems. This role of producers as well as the effects and consequences of supply management have made these systems quite controversial. Until now at least an ability to compromise has carried them through the controversy.

Two different structural models are used for supply management. These are the CDC for the industrial milk sector with members appointed by the federal government, and the NFPC and national agencies for the poultry and egg sectors with agency directors elected by the producers in each province. The latter model, in particular, has contributed to the provinces' view of supply management as a regional development program through market sharing. Both models require some form of power sharing or delegation agreement, and this is accomplished through the federal-provincial agreements.

At least three factors contributed to the creation of supply management: 1) market failure in terms of persistent overproduction of dairy and poultry products coupled with an uneven distribution of bargaining powers; 2) jurisdictional disputes between federal and provincial levels of governments and among provinces; and 3) rent-seeking by industry stakeholders promoting protection from imports and domestic production control.

This chapter suggests two scenarios and/or hypotheses for further research. The first scenario is that some provincial boards may start leaving the supply management system, especially if exit costs are perceived to be less than the loss of producers' surplus from their internal adjustments to the new international trade rules (Saint-Louis, 1994). The second scenario is that federal regulatory agencies will conclude that desirable, structural adjustments in the whole food and fiber industry in Canada must take into account issues of competitiveness. The agencies must justify preemption of such organized price-leveling mechanisms as those tolerated in the past, especially when the possibility of rent-seeking by provincial stakeholders is greater in some provinces than in others.

Supply management is about power-sharing in a national market protected by import quotas or tariffs. In the context of trade liberalization, governments and business will be less able to exercise market power in the future. Planned adjustments are needed to meet increased international competition and to prevent the destruction of supply management systems from within.

Notes

1. Acknowledgments: Many thanks are due to Gilles Froment, graduate student at McGill, for his help in data analysis, literature review, and general preparation of this chapter. Thanks are also due to David Kirk, former Secretary General of the Canadian Federation of Agriculture, François Lemieux, legal counsel to CEMA and CCMA, the Hon. Senator H. A. Olson and the Hon. E. F. Whelan both Former Ministers of Agriculture, for sharing their knowledge and insight on the subject. Finally, the authors appreciate the helpful input and comments of Glenn Flaten, former member of the National Farm Products Council, and Dr. Yvon Proulx, economist at the Union des Producteurs Agricoles du Québec.

2. The task force members and their affiliations were as follows: D. R. Campbell, University of Toronto; P. Comtois, Belanger, Sirois, Saint-Jacques, Comtois and Company, Chartered Accountants; J. C. Gilson, Dean of Graduate Studies and Chairman of Agricultural Economics, University of Manitoba; D. L. MacFarlane, Head, Agricultural Economics, McGill University; and D. H. Thain, School of Business, University of Western Ontario.

3. The use of the term "quota" to describe both production and import constraints has been commonplace in Canada but may lead to confusion in other countries. In the current GATT context, of course, the market access provision is defined by tariff-rate-quota or "contingent d' importation."

4. The United States has continued to press for freer trade of both dairy and poultry products, claiming that the high tariffs posted by Canada under GATT are not consistent with the principals of the North American Free Trade Agreement (NAFTA).

5. François Lemieux, personal communication. Mr. Lemieux acts as legal counsel for the CCMA and the CEMA.

6. Commonly known as the Ontario Egg Reference Case, this judgment (Agricultural Products Marketing Act, 1978, 84 D. L. R. 3rd 257) is cited by Green (1983) as an important constitutional test of the FPMAA of 1972.

7. This authority is now covered by a new regulation under the CDCA as a result of a recent court decision in British Columbia.

8. Eggs were priced through a nationally administered, weighted average COP formula for about the first 20 years of operation under supply management. This is not the case for all commodities.

9. This is not a sure thing. The most recent attempt to establish national supply management was by apple producers from five provinces that had the majority of production. Having eventually cleared all hurdles up to the point of the national plebiscite during the period 1988–92, the plan was eventually rejected by producers themselves when the vote was held.

10. The number of signatories is dependent on the number of provinces that wish to join the plan. In principle, the signatories represent the federal and provincial ministers and the producers in each province.

11. In the parlance of Rausser (1982), supply management could be viewed as a PERT (political economic resource transaction) operating as it does at arms length from government. Though some will argue that it has become a PEST (political economic seeking transfer), and there are good arguments to support that view, there is still some under-explored evidence of PERT qualities such as the effect of risk-reduction.

12. For example, the NPMAA passed by the federal government in 1934 was struck down in 1937 on the grounds of interference with intraprovincial trade.

13. The proliferation of producer marketing boards at the provincial and regional level were part of a global response to the crisis in agricultural markets in the 1930s. Le Heron (1993) cites several authors in describing what he calls "a remarkable commonality of response" by governments of many countries in granting producer marketing boards "statuary powers to restore order to depressed markets."

14. Senator H. A. Olson, personal communication, June 9, 1994. This level of egg prices was independently corroborated by the Hon. Carol Teichrob, MLA Saskatoon, who recalls the value of eggs being less than the cost of shipping the empty crate back to the farmers.

15. Even without a nationally administered price in the case of chicken, however, Coffin et al., (1989) observed a certain convergence of provincial chicken prices following the introduction of supply management.

16. Glenn Flaten, former member of the NFPC, personal communication.

17. Apparently, processors were not very sympathetic to the farmers plight at that time. According to personal communication from Jack Stueck, founding president of the Saskatchewan Producers Marketing Board, Canadian processors awarded a trophy to the plant that achieved the highest gross margin each year. As one might imagine, this did not do much for producer prices, especially in areas with little competition for the local processor.

18. Glenn Flaten, personal communication.

19. In 1995 dollars, this would be equivalent to over $600 million. Current expenditures on dairy subsidy are less than half that amount.

20. Initially, only Québec and Ontario joined the remaining provinces signing on over the next four years.

21. David Kirk, former Secretary General, Canadian Federation of Agriculture, personal communication, March 10, 1994.

22. The Hon. Ralph Ferguson, also former Minister of Agriculture in Ottawa, additionally gives credit to the Hon. David Stupich, then Minister of Agriculture in British Columbia, for influencing his fellow agricultural ministers in this regard.

23. The issue of corporate control was noted by Kidd and Hiscock (1968) in discussing alternative models for national marketing plans.

24. David Kirk, personal communication, March 10, 1994.

25. The quest for protection from imports and regional preservation may be viewed as a form of rent-seeking by provincial governments as well as by leading producers and processors.

26. The Food Prices Review Board was a federal agency created during the early 1970s to investigate the sources of inflation in food prices. Under the chairmanship of Beryl Plumbtree, the agency published numerous reports and was generally critical of supply management, generating an ongoing debate with the Hon. E. F. Whelan, then Minister of Agriculture and a strong supporter of supply management marketing boards.

27. While Schmitz and Schmitz (1994) deal mainly with the academic literature, Trant (1994) reminds us of a considerable body of reports from task forces, steering committees and consulting studies that have been dedicated "to make regulated markets behave more like unregulated or free and competitive markets."

28. R. Saint-Louis, one of the authors of this chapter, was one of the five members of this fact-finding commission.

References

Arcus, P. L. 1981. "Broilers and Eggs." Technical Report No. E13, Economic Council of Canada, Ottawa, Ontario, Canada.

Balcaen, L., R. Doyle, K. Matte, R. Morin, and W. Sherwood. 1992. *A Vision for the Future of the Canadian Dairy Industry*. Report of the Consultative Committee on the Future of the Dairy Industry, Dairy Farmers of Canada (December).

Barichello, R. 1981. "The Economics of Canadian Dairy Regulation." Technical Report No. E12, Economic Council of Canada, Ottawa, Ontario, Canada.

Campbell, D. R., P. Comtois, J. C. Gilson, D. L. MacFarlane, and D. H. Thain. 1969.

"Canadian Agriculture in the Seventies." Report of the Federal Task Force on Agriculture, Catalog No. A21–15. Ottawa, Ontario, Canada: The Queen's Printer.

Canada, Government of. 1972. National Farm Products Marketing Agencies Act (NFPMAA). Ottawa, Ontario, Canada.

Coffin, H. G., R. F. Romain, and R. Douglas. 1989. "Performance of the Canadian Poultry System Under Supply Management." Department of Agricultural Economics, Macdonald College of McGill University, Ste-Anne-de-Bellevue, Québec, Canada and Departement d' Economie Rurale, Université Laval, Québec City, Québec, Canada.

Federal-Provincial Agricultural Trade Policy Committee. 1988. *Interprovincial Barriers to Trade in Agricultural and Food Products.* (June).

Gilson. J. C., and R. Saint-Louis. 1986. "The Constitutional and Economic Case for a Common Agricultural Market in Canada." *Policy Issues and Alternatives Facing the Canadian Hog Industry,* Chapter 15. Study sponsored by the Canadian Pork Council and Agriculture Canada.

Green, C. 1983. "Agricultural Marketing Boards in Canada: An Economic and Legal Analysis." *University of Toronto Law Journal* 33: 407–433.

_____ . 1989. *Canadian Industrial Organization and Policy.* 3rd ed. P. 471. Toronto, Ontario, Canada: McGraw-Hill Ryerson Ltd.

Hill, J. T. 1963. "The Problem Involved in Planning Broiler Marketing: An Analysis of Planned Supply as the Answer to Overproduction." *The Economic Annalist* 33 (1): 12–16. Government of Canada (February).

Hiscock, G., and H. V. Walker. 1969. "A Report on Marketing Boards in Canada." Preliminary draft for the Task Force on Canadian Agriculture. Pp. 204–209. (22 January).

Josling, T. 1981. *Intervention and Regulation in Canadian Agriculture: A Comparison of Costs and Benefits Among Sectors.* Economic Council of Canada and the Institute of Research on Public Policy, Ottawa, Ontario, Canada.

Kidd, J. F. D., and G. A. Hiscock. 1968. "National Farm Product Marketing Legislation and Organization." *Canadian Farm Economics* 3 (4): 18–21. Canadian Department of Agriculture.

Lanoie, C. 1985. "Comparison of the Canadian and United States Food and Beverage Industries." *Food Market Commentary* 7 (3): 41–59. Agriculture Canada.

Le Heron, R. 1993. *Globalized Agriculture: Political Choice.* P. 54. New York, NY: Pergamon Press.

Mestern, H. J. 1972. "The Evolution of Supply Management in the Canadian Dairy Industry." *Canadian Farm Economics* 7 (5): 12–16.

National Poultry Task Force. 1991. *Towards the Development of a Second Generation of Poultry Supply Management Systems.* Ottawa, Ontario, Canada (15 March).

NFPMC. 1976. *Accord Fédéral-Provincial Relatif à la Revision et à la Consolidation du Système Global de Commercialisation par la Réglementation de la Commercialisation des Oeufs au Canada,* Version Finale. (29 July).

Proulx, Y., D-M. Gouin, and R. Saint-Louis. 1992. "Supply Management: A Performance Analysis." GREPA, Université Laval, Québec City, Québec, Canada (January).

Rausser, G. C. 1982. "Political Economic Markets: PERTs and PESTs in Food and

Agriculture." *American Journal Agricultural Economics* 64 (5): 821–833. (December).

Rosaasen, K. A., J. S. Lokken, and T. Richards. 1996. "Provincialism: Problems for the Regulators and the Regulated. " This volume.

Ruppel, F. J., F. O. Boader, and E. W. F. Peterson. 1991. "Federalism, Opportunism, and Multilateral Trade Negotiation in Agriculture." *American Journal of Agricultural Economics* 73: 1009–1019.

Saint-Louis, R., D. Ivison, M. Brechin, A. Vielfaure, and S. Williams. 1982. *Report of the Fact Finding Inquiry into Egg Production Costs to the National Farm Products Marketing Council.* (27 September).

Saint-Louis, R. 1994. "Les Plans Conjoints ne sont pas Morts." *Le Bulletin des Agriculteurs.* Pp. 11–13 (March).

Schmitz, A. 1983. "Supply Management in Canadian Agriculture: An Assessment of the Economic Effects." *Canadian Journal of Agricultural Economics* 30: 135–152.

Schmitz, A., and T. G. Schmitz. 1994. "Supply Management: The Past and Future." *Canadian Journal of Agricultural Economics* 42: 125–148.

Skogstad, G. 1993. "Policy Under Siege: Supply Management in Agricultural Marketing." University of Toronto, Toronto, Ontario, Canada.

Task Force on National Dairy Policy. 1990. Consultative Document, Agriculture Canada, Ottawa, Ontario, Canada (July).

Trant, G. I. 1994. "Scenarios for Regulated Products." Working paper 3–94. Policy Branch, Agriculture and Agri-Food Canada, Ottawa, Ontario, Canada.

Touche, Ross & Partners. 1981. "An Analysis of Provincial Variations in Egg Production Costs in Canada." (October).

Veeman, M. M. 1975. "Canadian Dairy Policies." Dairy Policies and Programs Discussion Paper AEEE/75/9. School of Agricultural Economics and Extension Education, University of Guelph, Guelph, Ontario, Canada.

_____ . 1982. "Social Costs of Supply-Restricting Marketing Boards." *Canadian Journal of Agricultural Economics* 30: 21–36.

Whelan, E. 1977. "A Proposal for the Establishment of a Chicken Marketing Agency." Discussion Paper (17 May).

14

Power Relationships in the Political Process

G. Skogstad

Abstract

The elimination of Article XI of the General Agreement on Tariffs and Trade (GATT) with the implementation of the Uruguay Round Agreement in 1995 removes the legal compulsion for Canada to manage its dairy and poultry supplies domestically in order to restrict imports. Canadian dairy and poultry producers are no longer legally compelled to honor the production-restricting requirements that enable supply management to function effectively. Thus, the preservation of orderly marketing and formula-based pricing and the adjustment of provincial market shares in the face of changes in market structures and consumer preferences are now all matters of voluntary agreement. This chapter appraises the viability of Canadian supply management in this altered environment by focusing upon reform initiatives underway in the poultry and dairy sectors. It argues that the composition of key decision-making structures in poultry supply management systems and the rules under which they have functioned have stymied their adjustment in the past and undermined their legitimacy. The future survival and adaptation of poultry supply management rests upon reforming its institutions and rules. However, the reformist strategy in poultry supply management is flawed, and the private and public stakeholders in the sector are limited in terms of their ability to build a long-lasting consensus. By contrast, the existing institutional framework and the ongoing reform process in dairy supply management are more likely to enable the consensus-building and compromises across private and public stakeholders that are necessary for its adaptability and survival.

Introduction

During the Uruguay Round, Canadian trade negotiators were unable to retain Article XI.2.c of GATT, which legally enabled Canada to protect supply-managed dairy and poultry sectors from foreign competition. Canada was successful, however, in negotiating "tariff equivalents" to replace the quantitative import restrictions. These tariff equivalents are sufficiently high to provide the industries with substantial protection from lower-cost imports in the medium term (National Farm Products Council, 1994).[1] As a result of these tariff levels, GATT poses little direct threat to Canadian dairy and poultry supply management. The North American Free Trade Agreement (NAFTA) may present a bigger threat, as the United States argues that Canadian tariffs imposed under GATT in January, 1995, are illegal under the terms of NAFTA, which takes precedence.[2]

The magnitude of the United States' threat is, as yet, indeterminate. However, the indirect challenge posed by NAFTA and GATT is apparent. The demise of Article XI.2.c injects a new calculus into supply management. It removes the legal compulsion for countries to manage their supplies domestically in order to restrict imports. Unlike import quotas, tariff equivalents do not compel producers to restrict their own production to preserve the domestic market for themselves. What has been a legal necessity in the past, if producers were to realize their self-interests of stable prices and profits, henceforth would be voluntary (National Farm Products Council, 1994). Will Canadian dairy and poultry sectors continue to have the necessary incentive to cooperate voluntarily in sharing Canada's consumer market, including honoring production-restricting regulations? Or will that incentive disappear, and will long-standing disputes and conflicts of interest over market share, pricing and decision-making rules between producers in different provinces prove insurmountable? The very survival of Canada's national supply management systems rests upon an affirmative answer to the first of these questions.

The history of Canadian supply management provides ambiguous guidelines in terms of assessing its continuing viability in the new environment. On the one hand, the troubled past of dairy and poultry marketing boards provides ample grounds for pessimism. Since their birth, Canadian poultry and, to a lesser extent, dairy supply management systems have been riddled with conflict. At the center of this conflict has been producers' disagreement about how to share growth in the domestic market and how to adjust provincial and producer market shares in the face of changes in market structure and consumer preferences. The cleavage has been between producers in provinces with growing consumer markets, who wish to expand their quota allocation and production to take advantage of new market opportunities, and producers in provinces with stagnant or declining consumer markets, who are anxious to retain their initial allocation of production share. Despite past periodic exhortations of the need for compromise and reform by Canada's agricultural ministers,[3] cleavages have persisted, and institutions of supply management have remained largely unchanged.

Beyond this, advocates of supply management have been unable to counter effectively two persistent criticisms. The first criticism is that supply management has been unfairly biased toward producer interests at the consumers' expense and contrary to some processors' concerns. The second criticism is that the original intent and goals of supply management — as a mechanism to enable producers in Canada's 10 provinces to share the domestic market in an orderly fashion — have been subverted as rigidity in market allocations, inefficiency in the dairy and poultry sectors and chronic squabbling among producers in different provinces have prevailed. The result is a legitimacy crisis and a serious questioning of the continued appropriateness of currently structured supply management systems.

On the other hand, if history offers grounds for doubting the future viability of Canadian supply management, it also suggests that an obituary is premature. First, Canadian supply management systems have survived repeated internal tensions in the past by developing internal mechanisms for brokering competing interests. Second, as economic analyses have demonstrated, there are significant economic gains that Canadian producers and processors realize from supply management. These gains are not likely to be foregone readily. Third, the "crisis" of supply management appears to be institutional rather than philosophical; the current methods of functioning rather than the principle of supply management are under attack. There appears to be a continuing, strong endorsement within the dairy and poultry policy communities for the principle of regulated marketing and the instruments of production controls and formula pricing.

Perhaps the most promising support for the future survival of Canadian supply management systems is the recognition of the need for system reforms by Canadian national and provincial governments and the dairy and poultry sectors. The national and provincial governments' unprecedented commitment to reforming supply management is now in its sixth year and has survived a change of government at the national level. The initiative toward comprehensive reform began in 1989 with the creation of task forces in the dairy and poultry sectors by the former Progressive Conservative Agriculture Minister, Don Mazankowski. Reform discussions among national and provincial governments and various components of the industry resumed in early 1994. The current Liberal government's Agriculture and Agri-Food Minister, Ralph Goodale, established the Federal-Provincial Task Force on Orderly Marketing immediately following the December, 1993 signing of GATT. Subsequently, ad hoc task forces have been created in each supply-managed sector. The reform initiative has already produced some significant operational and institutional changes in the chicken and dairy supply management systems.[4]

Governments' exhortations and their direct involvement in the redesign of supply management systems are the most significant new variables in the Canadian supply management equation. The Canadian government's message has been sympathetic but unrelenting: adapt and adopt new operational rules and institutional structures, or supply management systems will not survive in the future. Can the private and public stakeholders in dairy and poultry supply management systems agree on the reforms perceived to be necessary if the systems are to evolve and endure? The discussion that follows addresses this question by focusing on reform initiatives to date, including the proposed institutional reforms as well as the reform process itself.

It is argued that successful reform depends upon striking an equitable balance among competing private interests, between private and public interests and among competing public/governmental interests. The legitimacy and stability of future supply management systems hinges on correcting the current institutional biases, which privilege some private interests at the expense of other private interests, overrepresent some private interests at the expense of public interests and afford no mechanism, other than gridlock, for resolving conflicting public goals. An appraisal of the success of the reform processes entails determining whether the guiding principles that underline the reform process are likely to enable these imbalances, which are responsible for the current cleavages and tensions in supply management, to be overcome. Support for the appropriate redesign of supply management depends vitally, as well, upon the will and institutional capacity of the governments and private parties to the reforms to engage in consensus-building and compromises. An examination of the record to date suggests important differences between the contours of the reform process and its likelihood of success in the dairy and poultry sectors. A stronger role for government officials, and organizational structures, in the dairy industry able to build consensus and strike compromises, bodes well for the success of the dairy reform initiative. In contrast, the "back seat" role of governments in the poultry sector and the much weaker capacity of poultry sector organizations to build a national consensus pose significant hurdles to enduring reforms in poultry supply management. In the chicken sector, in particular, there appears to be an unwillingness to deal explicitly with competing governmental conceptions of provincial equity, and proposed reforms run a high risk of creating new imbalances among private interests. If this prevails, supply management reforms in the poultry sector — and especially in chicken marketing — will be unstable and unsustainable in the medium to long term.

At the heart of the reform exercise is the endeavor to establish new rules about how to operate supply management systems, including how to allocate production shares among producers and provinces. New decision-making institutions with new rule-making powers and new authority to enforce and adjust operational rules and substantive decisions are also being sought. The focus on rules and institutions recognizes that the current institutional structure of supply management systems has been flawed in failing to strike the balance among competing private and public interests that is necessary to the ongoing stability and legitimacy of dairy and poultry marketing boards. Thus, an appraisal of the capacity of the current reform initiative to overcome these institutional flaws and their attendant tensions is best achieved by first appreciating the origins and reasons for these biases. A look backward at the political process and calculus which prevailed at the creation and implementation of dairy and poultry supply management is therefore instructive if reform efforts are to secure a better equilibrium of competing interests.

Supply Management as Political Exchange Relationships

Unbalanced Political Exchange Relationships

Producers' needs and interests loomed large in governments' decisions to establish supply management systems in the dairy and poultry sectors between 1966 and 1978. Joint government action grew out of agreement at both levels of government that supply management was the appropriate policy instrument to solve economic problems of depressed producer prices and conflict over market share in the dairy and poultry sectors. However, the design of the dairy and poultry national marketing agencies differed significantly in terms of the respective role accorded public officials (provincial and national governments) and private officials (producers and processors) to formulate and implement supply management policies.

The existing decision-making structures and rules of poultry and dairy supply management systems reflect the nature of the political exchange that occurred among Canada's governments and between agricultural producers and governments at the time of the creation of national marketing agencies in the dairy, egg, chicken, turkey and broiler hatching sectors. The political exchange[5] that occurred was one in which each party (governments, producers) gave up some autonomy or authority in return for being able to realize goals deemed otherwise unattainable.

Two political exchanges involving two different parties occurred at the establishment of supply management systems and have continued thereafter. The first exchange involved governments and private interests — governments and producers, to be precise. Governments granted private groups (producers) preferential rights of decision-making (usually regulatory power in the marketplace) in order to realize their own goals as governments/politicians. Governments conceded some of their decision-making autonomy, delegating it to or sharing it with private interests. In exchange for their enhanced bargaining power in the marketplace and/or delegated political authority, the private parties implicitly accepted joint responsibility for decisions and limits on their behavior. A vital part of the political exchange is the groups' guarantee of their members' adherence to decisions reached as part of the ongoing implementation of supply management. Among the most important of the decisions that must be adhered to in supply management are limits on individual rights to produce/market the regulated commodity.

Given the division of jurisdiction over marketing in Canada, supply management always entails a second relationship of political exchange — the relationship between provincial and national governments. Federal and provincial governments agreed to cooperate in the establishment of marketing agencies to regulate production on a national plane. They shared the mutual goal of raising and stabilizing producer incomes. Moreover, governments perceived the benefits that would flow to them from supply management. For the provinces, benefits included stronger rural and provincial economies; for the Government of Canada, savings in price stabilization program costs could be anticipated.

All the involved parties had reasons for entering into the political exchange and creating supply management systems. The institutional design of the national marketing agencies, however, reflects the uneven bargaining capacity of the parties to the political exchange. The provinces and the producers overwhelmed the Government of Canada in the bargaining over the design of poultry supply management systems. Provincial governments allied with their producers, seeing their interests to be one and the same.[6] For their part, provincial governments were anxious to preserve their own legal and administrative powers, which they deemed necessary to advance their local economy. The producer-provincial alliance was facilitated by the fact that provinces had moved ahead of the national government in poultry marketing regulation; provincial poultry marketing boards were functioning across the country.

Hard-pressed by provincial governments, opposition parties and a well-organized farm lobby, the Government of Canada made significant concessions that left unbalanced the political exchange relationships in poultry supply management. The imbalance was in terms of both the Canadian government's future role and the representation of nonproducer interests. The structures and rules of poultry supply management systems tilted toward provincial and producer interests in a number of ways.[7] First, producers secured overwhelming control of the primary decision-making bodies in poultry supply management. Representatives of provincial marketing boards comprise the boards of directors of the national marketing agencies, ensuring due regard for producer interests as provincially construed.

Second, the federal supervisory body, the National Products Marketing Council (subsequently renamed the National Farm Products Council), was not given legal authority to be an effective and authoritative oversight agency. An effective restraining and monitoring capacity for the Government of Canada required that it be able ultimately to intervene directly in the industry and to substitute its own actions for those of the self-regulatory bodies (Streeck and Schmitter, 1985). This would have included the right of the National Farm Products Council (National Council) or Agriculture Minister to veto decisions of supply management bodies and replace agency decisions with its/his own regulation. The bargaining process that surrounded passage of the legislation to enable poultry supply management deprived the National Council of these powers. Moreover, its producer-dominated structure robbed the National Council of the political legitimacy to be an effective government oversight agency.

And third, provinces were assured their interests as governments and those of their producers would be given due regard. Their consent, and indeed that of all signatories (provincial departments responsible for agriculture, provincial commissions or agencies responsible for supervising provincial marketing boards and provincial marketing boards), was required for any changes in the core features of supply management systems, including the level of the national production quota and each province's share.[8]

Thus, there are perceived and real imbalances among private interests, between public and private interests and among competing public interests in poultry supply management systems. The consequences have been problems of illegitimacy[9] and instability. Illegitimacy is a result of the inadequate balance between public and private interests: the absence of other private stakeholders, like consumers and further processors, and the insufficient authority vested in the national agency entrusted with upholding the public (consumer) interest in supply management. Instability follows from the institutionalized struggle among provinces to pursue their own distinctive conceptions of their public interest. Conflict ensues as some provinces seek to retain their market share and other provinces seek to expand their market allotment.

Dairy supply management institutions reflect a more even political exchange as regards both the intergovernmental relationship and the roles of public and private officials. The government presence is stronger.

Governments at both levels are actively involved in dairy policy, but one government — the Government of Canada — through the Canadian Dairy Commission (CDC), stands at the apex of the system. CDC commissioners are appointed by and responsible to the Government of Canada. Significant rule-making authority is entrusted to the Canadian Milk Supply Management Committee (CMSMC) in which all provinces that participate in national industrial milk supply management are represented. The chair of the CDC heads the CMSMC, thereby stabilizing the intergovernmental relationship of political exchange. The composition of the CDC and its rule-making authority keep the government/private interest relationship a balanced political exchange. Moreover, other interests besides producers have an opportunity to be represented on dairy supply management bodies — including the CDC and advisory committees. The presence of these other stakeholders makes possible an equitable balancing of private interests.[10]

The differing institutional design of dairy supply management can be traced to a number of factors. The stronger federal government presence and the willingness of provinces to delegate their interprovincial marketing powers to an agency of the national government reflect the fact that the federal government had staked out its political claim to dairy policy prior to the implementation of dairy supply management. It had provided dairy subsidies for several decades and had taken leadership in regulating interprovincial milk flow by creating the CDC in 1965. Unlike in the poultry industry, the concept of a national dairy market already existed; industrial milk was flowing interprovincially when the national marketing system was put in place.

Unstable and Conflictual Political Exchange Relationships

The relationships of political exchange in poultry supply management have been unstable and conflictual. Various provinces have repeatedly threatened to withdraw from the national marketing systems. Some provinces have followed through on their threats (Federal-Provincial Task Force on Orderly Marketing, 1994).[11] Episodes of brinkmanship have recurred as national poultry marketing agencies have flouted the supervisory agency's rulings. In contrast, and although not without periods of internal tension, greater stability and cooperation characterize dairy supply management. These differing histories of dairy and poultry supply management reflect the greater capacity of public and private officials to balance competing private and public interests in the dairy supply management systems.

As earlier noted, the principal cleavage in poultry supply management, and the one which has created perennial conflict, is that among the private parties to the political exchange — the poultry producers themselves. Producers in different provinces have been unable to agree on how the Canadian consumer market should be shared and how it should be readjusted among provinces in the face of changes in provinces' competitiveness or growth (decline) in consumer markets. The institutional design of national marketing agencies provides no incentive or compulsion for poultry producers to compromise on market share. The producer representatives on the national agencies are representatives of provincial producers, not producers across the country, and so lack the incentive to take a broader look at their industry. There are no national poultry producer associations that represent all Canadian egg, chicken or turkey producers. Hence, there are no organizational forums within which farm leaders with a national perspective can broker and make short-term compromises necessary to advance long-term interests of the industry. Poultry producers are thus widely perceived as having been unable and unwilling to uphold their end of the bargain, that is to make the necessary compromises to ensure competitiveness and efficiency in their industry in return for stable and adequate incomes.

Private interests in dairy supply management have been more able to meet their obligations in the political exchange relationship. The existence of monopolistic organizations to represent both dairy producers and dairy processors on a national level is of critical importance in enabling them to do so. Their CDC representatives can speak in the best interest of Canadian producers and processors, not for the parochial interests of any one province's milk producers or milk processors. Cooperative political exchange is therefore possible in dairy supply management. Such cooperation contrasts with the conflictual political exchange that appears endemic to poultry marketing agencies.

The interprovincial conflict that characterizes the private/producer relationship in poultry supply management has its roots in the skewed intergovernmental exchange. The provincialism that is embedded in poultry decision-making bodies and rules of operation is the source of much of the inflexibility that causes internal conflict and criticisms of inefficiency and lack of market responsiveness. The fact that even one provincial signatory can veto change contributes to what Scharpf calls "the joint decision-making trap," supply management systems in which the status quo reigns (1989). Even when policies are suboptimal by their original objectives, they cannot be changed "as long as they are still preferred by even a single member" (Scharpf, 1989). In short, the composition of national poultry marketing agencies and the unanimity decision-making rule merge and blur private and public interests at the provincial level. Producers simultaneously champion their private and provincial interests, whether the latter be in favor of preserving a provincial poultry industry or expanding it.

Why were those public officials entrusted with upholding the public interest — governments and oversight regulatory bodies — unwilling or ineffective in constraining such provincialism and self-interested behavior? An examination of the behavior of the National Council suggests that the answer to this question lies in a combination of the Council's uncertainty regarding its legal mandate and its double-edged mandate. In the early years, the Council felt caught between its perceived obligation to make supply management work and its duty "to promote a strong, efficient and competitive production and marketing industry."[12] It thus tended to overlook bargained outcomes which upheld provincialism. Although the National Council has recently been more aggressive in seeking to check parochialism that mitigates market responsiveness, it acknowledges that even its current role has been "much more advisory than supervisory" (Federal-Provincial Task Force on Orderly Marketing, 1994).

For their part, even in the face of chronic internal strife in poultry supply management, Canada's ministers of agriculture eschewed direct intervention in the sector. Rhetoric, rather than action, was their preferred option.[13] Responsible ministers issued warnings to the industry to address its internal problems, but all the while they were affirming their support for the principles of supply management (Whelan, 1986).[14] By delegating national supply management authority to a third party and thereby removing themselves from the ongoing implementation of national marketing plans, governments were able to avoid directly implicating each other in their admonitions to national marketing agencies to "shape up or else." Certainly, no federal Minister of Agriculture ever publicly challenged provinces' veto — and stranglehold — over national marketing agreements. Their failure to do so reflects the fact that intergovernmental relations in agriculture are a matter of not only legal niceties but also high politics. As in so many other policy areas, strategic calculations, relating to the place of Québec in Canada and regional economic development goals, loom large. When faced with the choice of intervening and prompting a major intergovernmental skirmish or backing off from radical reform of a supply management system that is limping along but is still on its feet, agriculture ministers opted for the latter and retreated to the sidelines.

By failing to check the built-in tendencies toward provincialism that were present at the outset of poultry supply management,[15] governments and their delegated regulatory supervisors implicitly sanctioned them. The fact that there has been no significant readjustment of base quota shares among provinces and only limited success at allocating over-base quota shares on the basis of provinces' economic efficiency and changing consumer demand has lent legitimacy to objectives of provincial economic development. Despite not being a designated quota-allocation criterion in the federal-provincial agreements that enable poultry supply management, the right of each province to have a viable poultry (or dairy) industry came to be construed, in some quarters, as legitimate as goals of efficiency and market-responsiveness. Thus, two quite different conceptions of provincial equity coexist. One is a conception that recognizes every province's right to the economic stability afforded by supply-managed industries. The other is a conception that recognizes the right of provinces to exploit fully "the growth and development potential" of their dairy and poultry sectors (Federal-Provincial Task Force on Orderly Marketing, 1994). It is necessary to deal with this legacy and find a way to balance economic efficiency and provincial equity if the reform effort is to succeed and supply management systems are to be stable in the future.

Redesigning Supply Management: Correcting the Imbalances

The review of supply management systems that has been underway for the past five years proposes significant changes in the institutional design and substantive goals of the systems. Recommended changes in the institutions, rules and goals of supply management speak to the balance among competing public and private interests that must be struck if supply management systems are to be perceived as fair and garner the public and private support to be stable.

Rebalancing Private Interests

The proposed 'new' goals of future supply management systems and, even more so, the reform process upon which the Canadian government has embarked demonstrates the high priority placed upon rebalancing private interests. To rectify the existing structural bias toward producer interests, enhanced representation of nonproducer interests (consumers, processors, further processors) in supply management decision-making bodies is proposed. In the first instance the reform process looks to private interests to strike such a balance. The parliamentary secretary to the Agriculture and Agri-Food Minister, and the individual responsible for conducting the 1994 Federal-Provincial Task Force on Orderly Marketing, has observed that "the Minister and the Ministry . . . want all of the stakeholders to come forward . . . and tell the task force, the minister, and the provincial ministers what they feel needs to be done . . . the Minister could have asked departments or provincial ministers to tell them [sic] what they think needs to be done. But we think the right way to do it is to go out to all the players in the industry and ask them what they think needs to be looked at." This same view of the government's role is echoed by the chair of the National Council. The chair has stated that the National Council's role in reform discussions is that of "a catalyst for change in poultry supply management" through "liaison and facilitation" (Vanclief, 1994).[16]

Having drawn a blueprint for reform in the 1989 document, *Growing Together*, the Canadian government subsequently created the institutional framework wherein industry participants develop sectoral perspectives for submission and approval by federal and provincial agriculture ministers. In stark contrast to the political process under which supply management systems were established and by which they have operated, nonproducer stakeholders have been members of the various task forces and consultative committees formed to expedite reform discussions. Beginning with the federal-provincial task forces in 1989 and including the subsequent consultative committees and current ad hoc committees, representatives of processors, further processors and consumers have been part of the debate. It is the full inclusion of the processing sector that is the highest priority, owing to "their position as risk takers and in recognition of the vital interdependence between producers and the processing sector" (Federal-Provincial Task Force on Orderly Marketing, 1994). This reform strategy of including a broader range of private stakeholders in the redesign of supply management and eventually on supply management decision-making structures is consistent with European corporatist models of political exchange. Corporatist models incorporate all potentially antagonistic stakeholders into regulatory decision-making structures where they share rule-making authority with government officials. By including processors, further processors, consumers and so on, in the reform discussions and giving them a seat on decision-making or advisory bodies, a better balance is created among private interests and between public and private interests. A more equitable balance results from greater transparency in decision-making, the possibility for cross-stakeholder communication and, therefore, an enhanced likelihood of consensus. Thus, the 1994 Federal-Provincial Task Force on Orderly Marketing argues that the input of marketers of supply-managed products in the supply management review will enable "ongoing discussions and exchanges between producers, processors and further processors, as well as with the trade." The latter, in turn, will facilitate the "strategic alliances" required within each industry (Federal-Provincial Task Force on Orderly Marketing, 1994).

Attempts to reform policy through a corporatist process are most likely to succeed when private interests meet two conditions. First, those who speak for private parties — producers, processors, consumers — must be representative of their constituency, able to speak for it and capable of making binding commitments on its behalf, even when those commitments entail short-term concessions to advance members' long-run interests. Second, corporatist models are most likely to succeed in achieving an equitable balance of private interests when those interests exercise even bargaining power. With each affected party vigorously pursuing its own goals, the idea is that subsequent policies and decisions will represent a compromise in the best interest of all society.

Are these conditions met in the dairy and poultry sectors? In short, how appropriate is the corporatist reform model in effecting long-lasting reforms and resolving the chronic internal tensions that have plagued supply management? The organization and leadership resources of dairy farmers and dairy processors suggest that the first corporatist assumption is more valid in the dairy sector than it is in the poultry industry where there is an absence of national organization of chicken, egg or turkey producers or processors. As regards the second corporatist assumption, are producer and other private interests equally able to bargain for their interests? Unlike the situation which prevailed in the prelude to supply management, dairy and poultry producers are no longer economically distressed. Canada's poultry farmers, followed by dairy producers, enjoy the highest net income of any agricultural sector and certainly, to date, the most secure incomes.[17] By contrast, other stakeholders in supply management, especially further processors, can credibly claim their vulnerability given their exposure to foreign competition under liberalizing trade agreements. Consolidation in the poultry processing sector, however, means that, at least in some provinces, processors have considerable bargaining power vis-à-vis producers.

While there are interprovincial variations in the degree of competitiveness in the processing sector, the example of Ontario is instructive. In 1991, the two largest chicken processors controlled 68% of market share with 90% of market share in the hands of only five processors. In 1976, prior to the creation of national chicken supply management, six chicken processors shared 68% of the market and as many as a dozen processors absorbed 90% of the market. In 1978, the two largest chicken processors controlled only 35% of the market.[18] It is clear, at least in this province, that producers now have far fewer processors to whom they can sell. The situation in the poultry sector varies from province to province. In Québec, where the majority of processors are also producer-owned cooperatives, one might expect a more equitable balancing of producer and processor interests. Dairy producers are also somewhat less vulnerable for the same reason; many of the largest dairy processors are cooperatively owned by producers.

The current political economy within which supply management reform is proceeding is one in which power in the marketplace has shifted. In the provincial context and in the absence of producers' ownership of a share of processing capacity, processors have greater leverage in the marketplace. While this is a significant development in itself, the inequity in the bargaining power of producers and processors across provinces is of greater political importance. To the extent that producers and processors in one province are not at odds but rather share a joint interest in retaining or maximizing their existing provincial share of the national poultry market, the balance that must exist is less between producers and processors within a province and more between producers and processors across provinces. In a trading regime where domestic supply management is voluntary and in a domestic context where goals of economic efficiency are touted as primary, the bargaining power flows to producers and processors who are currently in the best position to serve consumer markets at the least cost.

Corporatist models rest upon a third assumption. It is assumed that governments will be party to policy deliberations and will intervene, both to strike a suitable balance of public and private interests as well as to correct unequal bargaining power among private interests (Offe, 1981; Streeck and Schmitter, 1985). Here, the dairy reform process differs starkly from the reform that is underway in poultry supply management. The Canadian government has actively participated in the dairy discussions[19] through the CDC's direct involvement in ongoing talks. Governments' role in the poultry reform discussions has been essentially exhortatory.[20] Even while stating that "governments are key partners in the process of renewal," the Canadian government has endorsed the industry's preference that private stakeholders take the initiative in developing "workable elements of solution to key problems" (Federal-Provincial Task Force on Orderly Marketing, 1994). The role of an indirect catalyst for change also characterizes the behavior of the National Council, which implicitly and explicitly endorsed endeavors initiated by the private sector to force changes in chicken supply management throughout 1993–94 (National Farm Products Council, 1994).[21] In trumpeting market-oriented reforms, the Council is acting in accord with one element of its mandate: the promotion of an efficient and competitive agricultural industry. Nonetheless, in sanctioning such "private processor-producer deals," the Council has provoked considerable controversy, fueled complaints of unfair treatment interprovincially and perhaps unwittingly made compromise even more unlikely.

Rebalancing Public Interests

By far, the thorniest issue in redesigning supply management is ensuring that public interests maintain sway over private stakeholders' goals. Major stakeholders, including provincial governments, have different "public" interests in supply management. The governments of Ontario, British Columbia and Alberta concur with the Task Force on Orderly Marketing that the national interest should be equated with a competitive and fully market-responsive national dairy and poultry industry. By contrast, the Atlantic provinces tend to argue that the national interest lies in serving regional interests in a "broad based geographic production viability in all areas of Canada" (Federal-Provincial Task Force on Orderly Marketing, 1994).[22] In the presence of regional economies, regional and national economic-building goals do not necessarily coincide, and hence, agreement on what constitutes the most efficient structure of supply management is elusive.

The Canadian government's Action Plan for supply management reform recognizes that the redesign of supply management must "accommodate regional priorities and concerns" (Federal-Provincial Task Force on Orderly Marketing, 1994). Institutional and procedural reforms are proposed to address this interprovincial dispute about the appropriate substantive goals of supply management. The emphasis is twofold: first, strengthening the legitimacy and authority of the oversight body and second, blocking the institutional opportunities for impasses and instability. The proposed augmentation of the powers of the National Council and the broadening of its base of representation would redress the perceived present imbalances among private interests and between private (producer) and public (consumer) interests. As well, a binding dispute settlement mechanism is recommended to end standoffs between the national poultry marketing agencies and the National Council. Reforms also seek to reconfigure the intergovernmental relationship. The Action Plan endorsed by Canada's agriculture ministers in March, 1994, proposes ending the unanimity-amending formula for the federal-provincial agreement and substituting a majority voting rule in key decision-making structures and in some fundamental operational decisions. In addition, impasses at the boards of directors of the national marketing agencies would be subject to a binding dispute settlement mechanism.

The most radical reform proposal is to end the veto of each provincial government over changes to the federal-provincial agreements that establish the rules under which supply management systems function. By eliminating a veto for each province over fundamental features of supply management, while simultaneously reorienting the systems toward economic efficiency goals, the very real prospect that certain provinces will lose their supply-managed sectors or suffer major economic declines is raised. Most concerned about this prospect are provinces that have a minority of current production. To their credit, government and private parties to the reform process have recognized the need to accommodate such concerns. Thus, they propose to offset the loss of the veto with guarantees of fair interprovincial treatment through independent review panels and arbitration committees.

Ultimately, however, the success of the reform initiative in redesigning supply management so that it is stable in the long run and perceived to be fair requires taking seriously the goals of smaller (and usually poorer) provinces to maintain stable rural economies. What is probably needed are commitments by the Canadian government in the form of governmental research and development, credit and other stimulative programs and adjustment policies. And yet, such commitments appear increasingly unlikely given the imperative of fiscal deficit reduction.

Conclusion

The ability of Canada's dairy and poultry supply management systems to survive rests upon institutional and procedural reforms that enable the systems to adapt to changing market circumstances and simultaneously shore up their legitimacy. The current reform process appropriately seeks to enlarge the sphere of support for supply management by paying attention to the interests of nonproducers, including processors, further processors and consumers. The success of reform discussions hinges on an overly optimistic assessment of the consensus-building abilities of private interests — at least in poultry supply management — and their capacity to put long-term goals ahead of short-term concerns. Outcomes, which have the result of favoring the interests of producers and processors in some provinces at the expense of those in other provinces, are unlikely to be politically tenable. If the survival of supply management necessitates a greater emphasis on goals of efficiency and market responsiveness as appears to be the case, and if these priorities occasion sectoral restructuring (growth in some provinces, decline in others), then acceptance of these outcomes is likely to be problematic in the absence of compensation for those producers and provinces economically disadvantaged by reforms in this direction. The likelihood of such ancillary adjustment strategies appears slim given the Canadian government's trend toward reduction of its fiscal commitments to agriculture. In the absence of national government initiatives to offset the provincial adjustment costs of redesigned supply management systems, the fate of Canada's national supply management systems rests disproportionately in the hands of industry stakeholders. And that fate, at least in chicken supply management, is by no means assured.

Notes

1. This conclusion also applies to dairy supply management.
2. The United States has officially protested Canada's January, 1995 implementation of tariffs on imports of dairy and poultry products, including

yogurt and ice cream, at levels consistent with those allowed by GATT. The United States argues that such tariffs are illegal under the terms of NAFTA. The NAFTA, like the Canada–United States Free Trade Agreement (CUSTA), restricts the signatory countries from imposing new tariffs on bilateral trade without the consent of the other parties to the agreement and commits Canada to eliminating existing tariffs. Canada argues that import quotas were preserved under NAFTA and that it is therefore legal to convert them to tariff equivalents. With specific reference to ice cream and yogurt, the United States argues that converting ice cream and yogurt import quotas to tariffs is illegal given that a 1989 GATT panel ruled that these import quotas were inconsistent with Canada's GATT obligations.

3. In 1983, Canadian Agriculture Minister Eugene Whelan responded to an insistent Ontario Minister of Agriculture and Food, Dennis Timbrell, to convene a joint meeting of federal and provincial agricultural ministers and signatories to all three poultry plans. In 1987, federal and provincial agricultural ministers joined forces to commission an external examination of supply management. However, neither the consultants' report nor the internal Agriculture Canada recommendations that buttressed it produced substantive changes in supply management.

4. Consumers and processors have been given representation on the Canadian Dairy Commission (CDC) and on the Canadian Milk Supply Management Committee (CMSMC), and the Consultative Committee membership has been broadened to include representatives of further processors.

5. The concept of political exchange is developed by Martin J. Bull (1992).

6. During the negotiations that led to the federal-provincial agreements underlying the poultry marketing agencies, the agriculture ministers in Ontario and Québec made it clear that they wanted the agencies run by the producers, not the federal government. Provincial, and not national, pricing was written into the Canadian Chicken Marketing Agency (CCMA) at the request of the provinces of Ontario, Québec, British Columbia and Alberta.

7. For an extended discussion, see Grace Skogstad (1980).

8. It should be noted that unanimous consent of signatories was not a legal feature of the first two poultry supply management systems, eggs and turkeys, but was written into the federal-provincial agreement that created the national marketing agencies in the chicken and broiler hatching sectors.

9. For an extended discussion of the legitimation problems of supply management, see Grace Skogstad (1993).

10. Although without a vote, processors have an advisory role on the CMSMC. The Consultative Committee that advises the CDC on pricing has recently been recomposed to reflect broader representation of stakeholders.

11. In dairy, British Columbia left the national dairy plan for 15 months in 1983-84. Alberta did not join between 1978–83 and then had only a contractual relationship with the CDC between 1983–89. In chicken marketing, British Columbia left the CCMA in 1990 and remains outside today. In 1993, Ontario served notice it would withdraw from the CCMA in 1993, a notice later withdrawn in 1994. Ontario and Québec have been operating their own surplus removal programs for eggs since 1992.

12. Section 22 of the Farm Products Marketing Act, 1972. The perceived obligation to make the marketing plans work arises because the National Council is a signatory to the plan.

13. As a rule in the past, when public criticisms of supply management and/or interprovincial tensions merited federal and provincial agriculture ministers' involvement, the latter's attention tended to be directed toward particular operational issues, such as allocation, rather than on major reforms to the institutions of supply management. See note 2 above and the ministerial action in 1983 and 1987.

14. Indeed, Eugene Whelan, Federal Minister of Agriculture from 1974–79, denied the appropriateness of ministerial involvement in national agencies' affairs. "The agriculture minister's involvement with a new marketing board," he said, "should pretty well end once the enabling legislation is passed."

15. Provincial self-sufficiency goals are implicitly addressed in the federal-provincial agreements that established the chicken and turkey agencies. Section 7 of the CCMA Proclamation stipulates that "total market requirement within each market area" and "the proportion of market demand in a province that is met by production in that province" shall figure into over-base allocations. Alterations in turkey quotas shall take into account "the existing production and storage facilities in each province."

16. Responsible officials have described their role in these terms. See the comments of Dr. Cliff McIsaac, Chair of the National Farm Products Council (1994).

17. Agriculture Canada (1993) reported the average net operating income of poultry and egg farms to be the highest of all farm types at $46,301, followed by dairy farms at $39,795. A subsequent Agriculture Canada document (1994) reported dairy farm net operating income in 1992 to be ahead at $42,424, followed by poultry and eggs at $40,375.

18. Figures obtained from the Ontario Chicken Marketing Board.

19. The Chair of the CDC chaired the dairy consultative committee, whose members included representatives of the Dairy Farmers, Dairy Bureau, National Dairy Council and the CDC. Its discussions about the future of the industry and its review of the pricing structure resulted in a joint statement, subsequently discussed by the provinces. Other consultative forums included a subcommittee of the CMSMC and a federal/provincial deputy minister committee that also had producer, consumer and processor representatives, and was meant to be a follow-up to the dairy and poultry task forces. It should be noted that the Deputy Minister's Steering Committee, created at the initiative of Agriculture Canada, failed in its attempt to assume leadership of the review process.

20. The 1994 discussions include one federal government bureaucrat as the chair of each of the ad hoc task forces in dairy, chickens, eggs, turkeys and broiler hatching. Provincial representatives also sit on these sectoral groups.

21. The National Council has supported initiatives by Ontario chicken producers and processors to obtain a greater share of the national chicken market and to inject changes designed to ensure greater market responsiveness and efficiency. The Council's role has generated much controversy in the CCMA and was challenged by Québec, Nova Scotia and Newfoundland. See the National Farm Products Council's *Inquiry into the complaint by la fédération des producteurs de volailles du Québec, the Nova Scotia Chicken Producers Board and the Newfoundland Chicken Marketing Board against the decision of the Canadian Chicken Marketing Agency respecting the third period 1994 quota allocations* (April, 1994).

22. The Government of New Brunswick believes, "The existing system was put in place based on a fundamental principle that each province should have a share of national production. We are concerned that provinces with larger treasuries would assist their producers directly or indirectly in acquiring quota from other provinces."

References

Agriculture Canada. 1993. *An Economic Overview of Selected Types, Canada 1991.* (November).

_____ . 1994. *Farm Income Financial Conditions and Governmental Expenditures, Data Book.* P. 14. (August).

Bull, M. J. 1992. "The Corporatist Ideal-Type and Political Exchange." *Political Studies* XL: 255–272.

Canada, Government of. 1989. *Growing Together.* Ottawa, Ontario, Canada.

Federal-Provincial Task Force on Orderly Marketing. 1994. *Action Plan Towards the Implementation of Sustainable Orderly Marketing Systems in the Canadian Dairy, Poultry and Egg Industries.* Ottawa, Ontario, Canada (23 March).

National Farm Products Council. 1994. *Annual Review 1993–1994.* P. 3, 19–23. Ottawa, Ontario, Canada.

_____ . 1994. Inquiry into the complaint by la fédération des producteurs de volailles du Québec, the Nova Scotia Chicken Producers Board and the Newfoundland Chicken Marketing Board against the decision of the CCMA respecting the third period 1994 quota allocations (April).

Offe, C. 1981. "The Attribution of Public Status to Interest Groups: Observations on the West German Case," in S. Berger, ed., *Organizing Interests in Western Europe: Pluralism, Corporatism and the Transformation of Politics.* Pp. 123–158. Cambridge, MA: Cambridge University Press.

Scharpf, F. W. 1989. "Decision Rules, Decision Styles and Policy Choices." Journal of Theoretical Politics 1(2): 197.

Skogstad, G. 1980. "The Farm Products Marketing Agencies Act: A Case Study of Agricultural Policy." *Canadian Public Policy.* Winter: 105–112.

_____ . 1993. "Policy Under Siege: Supply Management in Transition." Canadian Public Administration 36(1): 1–23.

Streeck, W., and P. C. Schmitter. 1985. "Community, Market, State — and Associations? The Prospective Contribution of Interest Governance to Social Order," in W. Streeck and P. C. Schmitter, eds., *Private Interest Government: Beyond Market and State.* P. 6. Beverly Hills, CA: Sage.

Vanclief, L. 1994. Testimony before the House of Commons Standing Committee on Agriculture and Agri-Food. Pp. 2–31. (22 March).

Whelan, E. with R. Archbold. 1986. *Whelan: The Man with the Green Stetson.* P. 149. Toronto, Ontario, Canada: Irwin.

15

Provincialism: Problems for the Regulators and the Regulated

K. A. Rosaasen, J. S. Lokken, and T. J. Richards

Abstract

It is no longer "business as usual" in Canadian supply-managed industries since Canada signed the General Agreement on Tariffs and Trade (GATT). The tariffs, which replace import controls, effectively maintain international barriers, but they are not contingent on the maintenance of supply management in Canada as were import controls. This difference changes the dynamics of interprovincial negotiation for market share within Canada. Formerly, political power was extremely important, but economics may soon become the key factor. The current disarray in the sector arises from provincial jockeying for position in the new "game," including an attempt by some provinces to have current shares legislated into the future.

This chapter provides examples of allocational problems in the "regulatory playing field." It dispels the myth that current quota values and industry size and distribution accurately reflect relative regional profitability, and it outlines the problems of political economic-seeking transfers (PESTS) in regulation, including the capture of regulators by the industry. Divisive issues such as import quota rights, interprovincial quota trading and provincial quota allocation are also discussed. This chapter leads to the conclusions that there is a need for the supply-managed system to define goals and priorities and that regulators need to adjust the system to make it more consistent with economic criteria. This may prevent system collapse and high adjustment costs. A move to include more economic principles in the regulation of supply management is a prerequisite for system survival.

Introduction

Changes in international trading rules brought about by GATT have created a new "playing field" for Canadian supply-managed industries.[1] It is no longer "business as usual" for these industries despite calm pronouncements to that effect by a number of observers of and participants in supply management. These commentators correctly see tariffication as maintaining extensive protection from foreign imports for a long time to come, but they do not understand the implications within Canada for supply management behind the border protection. Tariffication at the international level has changed the balance of power in interprovincial supply management negotiations.

Provinces are already beginning to jockey for position in the face of the new reality. Some provinces, especially the largest, are trying to ensure the maintenance or enhancement of their current position before the status quo changes while others are tentatively examining new options arising from changes in production and trading environments. Such maneuvering has left the supply-managed industries in a state of disarray. Confusion arises during this period of transition from an inability to predict wherein power lies — in political muscle or in cost-effectiveness of production, processing and distribution.

This chapter first explains why international tariffication changes the dynamics of provincial negotiation within the Canadian supply management system. It then puts into context the changes facing supply management by 1) providing examples of the regulatory process in theory and as it is practiced in Canada, 2) by dispelling some common myths regarding industry location and profitability and 3) by describing a number of problems and issues in Canadian supply management.[2] The conclusion is drawn that benefits provided by supply management for the producer sector can be maintained within Canada, even under GATT and the North American Free Trade Agreement (NAFTA), but a redistribution of these benefits must occur among provinces and among the producer, processor, retailer, importer and consumer sectors.

Supply Management Post-GATT:
The End of "Business As Usual"

The most important short-term impact of GATT on supply management is the alteration of the dynamics of negotiations among the Canadian provinces. Tariffication replaces quantitative restrictions thereby changing the terms of international protection afforded the Canadian supply-managed sector. Under Article XI of the previous GATT agreement, foreign imports were limited as long as supply management remained in place. A supply management system was said to be in place if over 80–85% of a country's total production of a commodity was regulated by production quotas. Under this restriction, if a large province, that is Ontario, Québec or several small provinces together, was to opt out of supply management, the national system would not control sufficient product to maintain the international import controls. Large provinces have had strength, owing to their ability to destroy, not necessarily their ability to produce. Borders could have been opened to cheap foreign products in order to swamp any unbridled expansion by a cost-efficient dissenter. During such an expansionary stage when investments are new and outstanding debts are large, an industry is vulnerable. This threat has kept provinces operating within the system, with a few notable exceptions that will be discussed later in this chapter. This exercise of political power in supply management has altered production patterns from those that would occur in a more open market.

The game has changed. Under the new GATT agreement, the maintenance of tariff protection for a sector is not dependent upon the maintenance of a supply management system. The external threat has ended. In the short run, tariffs remain in effect regardless of what happens internally, and a single province cannot threaten to open the Canadian border to a flood of cheap foreign products (Agriculture Canada, 1994). If a breakdown in supply management occurs, there could be an internal fight over market share among the Canadian provinces. Provinces need only be competitive within Canada (rather than being competitive on a world basis).

Bargaining power will eventually shift toward the most cost-efficient province.[3] This shift, however, could be protracted and painful if provincial treasuries become involved in the fight over market share. Still, under these new regulations, radical evolution of supply management within Canada is expected to occur long before any real pressure from international trade is felt. At a minimum, a realignment of provincial quota allocation, closer to the dictates of economic forces, appears to be in the direction for the future.

The transition costs for the Canadian supply-managed sectors under various policies and strategies are drastically different across provinces and across sectors. A slow and planned adjustment will reduce these transition costs. Conversely, an economic war for market shares will result in "red-ink" for many businesses and provinces. Thus, many fixed facilities will be closed in various regions before they become obsolete or fully depreciate.

The strategies traditionally used in supply management negotiations must change. The current situation can be compared to a poker game in which deuces are initially wild, but rules are switched in the middle of the game to those of straight poker with no wild cards! Under the old rules, the wild deuces might be equated to the ability of a large provincial player to dismantle border protection. Rule changes are dictated by the dealer, and the GATT agreement dictates that tariffication will proceed. A change in the relative values of the cards under new rules might be likened to a change in the relative values of provincial resource endowments, which are reflected in feed costs, environmental capability, location, population density and size of land base. For example, new production technology allows the conversion of low-value feed into high-value products, such as skinless chicken breasts or specialty cheese, so that production can occur at a great distance from the market. The result is that economic forces, rather than political forces, now become dominant.

It is crucial that regulators work with industries to make informed policy choices. The regulators must facilitate needed adjustments not merely rubber stamp efforts that seek to maintain the status quo. Maintenance of the status quo implies almost certain chaos and very high adjustment costs. Provinces must understand the topography of the new playing field. The configuration, or even survival, of supply-managed industries in Canada may well depend more upon the outcome of political maneuvering, which transpires before the planned dismantling of border protection, than on whatever happens afterward.

The Regulation Process in Theory and in Canada

The process of regulation is traditionally explained using public interest theory, which holds that the desire of government is to create the greatest good for the greatest number.[4] An alternative explanation is the theory of economic regulation, which holds that private interests achieve favorable regulation because of lobby pressures that create a cost for others.[5] Regulators are captured by the regulated and/or tend to behave in their own self-interest.

Regulations in Theory: Capture Theory

Capture theory is used to explain why regulators' enforcement of legislation changes or softens over a period of time. Generally, the captured agencies are interlinked with the industry, and any industry problems affect the regulatory body or agency.

Industries can capture regulatory bodies directly or indirectly. The direct approach employs the political process in which demands for legislation by special interest and lobby groups receive favorable attention as a result of their representatives being elected or appointed to agencies. This is an explicit rent-seeking activity.

Indirect capture may evolve in several different ways as outlined by Scherer (1980):

Even when legislators have only the public interest at heart in passing regulatory laws, those who are regulated end up as important beneficiaries by "capturing" the agency regulating them. This happens *inter alia* because the regulated firms use their political influence to have friendly regulators appointed, because the regulated enterprise has superior technical knowledge upon which regulatory agency staffs come to depend, and because regulators, like most people, seek identification and approval and are more likely to find them by cultivating a community of interest with the well-organized firms they regulate than with the remote and unresponsive public.

In addition, the regulator may be captured because of a monetary constraint. Regulators working on limited budgets may be forced to rely on financing from the industry, and, therefore, their budgets are linked to industry returns. The result of this process is a "growing together" where profit levels in the industry tend to be above normal, and a community of interest develops among the industry and the regulators. Regulators become advocates of what is "good for the industry." Few ask what is good for consumers, the province or the country. Identification of issues in terms of the people and the industry being regulated becomes the norm. The focus is on the industry's profits, future and protection, rather than on the protection of public interest.

The body of acronyms and technical terms that develops becomes a language that only the "initiates," that is the regulator and the regulated, understand, making former regulators valuable in the regulated industries and members of the regulated industries valuable as regulators. Knowledge of the institutions and "the lingo," such as CMSMC, CCMA, CEMA, sleeve, overquota penalties, skim-off and other industry-specific terms,[6] allows these individuals to command higher prices or rents.

Regulations in Practice

The Evolutionary Process: "As Topsy Grew." The regulatory process for Canadian supply management is complicated by the existence of a dual regulatory structure.[7] Each province is largely responsible for regulation within its own borders. There is another regulatory level supervised by the National Farm Products Marketing Council (NFPMC), or the Canadian Dairy Commission (CDC) in the case of dairy, with the provincial participants being signatories to national agreements. The total system includes the national marketing agency, the federal regulatory agency, the federal Minister of Agriculture, as well as the provincial marketing boards, the provincial ministers of agriculture and the provincial regulatory authorities.[8]

The blend of political forces, regulation and economic factors that comprises supply management in Canada creates a complex, cumbersome system. Unanimity is required by some agencies at the national level. Other agencies give equal votes to each province while still others have weighted voting based on the relative size of provincial industries. Provincial boards continue to enforce some regulations that predate national plans. One example is the variance in exemption levels for unregulated egg production between the provinces — 0 layers, 199 layers or 500 layers, depending on the province. Some provinces have rules which discourage or curtail vertical integration; others do not. Some provinces restrict the size of quota-holding per entity. Size limits differ among provinces. Administration of quota, such as rules on the transferability of quota, also differs between provinces. In cases in which quota is easily transferable, quota values are generally higher than in regions in which quota has been tied to facilities or has had a transfer tax applied to it (Dawson, Dau and Associates, 1983).

Provinces or individual firms adopt strategies to maximize benefits under changing rules. Chicken import quota regulations serve as an example of such a strategy adoption. Chicken imports normally occur under global import allocations granted to firms at the outset of supply management and based on historical imports. Supplemental quotas are available when domestic sources of product are unable to meet market demand. When the Canada–United States Free Trade Agreement (CUSTA) was being negotiated, the United States wanted greater access to Canadian supply-managed markets. Canadian firms, including processors and retailers of these products, wanted ownership of any expanded import quotas that might be written into the agreement. Producers have alleged that during this period firms ordered large volumes of specific types and weights of chicken that could not be provided by Canadian suppliers. Permission to import product to meet this order was then granted to the firm as a supplemental import quota. Subsequently, these supplemental import quotas became global import quotas under CUSTA, conveying to these firms a permanent right to import the increased volume each year. This "race for base" was a very profitable undertaking because import quotas allowed the holders to buy at US prices and sell at Canadian prices. This change benefited the firms that were allocated these import quotas, but it lowered the domestic proportion of chicken production and processing for Canadian consumption. Who were the regulators and the Canadian government representing when they allowed this type of activity?

Many observers would suggest that the Canadian supply management systems have drifted away from their early principles and spirit. The original founders of marketing boards sought changes to the "free market" system because they were displeased with market outcomes (Rosaasen and Maley, 1986). Problems of unstable prices and low incomes were attributed to producers' weak bargaining position and an inability to manage supply. Recognition of common problems and a desire to cooperate in overcoming them were the driving forces that united producers across Canada in a highly-regulated system.[9]

This particular type of intervention on behalf of a specific group in an industry results in industries that maintain a horizontal structuring with visible groups of producers, processors and retailers. The horizontally organized groups sometimes behave as antagonistic independents rather than recognizing their interdependency. The product flow is vertical and requires coordination, but these vertical linkages in the private sector have not formed to the same degree in Canada as they have in the United States. The natural links and the potential to lower transaction costs from the input supplier through the producer, processor, wholesaler and retailer are not looked upon as opportunities for efficiency gains. Admittedly, there are differences between provinces with regard to how aggressively vertical integration is opposed, either implicitly or explicitly, through regulation.

From a collective producer perspective, Canadian supply management reduced price instability and improved incomes. However, individual greed, as reflected in the actions of producers and provinces, soon became visible. Provinces often pursued strategies which maximized returns for an individual producer group or province, not for the total Canadian sector or industry. Provinces that threatened withdrawal from the national system, that is British Columbia and Alberta, in chicken benefited by extracting a larger share of national quota. Precedents were set that rewarded dissenters.

The original founders of supply management used political intervention to alter an undesirable market outcome. Today, if the political outcome is undesirable from the perspective of some provinces, it must be expected that these provinces may seek to revert to an open market system because of their disadvantages in the current political/regulatory arena.

Undefined Goals and Objectives. A major ongoing problem in supply-managed sectors is a failure to prioritize goals and objectives. Adequate incomes for efficient producers, stable supplies and prices, safe and wholesome food for consumers, increased regional developments and a progressive and efficient system have been espoused as goals at various times. The weighting of these goals has determined the evolution of the system. If regional development is the primary goal as would appear by some provinces' actions, then one can argue for a policy to produce pineapples in Prince Edward Island and bananas at Brandon, Manitoba! Economic development policies, though legitimate, should attract footloose industries rather than supply- or market-oriented businesses. If economic criteria, that is cost-efficient producers, comparative advantage and trade gains, are not considered important, then it is simply the most powerful political force that dictates regional production patterns.

In the Canadian chicken industry, criteria for interprovincial quota allocation have been developed. However, the relative weighting of the criteria has been left open to the judgment or interpretation of the actors and regulators. This has created a loose operational arrangement in which regulatory decisions can reward those with power and the most effective bargaining stance. For example, theoretically, new quota is allocated on the basis of several factors, including comparative advantage, growth in population and whether the actual production filled the previous quota allocation. However, since these criteria are not defined or weighted, in reality quota allocation becomes a "horse trading session" in which the most politically powerful are the most successful at achieving gains for their province and their producers. The Molot Decision, a legal judgment on the turkey industry, states that a factor can be weighted as zero and still comply with the requirement of the national agreement on quota allocation, which requires that each of the factors be given a weight.

Equal Versus Fair. The Canadian government could develop an "equal" policy where all Canadians are allocated a sweater for the weather conditions from August 15 to November 15. One does not need to be an economist or a climatologist to realize that this equal treatment would leave people who are in Iqualuit (Arctic), Edmonton (Interior Plain), Victoria (Coastal) and Niagara Falls (Eastern) with relatively different levels of comfort. Similarly, there are problems operating a highly regulated national industry when many of the provinces have visible differences in resources, opportunity costs and other factors. When there is adherence to the principle of a single industry rule that treats all as equal even though circumstances are different, the end result is often drastically different for different provinces. For example, penalty sleeves for overproduction specify given ranges of variation from a production target. Conceptually, it is harder to get within 2% of an output target for a small province than it is for a large one. For instance, a fire in a single broiler hatching egg barn may disrupt the output plans of a small province but will have little influence on a province with a large output.[10] In this way, equal can mean nonequitable if it shows a lack of understanding of the conditions under which it is applied.

Regulations Can Influence Change. Regulations can curtail the ability to adjust or adopt to new technology or a new production alternative. When a new product is developed, such as the McDonald's McNugget, there is no potential in the Canadian system for its production to move to a processor located in a province that has a small production quota allocation. The size of processing plant needed to achieve production efficiencies would require an amount of raw product that is simply too large to be considered a reasonable level of quota increase for a small province.

When committee table decisions determine how markets function, how an industry evolves and the regional location of production, change can only occur on an incremental basis, and regional shifts are blocked. The largest economic costs to the supply management sector may not arise from the inefficiency of income transfers from consumers to producers but may instead arise from the misallocation of production on a regional basis across Canada. Regulators are largely unable to force adherence to economic factors, such as cost efficiency in quota allocation. In an open market, shifts in production from one area to another readily occur. The US broiler industry moved from Delaware and Maryland to Georgia, Arkansas, Alabama and North Carolina. Similarly, the potato industry was moved from Maine to Idaho. If the United States had adopted the current Canadian regulatory system in the 1950s, the broiler industry might still be in Delaware and Maryland. In the United States, chicken is now produced where the profits are, not where the people are![11] The introduction of modern technology has enabled chicken production and industrial milk production to be located at a distance from consumers.

Cost of Production (COP) Pricing. There is a tendency for firms in specialized industries to keep information tightly capped. For this reason COP surveys in the supply-managed industries cannot be used to accurately determine farm management alternatives, nor can they be used in making effective management decisions based on precise enterprise costs. In these industries, producers gain if reported costs are high because product prices are more likely to increase. COP pricing may also lead to X-inefficiencies since cost-cutting measures are neither sought nor adopted because of comfortable profit margins.[12]

The Quota Value Myth. Will Rogers once stated, "It isn't what we don't know that gives us trouble; it's what we know that ain't so" (Fitzhenry and Whiteside, 1981). Myths have developed surrounding the relative production efficiencies of various regions. These myths are based on the use of simplistic logic regarding visible factors such as quota value.

Economists predict that quota value should represent the difference between the costs for the firm and the market price capitalized forward, based on the expectations of the buyers and sellers. If all conditions are assumed to be similar in each region, then quota values should be highest in the regions where production is most profitable and producers' expectations are the highest. However, quota prices may have more to do with the mechanics of supply management than with relative efficiencies (Personal communication (1994) with Bruno Larue, Assistant Professor of Rural Economics, Université Laval, Québec City, Québec, Canada). Milk produced in excess of quota has a discounted or penalty price. Quota value may then reflect the difference between the penalty price of milk and the in-quota price of milk, rather than the difference between milk price and the firm's COP. This is especially important where quota can be traded as unfilled, giving the purchaser the right to deliver 150% of quota in the first year (as happens in some provinces).

If broiler quota values are $15 per bird in province A and $24 per bird in province B, the obvious implication is that chicken production (or dairy production with a similar quota value difference) in province B is more profitable, assuming that all other conditions are similar and that there is an efficient market where producers who purchase quota have similar information. These assumptions may be wrong. Consider these potential differences:

- Province A has not allowed quota to be traded or priced as a tradeable item. Rather, quota is tied to assets, and a sale may be overturned in which the quota is deemed as being paid for. In province B, quota trading has been facilitated by the establishment of quota exchanges that allow the trading of small blocks of quota and that openly publicize quota pricing information. (Manitoba and Saskatchewan maintained restrictive trading rules when many other provinces did not.)

- There are only 100 producers in the industry in province A, and few outside the industry realize the level of profit achieved by current participants. In province B, production is large with a high-profile industry. (The Prairie provinces have a large number of farms of which supply management enterprises make up only a small portion. For example, Saskatchewan has over 50,000 farms but only about 70 licensed chicken producers while Ontario has over 68,000 farms and 970 licensed chicken producers (Statistics Canada, 1993; CCMA, 1993).

- There are differing expectations concerning the longevity of supply management; the expectations in province A are for two years, and the expectations in province B are for 10 years.

- At specific pivotal times, there are or have been more alternatives for profitable enterprises in which farmers could engage in province A relative to province B. Cereals and oilseeds were very attractive enterprise alternatives on the prairies at the time supply management was established, and this conditioned a whole generation of farmers.

- In dairy, province A originally held a large proportion of quota as cream quota, which producers gave up over time. This was distributed to milk producers to keep barns filled as cow productivity increased. Province B had no cream quota, and quota had to be purchased as cow productivity increased. The prairie provinces historically had a large cream volume as a percentage of total output.

- The financial community accepts quota as security for between 50%

and 70% of current market value of the quota in province B where quota value is highest but considers quota to have a value of zero for security in province A. Dairy quota is accepted as security at 70% of market value in Québec by the Royal Bank and Farm Credit Corporation (FCC) but is valued at zero in Saskatchewan.[13]

Resources such as labor, management and capital do not flow rapidly between some provinces. Differences in language, culture, climate and perceptions between various regions limit the mobility of management and labor. Farmers' capital will usually only relocate under stress or over a very long term.[14] Quota values are, therefore, a weak indicator of relative regional enterprise profitability.[15] They capture a number of differing institutional arrangements among the provinces.

Divisive Issues/Proposals

Various issues/proposals currently being considered in the supply-managed industries have the potential to capture gains for certain producers or regions, often at unacceptable costs to others. Current disagreements in the supply-managed industries arise from producer and provincial greed and from misunderstanding among producers, industry officials, regulators and governments about the topography of the post-GATT playing field. Unless there is a better understanding of new bargaining positions, it appears that the current generation of farmers is about to relearn the hard lessons of the past, that is low prices that resulted from provincial disputes during the chicken and egg wars.

Import Quota Holdings

The issue of first receivership on import quotas creates rifts among various industry participants and provinces. The profits per unit of imported product may be much greater than the returns from processing the same unit within Canada.[16]

It is true that cost competitive regions that do not control import quota may be able over time to compete by lowering the Canadian product price relative to the US product price, thereby reducing import quota rents. In the meantime, however, processors and provinces who control import quota can dominate an industry. Such power may skew the development of an industry away from cost-effectiveness.

Import quota holdings are becoming more important because of the minimum access commitments in GATT.[17] Two issues must be addressed in regard to import quotas:

- Who should receive the new import quota?
- How should existing import quota rights be handled?

Import access in the GATT agreement is expressed as a percent of total Canadian consumption. Each province should receive a share of new import quota according to population, a proxy for consumption. An open system where tendering for import quota rights occurs on a frequent basis is preferred to an alternative system where there is no disclosure.[18] This would end the dumping of imports into the market of a competing province to injure its producers and processors. Import quotas, both new and existing, could become a positive force for the industry if auction proceeds were remitted to the provinces for research and market development. The quota auction could also specify that the product be delivered to each province based on its share of national consumption. The incentive for a province to disruptively collapse import quota rents (as a survival strategy) by increasing provincial output and lowering Canadian prices relative to US prices, would be significantly reduced.[19]

Failure to place all import quota on an open-bid basis could be, in our opinion, the wedge that leads to the early destruction of supply management at the national level, with the chicken sector as the first likely candidate.[20]

Interprovincial Quota Trading

It has been suggested that quota transfers be allowed between provinces. They are already occurring on a limited basis in the egg sector. The Coase theorem suggests that efficient economic solutions can be achieved by allowing property rights to be tradable (Tietenberg, 1988). For example, a neighbor's dispute in which one neighbor likes to play loud music and the other likes peace and quiet could be resolved by the allocation of property rights. If the right to silence is considered a property right, the neighbor who loves loud music will pay the neighbor who prefers silence an appropriate amount. Thus, an economically efficient solution to the dispute over sound levels is achieved. If the right to be noisy is a property right, then the neighbor who prefers silence will pay his/her music-loving neighbor to turn down the volume. A similar approach is sometimes considered for pollution; the pollution right can be given to the polluter or to the person adversely affected by pollution. The property right can be assigned to either side, and it will still bring about economic efficiency.[21] However, the choice of who will receive the initial allocation of the property right is a key factor in the determination of income distribution. Economic efficiency does not address this factor.

A major assumption in such analysis is that property rights have been properly defined. If quota is considered a producer right and individual producers are allowed to trade it, those producers will receive all the monetary gains from its sale. However, many externalities can be identified in such a transaction. Feed manufacturers, other input suppliers, truckers and processors are hurt by the transfer of quota from producers in their province to producers in another province. Early developers of the tobacco quota in the United States did not allow trade outside of small regions, probably in recognition of this factor (Rucker, Thurman and Borges, 1996). Similarly, at one time milk quota in Québec was traded on a subregional basis. A legal opinion indicates that, at least for some provinces, chicken quota is the right and property of the province and not the property of a producer.[22]

If cost-efficient provinces, such as Alberta or Saskatchewan,[23] were to endorse a scheme in which quota was made salable across provincial borders, then producers in these provinces would transfer wealth to producers in other provinces to gain the right to produce a product for which they would have a comparative advantage. In essence a tax would be paid by the efficient region, and the efficient producers would make payments to the less efficient producers in other regions, thereby helping them gain luxurious retirements by purchasing their quota. This would obviously increase the cost and slow the rate of adjustment of regional production toward cost-efficient regions.[24] Quota movement in any direction would imply that past regulations lagged behind economic changes, putting the industry out of touch with market realities, or that there had been recent and sudden shocks to a sector. Clearly, some provinces would receive undeserved windfall gains while others would incur major losses if the current distribution of production was given legal status as a property right.

A better option than interprovincial quota trading for any province which has low-cost feeds and other natural efficiencies is to simply opt out of the national program and increase output in a move toward a market solution. There would be no need for quota or for the purchase of the historical rights of others.

In an open market, production shifts to low-cost regions. Regions that cannot compete will suffer a loss in the value of fixed assets because they are shut down before they fully depreciate or become obsolete. This can be very costly in terms of social and short-term economic costs. Nationally, planned adjustment over time is a reasonable goal and would prevent high adjustment costs.[25] However, there is no reason to effectively patent chicken or dairy production, which would give current producers the right to charge new producers to enter the industry. Interprovincial quota trading is not a reasonable way to redistribute income within Canada. Income distribution policies should be explicit, not implicit.

Compensation for Erosion of Quota Value

Some observers have suggested that compensation be paid to producers for the loss in quota values caused by the changes initiated by GATT.

However, compensation by the federal government is not a logical demand. While it is true that international trade policy is a federal responsibility, GATT has not terminated supply management nor decreed its demise.

What may terminate supply management is an internal breakdown of discipline. Provincial actions will be the key determinant in any quota value meltdown. Provincial legislation created and developed marketing boards, and the commodities controlled by marketing boards are produced largely for domestic consumption in each province. The federal mechanism encourages coordination and cooperation, but it cannot stop interprovincial fighting. Perhaps the federal government should make it clear to the provinces that there will be no federal compensation given to producers for quota value erosion arising from market share disputes that expand output and lower prices. Such a position by the federal government might serve as an important "attitude adjustment" for provincial boards, regulators and departments of finance.

Any compensation for policy changes should be structured to facilitate adjustment in the sector and should seek to encourage greater economic efficiencies and market responsiveness. Provinces could be asked to compensate their own industries in areas where provincial legislation can be shown clearly to have retarded industry growth by limiting the size of firms or blocking firm expansion. In those provinces, some legitimate industry demands are: the repeal of growth-restricting regulations and assistance in finding and achieving efficiently sized operations, using state of the art technology.

Strategy Games

In a managed system such as supply management someone must take responsibility for the decisions of who, what, where and how much to produce.[26] Politicians and regulators are sometimes uncomfortable with openly making these decisions. As the bargaining powers of different provinces change because of revised GATT rules, various provinces can ignore the decrees that are made or can make it more difficult for regulators to decide what they will decree. Provinces are prone to seeking local advantages. They recognize the jobs and added value that are provided by local supply-managed production and processing, and they pursue strategies to persuade profitable value added industries to locate in their jurisdictions.

As described earlier, provincial regulators take a positive attitude toward these industries and begin to parrot the goals of the producer segment of the industry, often broadening them to include economic development objectives. This is much easier and is personally more rewarding for a regulator than is the alternative of seeking to protect the distant consumer. Provincial competition for market share becomes the norm. Different provinces tailor different regulations in response to pressure, and provincial systems evolve differently.

Given this situation, some might suggest that a producer group or province seek to maximize its position under the new trading rules. What framework should be used for this analysis? Perhaps the most rigid and simplified economic example of a single profit-maximizing individual should be used. Only dollars would be considered then. Human consequences or adjustment costs would be excluded explicitly by assuming they are outside of the area to be measured. This approach is exemplified by the caricature of *homo economicus* (Figure 15.1). The caricature portrays how some groups view economists and their decision process. *Homo economicus* is singularly dollar-driven — the only factor that matters is the monetary outcome for himself.

"Homo economicus"

FIGURE 15.1   "Homo economicus," How Some People View Economists
   Source: St. Hill (1974).

Using this framework, if Saskatchewan chicken producers sought to maximize benefits for themselves, a cold, hard recommendation might be that they support interprovincial quota trading with no restrictions on trade volume and no waiting period for approval of a quota transaction. Then the producers' strategy would be to sell all their chicken quota to producers in other provinces at one time otherwise those selling early might lower the price for those selling later. The strategy would have to be kept secret or farmers in other provinces would not buy the quota.[27] If such a move by the chicken producers took place, Saskatchewan, as a province, would also have the option of withdrawing from the CCMA by giving appropriate notice. Then after the one-year waiting period, Saskatchewan would be legally outside the agency, and Saskatchewan chicken producers could resume production on a larger scale, or the Saskatchewan government could invite an integrator to establish large-scale production.[28]

Given the current economic and political environment in the Saskatchewan chicken production sector, any new organizational form is likely to be resisted.[29] A complete dismantling of the existing industry would eliminate this problem. A decaying city is most likely to be successful if rebuilt from ashes! Human and other costs of a "scorched earth policy" preclude this from being a desirable alternative.

If this were a provincial strategy, Saskatchewan would be trying to maximize benefits for itself with no consideration of the costs for Canada and other provinces. Seeking a political/legal route to "beggar your neighbor" is not a responsible strategy.

Conflicts in a National System:
The Political Market Versus the Economic Market[30]

Regional Differences

Provincialism is not new in agriculture; neither is it unique to the supply-managed sector. In the past, problems with provincialism have arisen in the pork, beef and grains industries.[31] Some provinces would fare better if there was a move toward an economic playing field while other provinces prefer to remain in or move further into the political/regulatory realm.

Surplus Removal

In the supply-managed industries, conflicts over funding of surplus removal programs have occurred. Although surplus removal is usually considered a national responsibility, at times it has been considered a provincial activity.

The use of an overproduction sleeve to ensure an adequate supply for Canadian consumers has not been problem-free. Sometimes the overproduction becomes part of the planning process. For example, eggs have been priced at levels below cost (surplus removal price) because domestic egg breakers want continual access to underpriced product to keep their plants operating at high levels of capacity.

In the dairy sector, losses are incurred on exports of skim milk powder or butter powder relative to COP. Sometimes what appears to be a better deal is offered for the purpose of removing surplus. Recently, a firm proposed a chocolate crumb plant in Canada with the proviso of a guaranteed supply of dairy inputs at low prices. However, contracting a guaranteed volume at a low price is not surplus removal; it is planned production at prices below cost. It does not provide a flexible sleeve to smooth the operation of a market. This is simply bad economics.[32] There are gains for the plant, for the producers who supply the milk and for the province in which the plant is located. There are costs to all other provinces and producers who must pay levies to cover the discrepancy between the high milk price received by the milk producers who supply the plant and the price paid by the chocolate crumb plant. Surplus removal should be an *ex post* activity, not an *ex ante* activity! It should not be used to distribute benefits to the province that lobbies most effectively.

Trade Versus No Trade. There is considerable talk about free trade and capturing the gains from trade.[33] Trade can be stabilizing since it expands market size. It also can enable efficiency gains through specialization. However, trade can be destabilizing if border shocks or closures occur because of bogus health claims or other nontariff barriers. History indicates that borders are not immune to the tides of nationalism. Disruptive trade can trigger high adjustment costs.

Costs of Adjustment

Much of economics concentrates on how markets respond to random shocks and how they move toward new equilibriums. The curves bend nicely, and good economists can plot the new equilibrium point on these curves. This type of analysis says very little, however, about the costs of making that adjustment. Costs may be incurred in terms of fixed facilities rendered nonproductive before their useful life has expired, or there may be human costs of lost productivity and difficulty in adjusting to another level of employment, perhaps at a lower rate of remuneration. Despite the difficulty in obtaining empirical data, adjustment costs must be considered in economic analysis.

Canadian Unity and Canadian Regulators

Canada faces internal unity problems. Political decisions that are patently unfair on a regional basis are likely to encourage even greater rifts within Canada. A number of years ago, Thomas Stout (1977), an American, referred to Canada as a "lone green bean" of population along the US border. Stout suggested that Canada's natural geographic links are with US border states. Considerable barriers, both physical and cultural, exist in Canada on an east/west basis. If small provinces face additional legislated barriers to trade within their own country, it would be economically logical for them to look to another country for access to markets and the capture of gains from trade. One of the key economic responsibilities of the federal government in contributing to the preservation of Canada is to maintain the economic union by asserting its authority to ensure that provinces are trading fairly and capturing gains from trade within Canada. The focus of national supply management agencies should be the fostering of long-term development in regions where comparative advantage exists, thereby reducing political friction that results from managed (or mismanaged) markets.

The Challenge for Regulators

Adjustment is on the horizon for supply management in Canada. The supply-managed sector has historically had a powerful and effective producer lobby. Producer leaders are articulate, politically well-connected and experienced in handling political processes and in dealing with bureaucrats. This makes effective regulation extremely difficult. The producer lobby recently has become less united, less focused and less effective as other interest groups along the marketing chain intervene and visible inequities of the current system cause rifts among provinces and their respective producer sectors.

Unequal power relationships and a clouded political process taint the operation of supply management in Canada. Rent-seeking strategies are pursued not only by producers but also by processors, retailers, input suppliers and provinces with conflicting interests and regulatory demands. The different interests of each group create a dynamic playing field on which the new course for supply management must be plotted.

The key actors in this process will be the federal government and the regulatory authorities. Elected officials must clearly articulate the goals for supply management. The regulators, using the analysis they receive from their staff, will largely determine the action or inaction of the governments, the course of adjustment and, therefore, the costs of adjustment. Regulators will face tough decisions about what is fair, relative to what is currently politically expedient. Their decisions and recommendations to their political masters and the national agencies may well determine if the sector remains viable over the longer term or whether it seeks to maintain current economic rents at the risk of major dissension, discipline breakdown and a sectoral collapse.

Supply management entails a set of rules and a legal process. These define and enforce what is currently legal but not necessarily what is equitable or needed at the present time. Views and values of society change over time,[34] and at this point in time it seems clear that supply management regulations should be adjusted to allow cost-efficient regions to expand output.

Revisions in the supply management system should allow production to slowly shift over time to areas where the product can be profitably and efficiently produced and sold to consumers. Canadian regulations should facilitate managed change to avoid the costly adjustment process that could be triggered by chicken or milk wars. The regulators should not participate in the freezing of regional production patterns or in noneconomic proposals, such as the one that would create a flat price for milk across Canada (Balcaen, 1992). The regulators have a responsibility to allow trade within Canada and to foster an efficient and competitive industry, not to muzzle market signals. In our opinion, Canada is ill-prepared currently to drop all protection on supply-managed products immediately or even over a three-to five-year time span. If appropriate adjustments are not begun, the alternative may be international trade in supply-managed commodities where Canada is only an importer.

Conclusion

The benefits of supply management must be redistributed beyond the current quota holders if the system is to be preserved or considered worthy of preservation from a societal perspective. The Canadian regulatory agencies for supply management and the producers are given a grade of D for the exercise of their responsibility in managing the supply-managed sectors within Canada. Politically powerful provinces have been allowed to dominate politically weak provinces. An immediate return to the free market will be a much more viable option for some of the politically weaker provinces that are cost-efficient unless the federal regulatory agencies take immediate corrective action. The chartered banks and other financial lenders, including the FCC, provincial lenders and credit unions, should realize the repercussions of a quota value meltdown, which is the probable outcome for supply management on its current path.

The authors favor adjustment but do not argue for a "scorched earth policy." Planned and managed adjustment over time can accomplish the necessary shift and is expected to entail lower adjustment costs. The Saskatchewan position of seeking to correct the current course of action by offering constructive alternatives, viewed by some as stalling or foot dragging, is legitimate. The appropriate Saskatchewan and, indeed, Canadian strategy at this time is to move toward an outcome that is more consistent with market forces.

There are gains from trade! Political decision makers should seek to capture gains within Canada by fostering trade between provinces since, by definition, all of the gains from interprovincial trade accrue to Canada.

The provinces must begin to develop healthier trade relations, fostered by a revision in the regulatory process for the supply-managed sector in Canada. The major challenge for the federal government and the regulators is the development of a clear set of goals and objectives for supply management and the subsequent adjustment of the regulatory process for the achievement of these goals and objectives. In our opinion, Canadian supply management will be a failed historical experiment if changes are not initiated!

Notes

1. This chapter assumes that GATT takes precedence over CUSTA or NAFTA on the issue of new tariffs, that is tariffication of former quantitative restrictions on imports. If this is not the case, there will be very little time for any adjustment to US pressures by the Canadian supply-managed industries. Canadian law submits to GATT, so Canada's domestic policy will be revised to accommodate the new rules including tariffication. The border arrangements with the United States will remain uncertain if the United States treats the GATT rules as subject to US law rather than vice versa. Some legal opinions suggest that the United States is using this approach in their implementation legislations (1994 personal communication with Darren Eurich, University of Saskatchewan, Saskatoon, Saskatchewan, Canada).

2. The chapter arises from a Saskatchewan/Western Canadian perspective. However, the observations and recommendations in the chapter, which some contend favor Saskatchewan/Western Canada, are of long-run national interest in the authors' opinion.

3. Cost efficiency includes the entire industry from input supply to processing and distribution, not only the farm segment.

4. The literature on public interest theory and the rent-seeking behavior of groups is widely known and will not be cited here.

5. The rent-seeking activity is sometimes split into productive and nonproductive rent-seeking and is labeled political economic resource transactions (PERTs)and (PESTs), respectively (Rausser, 1982).

6. The acronyms and their meanings are as follows: CMSMC (Canadian Milk Supply Management Committee), CCMA (Canadian Chicken Marketing Agency) and CEMA (Canadian Egg Marketing Agency). The technical terms sleeve, overquota penalties and skim-off are used by some or all of the Canadian supply-managed sectors.

7. Supply management in Canada has both federal and provincial involvement because of the split jurisdiction in agriculture. The evolution of supply management and the method used by provincial agencies to eventually establish national agencies differ in timing and design by commodity. The federal and provincial supervisory role and some of the voting procedures also differ by agency. The federal/provincial legal structure is also important, but all are beyond the scope of this chapter. See the description regarding the division of federal and provincial powers in Chapter 13 of this volume.

8. The federal regulatory agency is not a signatory to some plans, but it is to others. The national marketing agencies are not generally signatories although CEMA is a signatory to the amended agreement and the Canadian Broiler Hatching Egg Agency (CBHEA) signed the document that created its existence (Personal communication (1994), NFPMC staff).

9. A personal interview was conducted with Jack Steuck, one of the original founders and former chairman of the Saskatchewan Chicken Producers Marketing Board and a former member of the NPMC in Saskatchewan, that bears this out. Similar responses were heard in conversations with early participants such as Glen Flaten; regulators such as Dr. Leo Kristjanson and Blair Backman; and a former manager of a provincial agency in Saskatchewan, Everitt Ritson. In the vernacular of economists, supply management began as a PERT but has become a PEST.

10. A recent fire at a single broiler hatching egg production unit in Saskatchewan destroyed facilities which produce 15.3% of the industry output (based on provincial-allocated quota).

11. This statement is not intended to provide an explicit or implicit endorsement of the vertically integrated US chicken industry. That sector has other shortcomings.

12. Liebenstein's X-inefficiency principle suggests that when a system faces adversity, new innovations result — innovations that may otherwise have gone undiscovered (Rosaasen and Schmitz, 1985).

13. Accepting quota as a security for a loan may not be a good lending policy. For example, in hindsight it was an error for both borrower and lender to value farmland in the Rosetown area of Saskatchewan at $1,200 per cultivated acre in the 1980s. Current expectations about quota value in the supply-managed industries may similarly be in error.

14. Saskatchewan has attracted new entrants into chicken production and dairy production from other provinces. A partial list of these is available from the authors.

15. Relative quota values in the same region between chicken, dairy and eggs may indicate the relative comparability of those enterprises within a region, compared to other regions or the national average. Simple commodity quota value comparisons across provinces are likely only to be useful when looking at changes in trends within a province.

16. For example, if a unit of chicken can be purchased in the United States for 90¢ and then sold in the Canadian market at $1.30 (common currency) and total transfer costs are 10¢, then there is a 30% profit. (Prices are hypothetical and only used to demonstrate the problem that import quota rights create.) Similar arrangements exist for cheese imports in the dairy industry. Various firms have conflicting interests; some firms want Canadian prices to remain stable or increase while others prefer a price decline.

17. The new GATT and NAFTA agreements require that by the end of the agreement Canada allows import access of 3% and increases access to 5% of the domestic market for a commodity group. Canadian chicken imports will be unaffected because they are already at about 7% of domestic consumption.

18. This was recommended by the Canadian International Trade Tribunal (1992).

19. An estimate of the import quota rents was $64.1 million annually for chicken imports into Canada from the United States (Deloitte and Touche, 1991).

20. Consistent with our view of reducing adjustment costs, existing import quota rights might be given benevolent treatment and phased out over a three-to five-year period on a proportionate basis. Import quota holders are at risk because of potential changes in regulations and financial markets, as they are "net long" the Canadian dollar. If the Canadian dollar declines to US $0.55, it is expected that the costs of chicken production will decline in Canada relative to the United States. The narrowing of the price spread erodes the value of the import quota. Import quota values fall to zero when Canadian prices decline to the US price plus transfer costs.

21. The granting of the pollution right to past sinners has been equated to the logic, "If you pay me . . . I'll quit pissing in your soup!" (Rosaasen and Lokken, 1993).

22. A legal opinion by Armitage indicated quota was not the property of chicken producers in Saskatchewan (1993).

23. Provincial and regional cost efficiencies can be debated, but the recent CCMA COP surveys support this statement at the farm level.

24. In the short run, interprovincial quota transfers might move in the direction of inefficient provinces if producers there are willing to pay high prices based on high existing quota values (see previous subsection, *The Quota Value Myth*) or if some provincial governments put their treasuries behind quota purchases in a bid to expand their share of production. In the 1970s Saskatchewan forfeited milk quota, which then moved to Québec, when Saskatchewan producers shifted to producing wheat instead of milk because of short-run market signals. In that case, no compensation was paid to Saskatchewan by other farmers or provinces.

25. A reasonable strategy is to allow a shift in regional production to occur, perhaps somewhat slower than the shift that might occur in an open market, in recognition of the adjustment costs and the logic of allowing some facilities to be utilized rather than abandoned before becoming obsolete or fully depreciated. In addition, a farmer who is four or five years from retirement can be more gently phased out of the sector.

26. There are basically only three ways to allocate resources: tradition, decree and the open market. All of them have some problems. Supply management is largely operated by decree.

27. This strategy cannot be defended as being fair or rational for the country, but if profit-maximizing Saskatchewan chicken producers can use it - good for them! They will have lobbied well and won in the political marketplace. "Maximize for me" is the key to this strategy. It is neither endorsed nor recommended by the authors. The processing segment in Saskatchewan could shut down the use of this strategy. Perhaps the chicken producers would buy the plant for 10¢ on the dollar.

28. The regional shift of the US broiler industry to low-cost areas was led by the vertical integrators.

29. The US hog industry is vertically integrating. It is moving to North

Carolina and away from Iowa which appears to have a feed-cost advantage. Will Iowa farmers tolerate the entry of vertical integrators? Probably not until economic conditions get much worse. Iowa limits the scale of operations and openly opposes corporate farming. The economic environment for chicken production in Saskatchewan appears similar to the hog situation in Iowa.

30. The regulatory changes proposed in the *Report of the Consultation Committee on the Future of the Dairy Industry* (Balcaen, 1992) are examples of problems encountered in the regulatory field in Canada.

31. Past conflicts in the red meat sector were over levels of provincial support for the sector, top-loading versus bottom-loading, and over how government actions may have distorted regional comparative advantage. (Top-loading provincial subsidies were those that supported outputs while bottom-loading provincial subsidies supported inputs). Central Canada and the Prairies also line up differently on issues such as grain transportation and feed grains policy.

32. For a period during 1994, Saskatchewan stood alone in opposing this waste of resources.

33. Perhaps the question of whether free trade is achievable in the real world of politics, special interests and nationalism, or if it is simply a "dream for economists," should be more closely examined.

34. Recall in British law that a woman was considered a chattel and was the property of her husband. It was the law of the day, but it is no longer the law! The views and values of society have changed on this issue.

References

Agriculture Canada. 1990. "Issues in Federal Legislation Governing Supply Management in Canada: Phase I Poultry and Egg Sectors." Working Paper, Policy Branch, Ottawa, Ontario, Canada (December).

_____ . 1994. *Tariff Equivalents*. Facsimile Transmittal Notice (April).

Armitage, D. 1994. Legal opinion submitted to K. A. Rosaasen upon request.

Balcaen, L., R. Doyle, K. Matte, R. Morin, and W. Sherwood. 1992. *A Vision for the Future of the Canadian Dairy Industry*. Report of the Consultation Committee on the Future of the Dairy Industry. Dairy Farmers of Canada (December).

Canadian International Trade Tribunal. 1992. *An Inquiry into the Allocation of Import Quotas*. Reference No. GC-91-001, Minister of Supply and Services, Ottawa, Ontario, Canada (October).

CCMA. 1993. *Data Handbook 1993*. Ottawa, Ontario, Canada.

Dairy Industry Strategic Planning Committee. 1994. *Various Scenarios for the Future of the Canadian Dairy Industry*. Canadian Milk Supply Management Committee (October).

Daly, H. E., and J. B. Cobb, Jr. 1989. *For the Common Good*. Boston, MA: Beacon Press.

Dawson, Dau and Associates. 1983. *A Research Study on the Management of Quota in Alberta*, Parts I, II, III. Under contract to the Agricultural Products Marketing Council, Alberta Agriculture, Edmonton, Alberta, Canada.

Deloitte and Touche, Management Consultants. 1991. *A Study of Options for Pricing*

Industrial Milk in Canada. Toronto, Ontario, Canada (15 February).

Fitzhenry, R. I., ed. 1991. *The Fitzhenry and Whiteside Book of Quotations*. P. 163. Toronto, Ontario, Canada: Fitzhenry and Whiteside.

National Farm Products Marketing Council. Various years. *Annual Reports*.

Rausser, G. C. 1982. "Political Economics Markets: PERTs and PESTs in Food and Agriculture." *American Journal of Agricultural Economics* 64 (5): 821–33.

Revenue Canada Taxation. 1990. *Income Tax Act: Patronage Dividends*. Interpretation Bulletin No. IT-362R (10 August).

Rosaasen, K. A., and J. S. Lokken. 1993. "Canadian Agricultural Policies and Other Initiatives and their Impacts on Prairie Agriculture." Plenary Papers for the *International Workshop on Sustainable Land Management for the 21st Century*, Volume 2. Pp. 343–68. University of Lethbridge, Lethbridge, Alberta, Canada (20–26 June).

Rosaasen, K. A., and D. Maley. 1986. "Forward Planning: An Alternative Marketing Mechanism." *Journal of Agricultural Economics* 33: 205–220.

Rosaasen, K. A., and A. Schmitz. 1985. *The Influence of Feed Grain Freight Weights on the Red Meat Industry in the Prairie Provinces*. Study prepared for the Hall Committee of Inquiry on Method of Payment (February).

Rucker, R. R., W. N. Thurman, and R. B. Borges. 1996. "GATT and the US Peanut Market." This volume.

Scherer, F. M. 1980. *Industrial Market Structure and Economic Performance*. 2d ed. Boston, MA: Houghton Mifflin Company.

Statistics Canada. 1993. *Production of Poultry and Eggs*. Catalogue 23–202 Annual. Industry, Science and Technology, Ottawa, Ontario, Canada.

St. Hill, T. N. 1974. *Thomas Nast: Cartoons and Illustrations*. New York, NY: Dover Publications, Inc.

Stout, T. T. 1977. "Canadians at the Meat-Grain Interface, "in D. G. Devine, and R. J. Sparling, eds., *Proceedings of the Meat-Grain Interface Project*, Volume 1. Saskatoon, Saskatchewan, Canada (March).

Tietenberg, T. 1988. *Environmental and Natural Resource Economics*. 2d ed. Glenview IL: Scott, Foresman and Co.

16

Provincial Versus Centralized Pricing: Protectionism and Institutional Design

G. Gartner

Abstract

This chapter describes and analyzes the political-economic dynamics of institutional change within the Canadian supply management systems for poultry and eggs, using the Canadian egg and chicken marketing models as case studies. The investigation traces the historical origins of the systems as producers searched for means of improving their bargaining position to countervail the power of the concentrated processing, wholesale and retail sectors. It describes the efforts to develop and refine the supply management and pricing models in order to improve their performance in the face of changing market and international trading circumstances. This chapter assesses the performance of the two models by evaluating their response to domestic producer and industry needs while at the same time putting in place the necessary adjustments to effectively respond to new external competitive forces.

Introduction

The primary purpose of this chapter is to describe in some detail the operation of the Canadian pricing models for chicken and eggs as they have developed over the years and to point to their future evolution resulting from changing market and trade circumstances. Their performance in achieving industry and market objectives for markets experiencing imperfect competition will also be evaluated.

The administered pricing arrangements established by the chicken and egg industries have long been at the center of controversy and debate. Moreover, they have given rise to a wealth of academic research and literature by those interested in assessing the performance of concentrated agricultural marketing systems. Although the pricing models differ as to how they have structured their operating procedures, they share a common heritage and basis for coming into existence. Prior to their formation, primary producers found themselves in a highly imbalanced negotiating position in the marketplace. At the local or provincial level, hundreds of producers faced the concentrated buying power of processors, graders and wholesalers for their perishable products. Nationally and throughout the food chain, ineffective market signals and communication often resulted in wild swings in production levels and prices. As "price takers" under these circumstances, primary producers were on the "cracking end of the whip" because of their lack of bargaining power in the marketplace. As a consequence, producers suffered the effects of chronically low prices and wide fluctuations in production volumes and prices. Efforts to redress the imbalance in market power were first attempted by provincial marketing boards, which had been sanctioned by some provincial governments. However, provincial boards were limited in their ability to raise and/or stabilize prices because of their confined geographical influence and the lack of production discipline throughout the industries. Some would argue that their presence aggravated the already unstable market situation through aggressive marketing strategies designed to gain market share in other provinces. This activity culminated in what became known as the "chicken and egg wars." This interim difficulty with central desk selling was overshadowed by the realization that countervailing market power could not be achieved nor could prices and incomes be stabilized unless market demand was synchronized with supply generated both nationally and externally.

Any examination of the pricing activities must therefore be put into the broader context of the overall objectives of the agencies with pricing being only one of several interrelated programs designed to achieve industry stability and enhanced producer returns. An analysis of pricing performance is complicated by external forces beyond the control of any industry participant, making comparisons, such as those involving US and Canadian prices, difficult.

An analysis would also be lacking if the market strategies of wholesalers and retailers were not factored into an assessment of the pricing models that examined performance from the standpoint of consumer benefits accruing from the system. Schmitz and Schmitz (1994) observed that little is really known about the interaction between supply management at the farm gate and other sectors in the food chain.

The Pricing Models: Market Settings

Having shared a common origin for their formation, the pricing models for chicken and eggs began to diverge upon the creation of the national agencies. By bringing with them national supply coordination and quantitative border restrictions, the systems then held the potential to genuinely exercise countervailing market power within limits on behalf of their producers. Two major factors determined the extent to which the agencies employed pricing strategies to enhance producer returns. First, the respective federal-provincial agreements provided the opportunity for producers to receive their cost of production (COP) plus a reasonable return on their investment, but no more. That constraint was stipulated from the outset by government signatories as a public safeguard. The National Farm Products Council was charged with the responsibility of monitoring this and other activities of the national agencies to ensure that the national marketing plans were respected. Second, the agencies were charged with the responsibility of meeting all Canadian needs for chicken and eggs over and above those amounts allowed to enter the market by way of historical imports enshrined at the time the agencies were created. Any deliberate attempt to short the market provided buyers with a case for requesting supplementary import permits to fill a genuine market that could not be met by domestic supplies. In the case of chicken, shortages can arise because of unexpected increases or shifts in demand that stem from many factors, including substitution with red meats. Domestic chicken shortages have been less predictable but have had a shorter response time than eggs. For example, egg shortages are more of a seasonal nature because of a limited ability to adjust production to fully coincide with the peak demand for eggs at Christmas and Easter.

The divergence in pricing structures that arose as the national agencies assumed their overall responsibilities for managing the systems can be attributed to a number of factors. The most important of these factors are discussed below.

Variability in Product Mix

Over its production cycle a laying hen will produce approximately 6% small, 21% medium, 47% large, 22% extra large and 4% off-grade eggs. Chicken weights are more readily managed by varying the length of the feeding period to serve different market needs. However, the egg size of the laying hen is genetically determined and can only be adjusted through long-term breeding programs.

Variability in Product Demand

Consumer grade preference (influenced to some extent by retail pricing strategies) rarely coincides with the biological size pattern of the laying hen. Moreover, grade preference varies significantly across Canada. While Québec consumers have a strong preference for medium grade eggs, the rest of Canada requires a higher percentage of large grade eggs than is normally needed by producers. To further complicate the matter, new flock placements do not perfectly coincide with peak demand periods, and they tend to be "lumpy" in nature because of climate and the need to complement the timing of other farm enterprises.

Consumer demand for chicken can also vary throughout the year but not of the magnitude experienced by eggs. On the other hand, further processor demands present a greater challenge to chicken producers because of the volumes and weights required. With the primary product being the bird itself, rather than the eggs, the chicken industry has the capacity to meet changing market conditions more rapidly and directly, compared to the egg-producing sector.

Growth in Demand

The chicken industry is a growth industry with production and marketing increasing at a rate of 4–5% annually since 1970. It is no surprise that the dominant policy issue of the chicken agency revolves around the appropriate criteria and mechanisms for determining and distributing production over and above base allocations.

Table egg consumption grew until the early 1980s to about 19 dozen per capita, and then demand steadily declined until recent years when a modest upturn in table consumption was seen. This decline was largely attributed to changing lifestyles and public concern about cholesterol and heart disease. The demand for eggs in processed form began to expand as a result of food-manufacturing growth, increased usage in food service and new international demand for egg components. However, this shift was not sufficient to offset the decline in table consumption, leaving overall per capita consumption at 15.4 dozen (13.0 table and 2.4 processed).

Product Perishability

Both fresh chicken and eggs lose quality and value over short time periods although improved storage and handling techniques have been successful in extending their shelf life. Chicken that is not required for the fresh market can be frozen although some loss in quality and value does occur. The only real option for extending the use of eggs is to contract with domestic processors (breakers) and/or to tender internationally since eggs in shell form cannot be preserved by freezing. When broken and pasteurized, the product can be stored in frozen or liquid form for extended periods of time. However, the value to the producer is sharply discounted by as much as 50% of the fresh table price.

Product Identification in Different Markets

A compelling reason for direct central intervention in eggs is that it is not visibly possible to distinguish between table versus processor eggs, making it difficult to segment the market. Since weights differ for various chicken markets, less concern exists for product identification and pricing in maintaining the integrity of efforts to segment the chicken market.

For the reasons outlined above, the agencies formulated pricing and marketing structures patterned to meet the particular needs and characteristics of their respective industries. An inherent feature of these arrangements is that they have been and continue to be in a constant state of evolution — changing and adapting to new circumstances as they arise.

The Egg Pricing Model

Although the Canadian Egg Marketing Agency (CEMA) was created in 1972, pricing authority was not transferred from the provinces until 1976 through amendments to the federal-provincial agreement on eggs. It was only after some hard lessons were learned from the experiences of the early 1970s that signatories recognized the need to establish more effective and direct national coordination of egg marketing. The lack of production and marketing discipline, which existed at the provincial level, left CEMA with the unenviable task of disposing of eggs not required in the table market at prices negotiated by provincial boards. At that time, the breaker sector was still in its infancy and could only absorb 2–3% of available supply. A central element in the marketing strategy of the day was a storage system to hold excess eggs off the market until such time as they were required — the objective being to stabilize supply and prices in the face of seasonal demand fluctuations. Unfortunately, the strategy met with limited success. Burdened with a chronically high level of supply, storage stocks mounted as the pace of disposal fell behind the pace of accumulation. Instances of spoilage occurred and were widely publicized by the media and by opponents of the system. At the same time some provincial boards actively pursued markets in other provinces in an effort to maintain prices and incomes for their producers.

In their assessment of the chaotic first years of the national system, it became clear to signatories that CEMA should not only possess central pricing authority, but it should also have other powers that, together with pricing, would restore order and discipline within the system. These provisions included the ability to operate a truly national surplus removal program in which CEMA directly bought eggs in "surplus regions" and sold them into deficit table markets or negotiated their sale to breakers. The CEMA was also granted authority to impose antidumping penalties on graders found selling interprovincially below local prices plus transportation.

Armed with greater authority to directly participate along with the provinces in managing the affairs of the system, CEMA set out to develop a marketing structure that consciously segmented the national egg market into table and processed product. With its allied surplus removal program and levy system acting as essential instruments for accomplishing its task, CEMA formulated and began the implementation of a classical two-price system for eggs — a higher price regime for the larger and protected domestic table market and lower negotiated prices for processed eggs. In order to effectively achieve an important objective of the federal-provincial agreement "that producers receive their costs of production on average over time," CEMA undertook the first national COP survey in 1975. A formula, which served two purposes, was developed as a result of that survey. First, the formula provided a national benchmark of financial "well-being" against which the system's performance could be measured by producers. Second, the formula provided the necessary focus for CEMA to target its pricing in the table market and the means by which it could determine the price it would pay for surplus table eggs sold as processed product.

Using the COP formula as a close guide, CEMA established the price for grade A large on all eggs entering grading stations. Although CEMA "suggested" appropriate price spreads for other grades, the provincial boards were free to set their own differentials to suit their unique market circumstances. Levies charged by CEMA were collected from the grading stations by the provincial boards. These levies were remitted to CEMA for use in offsetting any losses incurred in the lower-valued processed market. Graders and retailers were free to pass on levies charged to the consumer, or they could absorb them internally, depending on their particular merchandising strategy. This aspect of CEMA's pricing arrangement was acceptable at the outset because levies were low (3–4 cents per dozen). However, as the industrial use of eggs expanded (necessitating larger levy charges), this vehicle for financing the two-price system became and remains a center of controversy. The debate was further inflamed by CEMA's use of consumer levies to offset deep losses incurred on the sale of eggs into world markets not required by domestic breakers. As an incentive to keep supply more in line with domestic requirements, in the late 1970s the levy was divided into consumer and producer levies. The consumer levy served the purpose of financing any losses on egg sales into the domestic processing (breaker) market and could be passed on to the

Canadian consumer. Losses on eggs not required in either the domestic table or processed market were then financed by the producer levy and could not be passed on to the consumer.

The CEMA used a variation of "base point pricing" to emulate what might occur in an unregulated, imperfectly competitive market — one that was already evolving into a vertically integrated structure before it was arrested by the advent of the national systems. Trying also to cover COPs in each province, CEMA first established prices for Manitoba and Ontario (the lowest cost provinces) in accordance with their COPs. Since product historically had moved east from Ontario and west from Manitoba, transportation differentials were added to the base prices to calculate the prices for the other provinces. Price listings were made on a monthly basis, following the updates of the cash cost component of the COP formula (determined provincially). Prices were adjusted in accordance with any significant movement in cash costs during the previous month.

While at first glance this initial attempt at establishing a centrally managed two-price system appeared fairly straightforward. In operational terms it was fraught with difficulties. Internally, the transportation differentials from Manitoba west performed quite well in achieving the dual objectives of facilitating the movement of eggs from surplus to deficit areas (when required) and enabling provinces to achieve their COP. Such was rarely the case, however, from Ontario east. Transportation differentials established for the Atlantic provinces were frequently inadequate to establish prices that enabled these provinces to cover all of their costs on a consistent basis. This feature of the pricing formula became a chronic irritant and cause for debate as to the priority of pricing objectives. The CEMA was caught in a dilemma and became a prisoner of its own formula. The "solution" to the problem lay in adjusting Ontario prices upward so that the Atlantic provinces could realize greater returns in line with their costs. However, in so doing, CEMA incurred the wrath of consumer groups and the attention of its supervisory body. As a result and much to the chagrin of the Atlantic producers, CEMA was forced to retreat in the face of public concerns regarding producer overpayments.

A further difficulty, which arose during the early years of the centralized pricing system, was the lack of an effective price discovery mechanism for eggs sold to domestic breakers. In addition to the challenge of sourcing eggs nationally for breakers throughout the country, CEMA encountered considerable difficulty in establishing and negotiating sale prices that minimized their losses yet were competitive with eggs available to breakers from abroad. The larger processors operated in the international market arena, and foreign competitors, in order to ensure that they were not at a disadvantage insofar as egg prices were concerned, kept a watchful eye on their Canadian counterparts. After several years of chaotic negotiations and conflict, CEMA and domestic processors agreed on a formula for pricing eggs that were destined for breaking purposes. The formula required that the price of large table eggs be sharply discounted to conform with the rate judged by CEMA to be the price of eggs available to US breakers. This change in breaker pricing met with only limited success in reducing the conflict with Canadian breakers. Threats of boycotting Canadian eggs and the growing possibility of dumping charges from US breakers forced another examination of the pricing formula. It was finally agreed that the new reference price to be used in the formula would be the biweekly Urner-Barry breaking stock quotation for mid-west heavy ungraded eggs. When adjusted for exchange rates and transportation costs to southern Ontario, the formula generated the Canadian breaker price for plants in that area. In conjunction with the new pricing arrangement, supply contracts were negotiated in which breakers were required to commit to accepting certain volumes of product over given time periods. The CEMA, in turn, was obliged to source product commitments or to allow the entry of imports to satisfy the needs of the individual breaker operations. These two initiatives went some distance to stabilize an important and growing market while at the same time shielding the sector from the clear prospect of dumping charges from US competitors. This formula and approach to breaker egg pricing remains in effect to this day.

The egg marketing structure continued to change in the 1990s. For example, seasonal pricing was introduced in 1990 in an effort to make the rigid formula pricing system more responsive to the seasonal demand for table eggs. The objective of the pricing revision was to price below COP during periods of weak demand and above COP just prior to and during peak demand at Christmas and Easter. Instead of attempting to return producer COP on a monthly basis, seasonal pricing sought to accomplish it on an annual basis. Ironically, seasonal pricing received strong resistance from graders and retailers who liked the predictability and stability of the earlier pricing practice. Retailers were unaccustomed to changing their price lists before high demand periods and often placed their suppliers in a margin squeeze. After its initial introduction and the establishment of improved communications with CEMA, the trade became more accepting of this practice. The program also received some resistance from producers in certain provinces with particularly "lumpy" flock placement periods. Those producers, who saw the benefit of having their flocks reach peak production during one or both periods of higher prices, were quick to do so if their operating situation permitted it.

The CEMA introduced the practice of selling eggs on the basis of weight rather than grade in 1991 to reduce the cost of its market operations and to improve the flexibility of their use in either table or processing markets.

The CEMA often found itself purchasing graded products which had not been clearly intended for the table market. By purchasing those eggs in ungraded or "nest-run" form, CEMA experienced a significant reduction in the cost of operating its two-price system by avoiding unnecessary grading, handling and storage charges.

The centralized pricing system, conceived in the 1970s and refined in the 1980s, was destined to undergo a major restructuring in the 1990s. Those changes were not driven by the failure of the pricing model to achieve its objectives but by internal provincial and producer controversy regarding an appropriate levy policy for the system. This conflict intensified in the late 1980s and early 1990s as levy requirements for financing the losses on CEMA's breaker sales grew in proportion to the shift in product sold into this market. Some issues involved in this conflict were: Should the losses on breaker sales continue to be financed by a uniform levy imposed equally on all eggs? Should the financial burden arising from these losses be allocated on the basis of the percentage of eggs declared available to the breaker market? The desire to alter the uniform levy practice was naturally led by provinces who produced a lower percentage of eggs that found their way into the breaker market — Ontario and Québec. Other provinces, such as Manitoba with close to 50% of its production destined for the breaker market, strongly opposed any modification to the uniform levy policy. They knew that a full shift in the burden to their producers and consumers would mean an almost twofold increase in levies to their provinces. They also knew that it was not feasible to pass on all the potential levy increase to their consumers. Consequently, their producers would have to bear the brunt of any amended levy policy.

The issue was finally brought to a head when Ontario, and subsequently Québec, partially withdrew from the federal-provincial agreement in 1992 by choosing to operate their own surplus removal program and contracting with CEMA for the remission of levies. The extreme uncertainty, which was created by not knowing when or in what amounts levies would flow to CEMA from almost 60% of Canadian production, spurred negotiations in an effort to keep the entire system, including pricing arrangements, intact. The outcome was a compromise consisting of a base uniform levy to finance a minimum level of breaker product, plus provincial levies that would vary in proportion to the volume of breaker product declared surplus to provincial table needs. With this revision went CEMA's ability to direct price because of the many uncertainties created by the revised levy system.

As it now stands, CEMA's centralized pricing system exists in name only. Rather than setting the national price for grade A large eggs, CEMA merely sets the price at which it is prepared to buy eggs surplus to the table market for resale into the breaker market. The provincial boards now establish all prices in accordance with local circumstances with an eye to prices prevailing in neighboring provinces and CEMA's buy back price. The CEMA still maintains and updates the COP formula on behalf of the provinces, calculates the price for breaker eggs and conducts tenders for the export sales of surplus eggs to even the breaker market. Under the revised pricing arrangement, priority is now placed on maintaining the competitiveness of respective provincial producers within the Canadian market although achieving COP for their producers remains a stated priority.

The Chicken Pricing Model

The decentralized chicken pricing model does not possess the high public profile that the egg pricing model does largely because it is volume-driven. Priority within national discussions about chicken revolves around production levels and their allocation. Accordingly, price negotiations are heavily conditioned by what is required to clear the market. Important parts of that assessment are data on red meat prices and supply and retail merchandising strategies at the meat counter. The COP information serves a somewhat different purpose in the chicken marketing system than it does in the case of eggs. Rather than creating a formula to directly generate product prices, COP information is used as a valuable guide to provincial boards in the course of ongoing negotiations with local processors. Its use is more vital at the national level to calculate the comparative advantage criterion for over-base production allocation.

A factor taken into account by both producers and processors during price negotiations is the differential required to maintain competitiveness with neighboring domestic and foreign supplies. Given all these constraints, provincial boards have a very limited range in which they can exercise any extensive countervailing market power through pricing strategies alone. The real impact has come by way of supply determination — hence the preoccupation with this vital aspect of managing the chicken marketing system from a national producer perspective.

In operational terms the crucial negotiating forum has been Ontario throughout most of the history of the chicken pricing model. With its large low-cost production base and high processor concentration, Ontario has been the price setter with other provinces closely following the pricing changes that occur from time to time in that province. Recently, however, Québec has begun to play a more prominent price leadership role as the system moves into an era of greater supply liberalization. Obviously, price negotiations occurring in central Canada are the focus of attention by the national agency and the other provinces because they represent a key factor in price deliberations occurring elsewhere. Not surprisingly, impasses have often occurred during the negotiation process, particularly in crucial market areas. To prevent pricing vacuums from arising, binding arbitration has been used from time to time to arrive at final prices. In these instances, provincial boards have used COP information to support their cases with a high degree of success.

Unlike CEMA, the Canadian Chicken Marketing Agency (CCMA) was not empowered with national pricing authority because it was never considered crucial to the successful operation of the national system. Its influence on product pricing is more indirect by way of its central role in supply determination and the most appropriate means by which it allocates that supply to meet future market requirements.

After many years of criticism from the further processing and retail industry for "shorting" the Canadian market, the CCMA (led by Ontario and Québec) engaged in a major supply increase to test the ability of the processors and retailers to absorb the increased product. This action coincided with the General Agreement on Tariffs and Trade (GATT) decision to convert quantitative restrictions to tariff equivalents but was not directly attributable to it. The shift in supply management strategy was driven more by an attempt to realign market shares for over-base quota. As expected, wholesale prices declined as the increased supplies entered the marketplace, but they were not fully reflected in price reductions at the retail level.

During this period of great instability within the industry, CCMA initiated a fundamental review of the system's quota allocation process. Pressure had been mounting for some time from Central Canadian boards and further processors to shift from a centralized national quota allocation system to one that determined production requirements on a decentralized or "bottom up" basis. Again the catalyst for this change was generated by Ontario — the province with the heaviest concentration of processing and further processing capacity.

The basic approach and principle inherent in the proposal was seen by many industry interests and government policymakers as consistent with the need for the system to improve regional supply sensitivity and price competitiveness as it confronted the realities of the new trade regime. Through most of 1994 intense negotiations were held to forge a consensus on the structure and operating procedures associated with a decentralized quota allocation system. Agreement in principle was reached in August, 1994, with the new plan to be launched in January, 1995. However, much negotiating time was devoted to achieving accommodations for high-cost provinces, such as Newfoundland, and determining procedures for placing "caps" on periodic allocations to the provinces. The decentralized allocation did come into effect in January, 1995, despite the lack of support from all provincial boards.

The negotiations associated with bringing forward a more flexible and market-sensitive supply determination system resolved two outstanding issues that had been nagging CCMA for a number of years. First, British Columbia reentered the national system, convinced that an allocation system would be in place that would more properly reflect its high growth requirements within the Canadian market. Second, the new allocation system agreement contains a more acceptable over- and under-production penalty system, which on National Council's insistence, will be enforced by CCMA starting January, 1995.

The related pricing arrangements remain basically unchanged under the new decentralized quota allocation system. Québec and Ontario will set live prices in Central Canada through periodic negotiations with processors. In turn, all other provinces will set their live prices no lower than Central Canadian prices, less transportation costs. The transportation cost data will be calculated using the costs of shipping eviscerated product converted to a live weight basis.

At this point in time, it would be premature to pass judgement on the performance of the new decentralized quota allocation system. One can only say that, in light of the required market sensitivity for the system in the future, it is conceptually a step in the right direction. However, if the chaotic circumstances, which gave rise to the new system, should continue to prevail during its early years of operation, then not much hope can be held out that this system will be any more sustainable than the one it replaced. The test of time may well be determined by the intensity of external threats to the Canadian market and the resultant degree of unity that the system might instill among provincial interests.

Performance of the Pricing Models

How well have the two pricing models performed in achieving the objectives of the systems? Have they played their required role along with other program activities? Because of the close interrelationship between pricing and other system activities, it is quite difficult to isolate the direct influence of the pricing models. However, conclusions drawn with respect to egg pricing may have greater credibility because of their more distinct profile in that system.

One test of their performance could be the extent to which they have enhanced and stabilized producer returns. Using national and provincial COP calculations as a rough benchmark, the models have performed quite well in this regard. The nationally weighted average returns have been consistently above nationally weighted costs for both commodities throughout the years of their effective operation. Some would argue that the models have performed too well in this regard. For example, in the early 1980s CEMA was heavily criticized for pricing table product in comparison with what was justified by the COP formula. By closely monitoring upper price levels, the margin between aggregate costs and revenues declined from just over one cent per dozen to a small fraction of a cent.

From an individual farmer standpoint, the pricing models may not have been held in high regard depending on his/her geographical location and the relevant cost position vis-à-vis the COP. In the case of chicken, Coffin, Romain and Douglas (1989) demonstrated that increasingly uniform prices were generated under the chicken pricing model with real producer prices gravitating more closely toward those of the lowest cost province. In the case of eggs, the rigid formula approach destined some of the Atlantic provinces to receive returns that more often than not were one to two cents below their published COP. In the West, there were occasions when Saskatchewan received less than their COP. The more significant impact of the pricing models was felt by those who, for various reasons, incurred production costs above their provincial COPs. By its very nature, the COP represents an average from a sample drawn from a producer population. When that COP is used as a guide to pricing or to actually determine prices, those producers with costs above the COP must either make decisions to reduce costs, accept lower returns or eventually leave the industry. In effect the COP/price relationship forces producers to compete among themselves. The one most effective means of doing so has been the rapid adoption of technological improvements and other managerial practices that have generated significant productivity increases in both chicken and eggs. Be that as it may, both the chicken and egg industries have experienced substantial consolidation of farm units over the life of the pricing models. For example, the number of regulated egg producers declined from a high of over 3,000 in 1972 to just under 1,700 in 1994. A similar pace of consolidation has been experienced by the chicken production sector. Some comfort can be taken from the fact that the pricing models and allied programs made a significant contribution to achieving the regional development objectives of the system although they did not generate the level of returns ideally sought by higher cost areas.

A subject of intensive debate has been the impact of pricing and other supply management programs on consumers. The reduction in real producer prices over the life of the pricing models has been well-documented and speaks highly of primary product performance with regard to consumer responsibility. One of the side benefits of the COP formula, or reference pricing, has been the close relationship between product and input prices. Taken together with the productivity gains made in the period, the benefits have also been reaped by consumers if they were in fact passed through by processors and retailers. The extent of this pass-through effect is open to debate. Moreover, food processors have received eggs at the US equivalent price, thereby enhancing their ability to compete in international markets with products containing a significant amount of egg components.

It is dangerous to generalize on the performance of the entire pricing structure, largely because of wide variation in the retail pricing strategies in existence throughout Canada. Using eggs as an example, producer prices increased by about 16 cents per dozen between 1987 and 1992 with little variation by region. The same situation occurred at the grader (wholesale) level. However, such was not the case at the retail level. Depending on their location, consumers either enjoyed a reduction in retail price over the same time period (Toronto – 3 cents per dozen) or a substantial increase (Winnipeg + 22 cents per dozen). Similar retail behavior has been identified with respect to chicken merchandising with the added complexity of the influence of red meat prices on retail pricing practices (Coffin, Romain and Douglas, 1989).

An interesting question to ask is: Would the pricing process and the outcome be all that different in an unregulated environment? I would suggest to you — not by a great deal because by now the industries would still experience price discovery through opposing concentrations of power. The only difference would likely be the identity of the players at the primary production level. Instead of the bargaining power provided to many hundreds of farm units through central desk selling, one would see an association of vertically integrated production units using supply management — US style — to maintain power in the marketplace.

Future Directions

The pace of change in the operation of the pricing models has been quite modest and slow-moving over the last 15–20 years, compared to what is expected to occur from now until the turn of the century. Those changes will be made as part of the industries' overall efforts to adjust to the changing trade and market circumstances.

The elimination of quantitative border restrictions and their conversion to tariff equivalents set the stage for a major realignment of power in the marketplace for poultry and eggs. It is fortunate that the tariff equivalents are of a magnitude that will allow for an orderly transition to the new trade and market regime if signatories and industry stakeholders are of a mind to make such a transition. The key to achieving a successful transition lies in maintaining supply discipline within the Canadian production system.

As part of that transition, the linkage between price and COP will become less and less important, and market conditions will become the predominant consideration in price negotiations. Should the next round of GATT negotiations bring a further (if not full) reduction in border protection, product pricing will be closely tied to US prices. The stark consequences of this scenario are that Canadian egg and chicken producers will not only experience chronically lower prices, but those prices will likely fluctuate more widely in concert with the cyclical swings of world prices.

Supply-managed systems are at important crossroads in their economic and political lives. Government and industry leaders have critical decisions to make regarding the future purpose and functioning of the systems on matters such as production, allocation and pricing — all vital to maintaining viable Canadian industries.

References

Agriculture Canada. Various years. *Poultry Market Review, Marketing and Economics Branch*. Ottawa, Ontario, Canada.

CCMA. 1993. *Competitiveness, Pricing and Cost of Production Studies: The Producer Viewpoint*. Ottawa, Ontario, Canada (June).

CEMA. 1991. *The Canadian Egg Industry: Supply-Management Myths and Realities*. Ottawa, Ontario, Canada (January).

Coffin, H. G., R. F. Romain, and M. Douglas. 1989. *Performance of the Canadian Poultry System under Supply-Management*. Pp. 116–121. Department of Agricultural Economics, Faculty of Agriculture, Macdonald College, Ste-Anne-De-Bellevue, Québec, Canada (January).

Statistics Canada. Various issues. *Production of Poultry and Eggs*. Ottawa, Ontario, Canada.

Schmitz, A., and T. G. Schmitz. 1994. *Supply Management: The Past and Future*. Department of Agricultural and Resource Economics, University of California, Berkeley, CA. Forthcoming–*The Canadian Journal of Agricultural Economics*.

17

Venturing into the Political Market

M. M. Veeman and L. M. Arthur

Abstract

This chapter is concerned with observations and conclusions based on the authors' representation of consumers' interests on the national dairy and poultry task forces. Part of a broad policy review for agriculture and food, these bodies and the subsequent Deputy Ministers' Supply Management Steering Committee sought consensus of major stakeholders on changes to improve the costs, inflexibilities and lack of transparency of Canada's national supply management programs. In contrast to naive public interest theories of regulation, as in hypotheses that supply management has enabled producers to offset the market power of processors and distributors, the operations of these programs involve many areas of joint interests and benefits to producers, processors, wholesalers and importers of supply-managed products at the expense of Canadian consumers. Many of the major problems identified in the task force process have persisted, a situation that reflects the political influence of the supply-side interests in the current supply management system.

Introduction

The general economic criticism of supply management is based less on the observation that it transfers large sums of money from consumers to producers than on the assessment that these transfers are inefficiently transacted (Barichello, 1983; Beck, Hoskins and Mumey, 1994; Schmitz and Schmitz, 1994; Veeman, 1982). Indeed, the Royal Commission on the Economic Union and Development Prospects for Canada (1985) concluded that for each dollar transferred from consumers to producers, about 25 cents is wasted through efficiency losses. Since their inception, Canadian supply management programs have made little progress in recognition of and action on consumers' concerns regarding regulated marketing.

Recognizing the costs of supply management, the Royal Commission on Economic Union noted that a social goal of augmenting and stabilizing farm incomes could be better met by income stabilization programs. Failing outright replacement, substantial reform of supply management was urged as was more adequate representation of consumer groups on regulatory boards and agencies.

Recommendations for consumer representation in the regulatory processes that specify marketed supplies, prices or both for supply-managed commodities recognize the unusual powers that regulatory bodies have been granted to regulate supplies and prices. These bodies have also been exempted from the usual provisions of Canada's competition legislation. Invariably, the legislation establishing the supply management institutions has specified minimum numbers of producer representatives, typically as a majority. Despite the lack of recognition of consumers' interests in the national supply management programs and the lack of inclusion of consumer representatives in the bodies administering these plans, consumer interest groups were encouraged when, in 1989, the national association, which represents a broad-based range of consumer issues, the Consumers' Association of Canada, was asked to nominate possible representatives to the federal Minister of Agriculture's tasks forces on supply management. Together, the dairy and poultry task forces, concerned respectively with the dairy and poultry (chicken, eggs, turkey) industries, were one component of the Conservative government's national review of policy for agriculture and food, a process that was launched in 1989 by then Federal Minister of Agriculture, Don Mazankowski, and subsequently continued through 1991.

One noteworthy factor of the review that was of particular importance to the supply-managed sectors was the pressure for policy adjustment from negotiations to reduce barriers to international agricultural trade under the General Agreement on Tariffs and Trade (GATT). A more immediate pressure was the increased competition for Canadian food processors — and particularly for "further processors" (secondary processors manufacturing foods of which regulated dairy and poultry products are ingredients) — as Canadian and US tariffs on foods and other goods fell under the Canada–United States Free Trade Agreement. The framework, designed by the federal governments for the dairy and poultry task forces, specified a federal commitment to supply management but noted needs for greater flexibility, market responsiveness and efficiency throughout these systems (Agriculture Canada, 1989). There were high hopes from many sector participants that the task forces would lead to real reforms in supply management systems. The two bodies tabled their reports, which included numbers of recommended changes in these systems, in 1991. (National Dairy Task Force, 1990 and 1991; National Poultry Task Force, 1991).

The purpose of this chapter is to document our observations and conclusions as members appointed specifically to represent consumers' interests in the policy review of the national supply management programs. This activity gave each of us an opportunity, unusual for professors of agricultural economics, to observe and participate in the agricultural policy process.

The Effectiveness of Minority Interests in a Stakeholder Consultative Process

The use of stakeholder consultative processes to seek consensus among diverse interest groups is an increasing practice used in Canada at both federal and provincial levels. Like other such groups, the task forces operated on a consensus basis. The theory of consensus building suggests that this even incorporates minority interests (Fisher and Ury, 1983; Fisher and Brown, 1988). However, experience in forums and other consensus-building processes suggests that often minority interests are simply strong-armed into defeat in order to achieve consensus. This use of force is necessary because in consensus-building processes, the minority interests cannot simply be ignored as they are in voting processes such as quota setting by supply management committees.

The outcome of a stakeholders' forum can be expected to reflect the political power of each group and the nature of the political and economic pressures on them. These pressures influence the extent to which each participant or sector can compromise — their bottom line. Some parties set such stringent bottom lines that no compromise is possible, thereby undermining the negotiating process. Fisher and Ury (1983) recommend the strategy to minority interests who are having difficulties making themselves heard.

The weakness of the consumers' position in decision making forums is the difficulty in setting a bottom line that can be enforced by the majority of consumers. This arises from the large number of consumers and consequent high costs of information and fragmentation of interests relative to the much smaller number of producers or supply-side interests. Thus, supply-side interests face lower information and organization costs and receive much greater individual benefits from the programs in question. During the period of the two task forces, the supply-side for both the poultry and dairy industries generally included primary processors and wholesalers in addition to primary producers.[1] This can be explained by the interaction of supply and price controls under existing supply management systems. (See Hollander, 1993). The coincidence of the interests of producers, primary processors and wholesalers is discussed in more detail later. The disparity in distribution of benefits and

TABLE 17.1 Annual Transfers Associated with Canadian Agricultural Policy,[a] 1991 and Capital Values of Quota[b]

| Sector | Producer Subsidy Equivalent | | Quota Value | Consumer Subsidy (Tax) Equivalent | |
	Aggregate $M	Per Farm[b] $	Per Farm[c] $	Aggregate $M	Per Consumer $
Dairy	2,838	98,167	199,449	(2,135)	(78.21)
Poultry and Egg	744	176,680	314,110	(487)	(17.84)

[a]Based on Organization for Economic Cooperation and Development (1993) estimates of producer and consumer subsidy equivalents

[b]Farms of specified type with sales greater than $2,500 (Statistics Canada, 1992)

[c]Data from Federal Provincial Task Force on Orderly Marketing (March 11, 1994)

TABLE 17.2 Canadian Family Food Expenditures by Income Level, 1992[a]

Income Classes	Lowest Income Quintile	2nd Quintile	3rd Quintile	4th Quintile	Highest Income Quintile	Entire Group
Total Food Expenditure						
Weekly average (in dollars)	60.70	88.65	108.31	130.13	164.08	110.44
As % of income	27.8	19.2	14.9	12.4	9.3	13.1
Expenditure on Dairy Products						
As % of income	3.4	2.2	1.6	1.3	0.9	1.4
As % of food expenditure	12.1	11.4	10.8	10.4	9.3	10.5
Expenditure on Fresh or Frozen Poultry and Eggs						
As % of income	1.3	0.9	0.6	0.5	0.4	0.5
As % of food expenditure	4.6	4.4	4	3.9	3.8	4

[a] Before tax income; all households
Source: Statistics Canada (1992a)

costs from the supply management programs is illustrated by information in Table 17.1, which gives a general indication of the substantial transfers to the producer level from associative action and lobbying (rent-seeking).

The basis of the dispersed and fragmented nature of consumer interests in supply-managed poultry and dairy commodities is also demonstrated in Table 17.2. This information from Statistic Canada's 1992 survey of household food expenditures indicates that these expenditures represent only 13% of income before tax (only 0.5% for poultry products and 1.4% for dairy products) for the average household. These expenditures are less in absolute terms but are greater in relative terms for low income households since these are basic foods for most Canadians (Statistics Canada, 1992a). Thus, the consumer costs of supply management have a greater impact on lower income consumers, yet the political power of this group is small. Factors contributing to the substantial political influence of the supply management coalition are noted later in this chapter and are discussed in more detail elsewhere. (See Veeman, 1990).

On the poultry task force, the supply-side's bottom line required that market power not be lost; that is those representing the supply side interests appeared to be willing to add flexibility if they could maintain the majority in voting situations. Less flexibility was evident in the positions taken by the supply-side interests in the dairy case. The bottom line for further processors of dairy products was to achieve input cost reduction by whatever means necessary in order to reduce the squeeze provided by competing imports relative to the cost of high-priced domestic dairy ingredients. Their position was bolstered, especially for frozen pizza manufacturers, by the threat of moving their plants from Canada to the United States. Industry representatives on the demand side — retailers, the fast food industry, restaurants — had a weaker bottom line; they simply wanted to weaken the majority vote enough (to equal supply- and demand-side votes in the poultry case) to force more flexibility. The consumers' position had a much lower bottom line. If they "walked," no one would care; most consumers would continue to drink milk and eat chicken, eggs and cheese.

Consumers do have some other options besides participating in decisionmaking forums. Their primary option is to avoid participation in such forums and to opt instead to lobby decision makers, both those sitting at the forum tables and those who are not, such as ministers. Nevertheless, it is difficult to decline the opportunity for representation which provides several benefits. These benefits include: the receipt of a substantial amount of information, which is not readily available otherwise because of the complexity and opaqueness of the regulatory regime; direct contact with influential members of industry; and the potential to influence the decision, if only subtly, using the information and contacts yielded by the process or through a coalition. Consumer representation on the bodies associated with the national supply management plans does not involve representation on the agencies, that is the national boards, that are dominated by producers. Following the 1989–91 national review of agricultural policy during which consumers were explicitly viewed as a stakeholder group, actions were taken to effect consumer representation on major subcommittees of the national boards. Consumers had one vote on the Canadian Chicken Marketing Agency's (CCMA) supply management committee; they were given nonvoting representation on the Canadian Milk Supply Management Committee (CMSMC). They were also represented on the agencies' consultative or advisory committees, but not on supply management committees, for eggs and turkey. This order of representation provided some information for the consumers' group but had little concrete influence on the programs.

In the most recent and ongoing process, which was initiated by the federal government in 1994, to review national supply management programs and to adjust these to Uruguay Round commitments, the consumer group has not been recognized as a stakeholder, and the national agencies have moved to reduce the input from this group. During this period, funding that had been provided by the CCMA to support consumer representation was withdrawn; the Canadian Egg Marketing Agency (CEMA) moved to discontinue its consultative committee; and the consultative committees of the other national agencies became inactive. In general, from early 1994, consultation on post-Uruguay Round adjustments on supply management and representation on the bodies associated with supply management focused much more on supply-oriented groups and much less on demand-oriented groups, such as consumers, grocery distributors and restaurants.

The potential downsides of consumer representation on supply management boards or their subcommittees are not always immediately recognized. These downsides include: the appearance of sanctioning a process in which consumers have not truly participated; reduced credibility if the consumer representative can get neither support nor ammunition from the consumer interest (Skogstad, 1993); reduced opportunity to lobby outside the process (if participants treat the process fairly); and not the least important, substantial expenditures of time and resources, typically of volunteer representatives.

The most effective means for consumers to influence decisions is not always clear, but consumer groups must consider the merits and costs of joining processes if these provide little or no potential for influence. The consumer interest side can only play a meaningful role in these decisionmaking processes if it is organized — and perhaps a limited role is likely even then (Sarker, Meilke and Hoyle, 1993). In some arenas, consumer organization has swayed political opinion and prompted changes, normally in coalition with other interests. For example, consumers were effective in getting provinces on-side in lobbying against Labatts "Maximum Ice" high alcohol beer (Chamberlain, 1993). In the area of food products, however, consumers are poorly organized and influence industry primarily through their aggregate purchasing patterns. The food basket is so diverse and complex that items are rarely singled out for special attention, but it is avoidance or boycotting behavior that speaks loudest to the food industry. Memorable examples are cases of known or imagined health risks in which consumers' actions have forced major food industry changes — pork and trichinosis; beef and high-fat content; eggs and cholesterol; and low-fat and high-fat fluid milks and ice cream.

Conflicting Versus Coincident Interests of Stakeholders

As consumer representatives in the task force processes, we sometimes attempted to strengthen our position by siding with the other members of the demand side. A primary strategy of the supply side was to isolate us from other demand-side constituents by highlighting the retail pricing powers of food stores, restaurants and fast food outlets and their unwillingness to reveal margins. It was implied that retail margins are wide and that the retail sector was responsible for consumer gouging. Unfortunately, competition in the retailing sectors does result in reluctance to reveal margins, reducing the information for consumers to use in developing a bottom line or backing their position.

Some members of the primary processing sector sought our support on specific issues although we observed their interests to be largely and ultimately on the supply side. For example, in dairy the coincidence of producer and processor interests comes only in part from the extensive vertical integration of producers' groups into processing and arises primarily because of the nature of regulation in the industry. One facet of the coincidence in interests induced by regulation is the nature of support for industrial milk prices that provides floor prices for processed dairy products. These prices specify a processor margin as well as providing "target returns" for producers. Each contains components that are favorable to the respective group, providing a basis for acquiescence to these prices by both groups. A telling comment was vehemently expressed by a processor in a dairy task force meeting who responded to a producer's query of the processor markup by saying, "We didn't object to the components in your part of the price, and we expect you not to object to ours." These incentives have contributed to the maintenance of an industrial milk price floor based on process costing factors that are outdated and farm costing procedures that can be defended only on distributional grounds. The expense, in both cases, is borne by consumers. Coincidence of the interests of dairy producers, processors and wholesalers has also been fostered by a degree of planned overproduction, the 4% sleeve, that adds to the cost of market support operations and subsidized exports. These benefit producers and processors but not Canadian consumers.

The regulation-induced coincidence of interests in poultry between producers, primary processors, egg graders and wholesalers differs more in detail than in effect. Thus, the level of protection that has been provided for the poultry primary processing and wholesaling level by import quotas has been significantly higher than the level of protection provided for the producer level, as judged by measures of the effective rate of protection assessed at both of these levels of the marketing chain (Cymbal and Veeman, 1994). One striking illustration of the coincidence of producer and other industry interests in the maintenance of a system of high internal prices, through limitations on imports, is seen in the substantial economic benefits that accrue to importers of regulated products, especially importers of chicken and cheese.[2] Estimates of the annual net benefits (annual rents) accruing to those fortunate to hold dairy and poultry import quota are about $100 million per year, a potent incentive for coalitions of importers/wholesalers with supply-side interests to maintain the cartel.

One apparent outcome of the dairy task force has been the strengthened coalition of dairy producers and processors as milk producers have become convinced that explicit cooperation with processors can bolster, rather than weaken, their position and power. The strategic alliance between these two interests allows issues of conflict and concern for both groups to be addressed, but unless there are changes in regulated institutions and administered pricing procedures, we consider that this alliance will, at best, be neutral to consumer interests. It is already operating to the detriment of consumers. A specific example of this is the weakening of the effectiveness of consumer representation on the consultative committee of the Canadian Dairy Commission (CDC) as significant policy issues are diverted to subcommittees composed only of producers and processors. Even more fundamental is the current government's implicit rejection of the dairy task force's consensus recommendation for explicit consumer representation by one of three CDC members. Another dairy example is the dairy supply coalition's refusal to share in the costs of a benchmarking approach to cost-of-production procedures, an attempt to address competitiveness in the process of regulated pricing. Other examples are the coalition's self-serving proposals regarding the administration of post-GATT tariff-rate import quotas and the moves to react to restrictions on export subsidization of Canadian dairy exports that have arisen from GATT by extending the system of price discrimination to include user class of milk and price pooling. All of these procedures are detrimental to Canadian consumers.

Numbers of advocates of supply management have based their support on concerns that concentrated and powerful food processing and distributing sectors can exert power against producers. This concern was certainly one of the motivating influences in the early development of agricultural marketing boards (Veeman, 1987). Our observation of supply management programs today is that their strong regulatory powers now provide incentives for collaboration between producers and processors in joint exertion of market power against consumers.

Even with coalitions on the demand side, countervailing coalitions that encompass consumers' interests are unlikely to be instruments of major change in the supply-managed sectors. Schmitz and Schmitz (1994) suggest that change may be brought about by infighting within the producer coalition. Infighting is occurring in the chicken and egg producer coalitions; chicken markets are dramatically increasing, and egg markets are decreasing with the respective growth and decline concentrated in specific provinces, for example, the growth in British Columbia and Ontario. The need for reallocation of quota and the varying needs for change, even in global quota, divide the supply coalition and weaken the supply management system. At various points in time, Ontario, British Columbia and Alberta have threatened to leave or have actually left the poultry supply management system. This view of the potential for change suggests that consumers might serve their interests better by backing components of the producers' groups who are pushing for change by altering global quota or reallocating existing quota.

The difficulty in solving infighting on the poultry supply side arises from a defect in the institutional arrangements of poultry supply management, which currently requires unanimity among all signatories for changes in marketing plans, including changes in the regional allocation of quota. Since those provincial governments and boards that expect to lose quota have had neither political nor economic incentives to agree to such loss, their unwillingness to cooperate is not surprising.

The unwillingness of provincial governments to agree to changes that might potentially reduce levels of provincial quota was a major cause of the lack of progress experienced by the supply management steering committee, a 1991–92 working group of the deputy ministers of agriculture and a small number of appointed stakeholder representatives. The purpose of this group was to seek adoption of the recommendations of the two supply management task forces. The lack of progress from this endeavor illustrates several characteristics of current supply management programs: their use by provinces as regional development programs; their considerable profit room as well as the toleration of production and distribution inefficiencies; and the leverage from producers' ability to lobby at two levels of government, which arises from the shared jurisdiction of provincial and federal governments over agriculture. However, the economic losses to producers and processors from failure of the cartel are appreciable. Thus, the supply-side coalitions embodied in the subsequent and current task force on orderly marketing[3] can be expected eventually to cooperate in changing the institutional framework to reduce infighting.

The incentives for cooperation are highlighted by the capital value of producers' marketing quota. This value totaled about $7.6 billion in 1992 according to the financial survey of Canadian farms in that year (Farm Credit Corporation, 1992). Quota was valued at $199,449 on average for dairy farms in 1991, relative to total assets of $749,992 and a net worth of $580,156. For the average poultry and egg farm, quota was valued at $314,140 while assets and net worth averaged $1,029,775 and $798,561, respectively, in 1991 (Federal Provincial Task Force on Orderly Marketing, March 11, 1994). Despite current strategic positioning by provincial boards, we expect that the possibility of economic losses of this magnitude will eventually encourage cooperation of provincial boards and governments in order to maintain the cartel.

Relative Influence of Political and Economic Pressures

One aspect of the task force processes that was particularly frustrating to economists was the virtual irrelevance of economic pressures in determining outcomes. As noted above, at times we strengthened our position as consumer representatives by siding with other parties with major demand-side exposure for whom economic forces were absolutely fundamental. Most of the demand-side sectors operate in a highly competitive environment; they are, therefore, extremely sensitive to and well-informed of the pricing problems, lost marketing opportunities and other market problems resulting from the failure of the supply management system to accommodate the changing market. These kinds of economic arguments and related economic solutions appeared to be completely irrelevant to the majority of supply-side interests at the tables as was the fact that prices for the regulated commodities well exceeded the competitive price as indicated by world or US prices. Basically, dominant supply-side interests were not prepared to recognize or accept any means for determining prices other than cost of production formulae or any means for determining demand-side interests other than voting on quota. The only acceptable solutions appeared to be administrative ones for which potential progress was not only extremely slow but was also easily diverted or reversed.[4]

The lack of success in appreciable reform of Canadian supply management programs from the 1989–91 agricultural policy review is reflected in the very considerable political influence of that sector, particularly the dairy sector. This influence was also evident in Canada's dichotomous and inconsistent position in the Uruguay Round negotiations on agricultural trade that led Canada to press for retention and strengthening of GATT Article XI provisions justifying import protection for supply-managed products. At the same time Canada sought lower trade barriers and distortions for other farm products. The importance of supply-managed sectors in Québec and Ontario, the dominant influence in national politics of these two large provinces and concerns regarding Québec separation all contribute to this influence.

Conclusions

The government-sanctioned regulation embodied in the national supply management programs does not arise from market failure. It is, rather, a powerful demonstration of Stigler's (1971) political/economic theory of regulation that explains the provision of regulation by politicians in return for political support. This provision responds to the demand for regulation by producers who receive economic benefits from the restriction of entry and limitation of import competition (Stigler, 1971). One form of the market-failure hypothesis of regulation, termed by Green (1990) the naive-public interest theory, has sometimes been argued to apply to agriculture. This is the argument that supply management provides offsetting power for relatively small producers who are faced by more powerful purchasers, processors and distributors. The argument is not borne out by the evidence for these sectors as was illustrated by information in Table 17.1 and by the nature of the programs which confer on supply-side interests a degree of market power that is considerably greater than necessary for producers to organize effective bargaining associations.

Our observation is that the nature of the regulatory interventions involved in the national supply management programs provides many instances of appreciable joint benefits to processors, wholesalers and importers of the regulated products and to established producers. These benefits are achieved at the expense of Canadian consumers. Some examples of these joint interests were provided in the chapter; these basically stem from the protection from imports and other profit room, which arises from administered pricing in a market in which domestic supply is also restricted. The supply management task forces sought decisionmaking structures to provide more balance between stakeholders' interests. They also proposed some changes to reduce excessive costs of the programs. The nature of that process, involving consensus among a broad group of stakeholders, led to recommendations for relatively modest changes in the existing systems. Overall, minor program changes, reflecting the relative political importance of groups that oppose substantive changes in the existing systems, have been made to date. Considerable room for economic and administrative reform still exists.

In the case of poultry industries, Skogstad has argued that the failure to make reforms is primarily the fault of Canada's first generation of marketing agencies and the associated governments that have failed to require "that producers think of their industry as a Canada-wide one, and enable their representatives in the private interest governments that implement supply management policies to set aside provincial concerns and promote the well-being of the Canadian industry as a whole" (1993). Nevertheless, consumers need more credible representation, "the type that will only come with better public funding of consumer advocates, either by government directly or by special levies of the marketing agencies" (Skogstad, 1993), as well as much patience if they are going to play a role in encouraging that change.

Notes

1. However, the relationship is not invariant and may change according to the issue or over time as adequacy of supply changes. An illustration was the battle during 1991–92 of the Ontario Chicken Producers Marketing Board and the Ontario processors over pricing policy before the Ontario Farm Products Appeal Tribunal. On the poultry task force itself, the primary processor representative chaired the committee, weakening this group's opportunity for input.

2. Also relevant in this context are the interests and efforts of current dairy importers to have their considerable potential financial stake maintained through their efforts to achieve legal ownership of post-GATT tariff-rate import quota. These efforts are supported by the supply-side coalition which currently encompasses this group. This coalition opposed allocation of import quota by auction, the method recommended for dairy products and eggs by the Canadian International Trade Tribunal (CITT) following the inquiry into the allocation of import quota (CITT, 1992). With auction allocation, import rents would be captured by the government, rather than by the importers. The supply-side dairy coalition also unsuccessfully proposed that new dairy tariff-rate import quota be allocated to the Canadian Dairy Commission (CDC) and that the CDC operate a private treaty quota exchange.

3. Consumer representation was excluded in the subsequent and latest generation of supply management task forces, the Federal-Provincial Task Force on Orderly Marketing, or Vanclief task force, named after its chair, the parliamentary secretary to the Liberal Minister of Agriculture, Ralph Goodale, who appointed this group in early 1994 to guide post-GATT policy for the supply-managed sectors. This body stated, "changes to existing structures should be tied to the level of risk assumed by the participants" (Federal-Provincial Task Force on Orderly Marketing, 1994). This view ignores the considerable consumer interests in and costs of the existing structures and reflects the supply-side composition of this task force.

4. A classic diversionary tactic is seen in the follow-up by CDC to the dairy task force consensus recommendation that the CDC provide problem analysis and develop an action plan to improve industry competitiveness in the context of a broad range of cited issues or areas. The response, to commission another study, focused on narrow issues and strategies, such as extension initiatives, rather than the broader issues of rationalization or regulatory changes cited in the task force recommendation.

References

Agriculture Canada. 1989. *Growing Together: A Vision for Canada's Agri-Food Industry*. Publication No. 5269/E and "Summary," Publication No. 5270/E, Minister of Supply and Services, Ottawa, Ontario, Canada.

Barichello, R. R. 1983. "Government Policies in Support of Canadian Agriculture: Their Costs." Discussion Paper No. 83–04, Department of Agricultural Economics, University of British Columbia, Vancouver, British Columbia, Canada.

Beck, R., C. Hoskins, and G. Mumey. 1994. "The Social Welfare Loss from Egg and Poultry Marketing Boards, Revisited." *Canadian Journal of Agricultural Economics* 42: 149–58.

CITT. 1992. *Analytic Staff Report, An Inquiry into the Allocation of Import Quotas*. Reference No. GC 91–001. Ottawa, Ontario, Canada.

Chamberlain, A. 1993. "Labatt boosts price, pulls ad on high alcohol maximum ice." *Toronto Star*. Section C: 1. Toronto, Ontario, Canada (23 November).

Cymbal, W., and M. Veeman. 1994. "Canadian Agriculture and Article XI: An Economic Analysis of Tariffication for Poultry Products." *World Agriculture in a Post-GATT Environment: New Rules, New Strategies*. Proceedings of June 13–15, 1994 conference, University of Saskatchewan, Saskatoon, Saskatchewan, Canada (in press).

Farm Credit Corporation. 1992. *Farm Survey 1992*. Ottawa, Ontario, Canada.

Federal-Provincial Task Force on Orderly Marketing. 1994. "Discussion Guide Toward an Action Plan for the Renewal of the Dairy, Poultry and Egg Industries." (28 January).

_____ . 1994. "Preliminary Draft Action Plan Towards the Implementation of Sustainable Orderly Marketing Systems in Canadian Dairy, Poultry and Egg Industries." (11 March).

Fisher, R., and S. Brown. 1988. *Getting Together*. Boston, MA: Houghton Mifflin Co.

Fisher, R., and W. Ury. 1983. *Getting to Yes*. New York, NY: Penguin Books.

Green, C. 1990. *Canadian Industrial Organization and Policy*, 3rd ed. Toronto, Ontario, Canada: McGraw-Hill Ryerson Limited.

Hollander, A. 1993. "Restricting Intra-industry Quota Transfers in Agriculture: Who Gains, Who Loses?" *Canadian Journal of Economics* 26: 969–975.

National Dairy Task Force. 1990. *Consultative Document*. (10 July).

_____ . 1991. *Report of the Task Force: Evolution of the Canadian Dairy Industry*. (31 May).

National Poultry Task Force. 1991. *Report of the Task Force: Towards the Development of a Second Generation of Poultry Supply Management Systems.* (15 March).

Organization for Economic Cooperation and Development. 1993. *Agricultural Policies, Markets and Trade: Monitoring and Outlook.* Paris, France.

Royal Commission on the Economic Union and Development Prospects for Canada. 1985. *Report of the Royal Commission, Vol. II.* Minister of Supply and Services, Ottawa, Ontario, Canada.

Sarker, R., K. Meilke, and M. Hoy. 1993. "The Political Economy of Systematic Government Intervention in Agriculture." *Canadian Journal of Agricultural Economics* 41(3): 289–309.

Schmitz, A., and T. G. Schmitz. 1994. "Supply Management: The Past and Future." *Canadian Journal of Agricultural Economics* 42: 125–48.

Skogstad, G. 1993. "Policy Under Siege: Supply Management in Agricultural Marketing." *Canadian Journal of Public Administration* 36(1): 1–25.

Statistics Canada. 1992. *Census, Overview of Canadian Agriculture: 1971–91.* Catalog No. 93–348.

_____ . 1992a. *Family Food Expenditures in Canada.* Catalog No. 62–554.

Stigler, G. 1971. "The Theory of Economic Regulation." *Bell Journal of Economics and Management Science*: 3–21.

Veeman, M. M. 1982. "Social Costs of Supply Restricting Marketing Boards: Reply." *Canadian Journal of Agricultural Economics* 30: 373–6.

_____ . 1987. "Marketing Boards: The Canadian Experience." *American Journal of Agricultural Economics* 69: 992–1000.

_____ . 1990. "The Political Economy of Agricultural Policy." *Canadian Journal of Agricultural Economics* 38: 365–83.

18

Vertical and Horizontal Coordination[1]

M. E. Fulton, M. Katz, and J. A. Vercammen

Abstract

This chapter develops a conceptual model of how prices and output are determined under supply management. The model explicitly incorporates the role of the retailing, processing and farm sectors and the bargaining or market power that the players in these sectors possess. The model also incorporates the regional (or horizontal) structure of the Canadian market. Since the mid 1980s, there has been a substantial increase in the interprovincial chicken trade because of new products, lower transportation costs and new market arrangements. Although marketing boards have historically been accustomed to setting price and quantity separately, these changes mean the separation of price and output decisions is no longer possible. The result is greater provincial rivalry among marketing boards and among processors as different companies and regions seek to expand their production base and increase pressure for vertical integration and processors attempt to provide themselves with production and price guarantees. Considerable strain and antagonism are also apparent in political decisions regarding the direction that supply management should take in the future.

Introduction

The determination of prices and output in industries that operate under supply management depends on the behavior of all sectors within these industries. Sectors that are likely to have an important impact on price and output determination are the retailing, processing and farm sectors. As Schmitz (1993) points out, while substantial attention has been directed toward the farm sector, little is known about the processing and retail sectors' roles in determining prices and output.

In considering supply management, it is important to understand not only the vertical relationships that exist between various sectors but also the horizontal relationships that exist within a sector. Canada is not a single market. Production, processing and retailing occur in many different regions of the country; products are traded among the different regions; and firms in one region compete not only with firms in their own region but with firms in other regions. This regional (or horizontal) structure is likely to have a substantial impact on prices, quantities and the location of production.

This chapter develops a conceptual framework that incorporates both the vertical and horizontal relationships that exist within Canadian supply management. Although this framework can be applied potentially to any of the supply-managed industries, particular attention is paid to the chicken industry. We begin by examining the vertical relationships that exist in supply management. We argue that until sometime around the mid-1980s vertical relationships played a dominant role in price determination in the chicken industry. Since the mid-1980s, however, we argue that the regional or horizontal relationships have played a much more important role. As a case in point, we argue that Ontario's challenge to supply management in late 1993 and early 1994 and the subsequent agreement by members of the Canadian Chicken Marketing Association (CCMA) to a new allocation and pricing mechanism can be explained in part by the changing nature of the horizontal relationships in supply management.

A Vertical Model of Supply Management

In this section of the chapter we develop a model to examine how prices and quantities are determined as the product moves from the farm production sector to the processing sector to the retailing sector and finally to consumers. By focusing on these sectors, the main elements of a product's transformation are captured. Although other sectors, such as further processing and the food service sector, are very important, they are not explicitly considered here since to do so would greatly complicate the analysis.

The model is constructed by first considering consumer demand, curve D in Figure 18.1, for the commodity. Consumer demand is met by a retailing sector that purchases a product from a processing sector and combines it with a retailing service. One unit of the processed product and one unit of retailing services are required to produce one unit of the retail product, that is fixed proportions production technology is assumed. A unit of retailing service can be purchased at price c. The resulting product is sold to consumers. The price of the final retail product is denoted as r.

Curve d represents the derived demand by retailers for a product produced by the processing sector. The processing sector purchases a product from the farm level, combines the farm product with a processing service and produces a product that is sold to the retail sector. It is assumed that one unit of the farm product and one unit of processing services are required to produce one unit of the processed product. A unit of the processing service can be purchased at price k. The price of the processed product is denoted as p. The price of the farm product is denoted as w, and the supply curve of the farm-level product is upward-sloping. The marginal cost at the farm level is denoted as m.

The price determination mechanism depends on the stage of the marketing chain being examined. The farm-level price is determined through implicit or explicit bilateral bargaining between the marketing board and processors. In contrast, the processor price is modeled as being determined through a more explicit market mechanism.

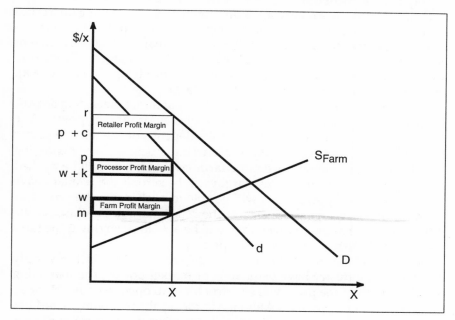

FIGURE 18.1   A Model of Price Determination under Supply Management

Figure 18.1 shows how prices in the vertical marketing chain are modeled. The marketing board sets output at level x which in turn determines the retail price, r. On the basis of their derived demand, d, retailers are willing to pay price, p, in order to purchase output x from the processors.[2] Since r > p + c, retailers earn a profit margin equal to the area labeled "Retailer Profit Margin."

The difference between p and m must cover the cost of processing, k, plus any profit margins earned by the processors and producers. Distribution of the profit margins between processors and producers is determined through bilateral bargaining between the marketing board and the processors over the farm output price, w. Prior to May, 1994, provincial production quotas were determined more or less as a fixed share of the national quota set by the national supply management agency. Given these provincial quotas, the producer price was negotiated on a bilateral basis between the producer marketing board and the processors or their association. Recent changes in the Canadian chicken industry have resulted in a dramatic change in this bargaining framework, so that negotiations are now about price and quantity. As will be examined later, this new bargaining relationship has important implications for how supply management operates.

Figure 18.1 is drawn on the assumption that the marketing board has some bargaining power and is able to negotiate the farm-level price, w, to raise it above the marginal cost, m. The result is that producers earn a rent equal to w - m, that is the area marked "Farm Profit Margin." Processors are also depicted as earning a positive per unit profit equal to p - (w + k), that is the area labeled "Processor Profit Margin."

The pricing model outlined above explicitly assumes marketing boards, processors and retailers all have some degree of market power. The distribution of profits among these sectors depends on the relative power held by retailers, processors and the marketing board. If the marketing board has no bargaining power, the farm-level price, w, will be pushed down to m, and the farm-level margin will be zero. If the retailers have significant market power, they will be able to push down the processor price, p, thus increasing the size of the retail margin. Similarly, if the processors have market power, they will be able to push down the farm price, w, and push up the processor price, p.

There is substantial evidence that different sectors in the supply management industry have bargaining or market power. The marketing boards in each of the provinces are structured to operate on behalf of all the producers in a province. Although in theory the marketing board has the power to unilaterally set the farm-level price, in practice the price is set via an implicit or explicit bargaining process. The board is often challenged by the processors, who themselves have market power vis-à-vis the marketing boards. As Coffin, Romain and Douglas (1989) point out, only one or two processors are operating in a number of the provinces while in other provinces a few large processors dominate the industry.

The Canadian retail industry is relatively concentrated, which is especially apparent when the country is examined on a regional basis. For example, in Ontario the top four food distributors own or sponsor 80%

of the corporate supermarkets and sponsored stores in the province. Similar percentages hold in other parts of Canada (Ferguson, 1992). The food service sector is also relatively concentrated, with McDonald's and Kentucky Fried Chicken representing a substantial proportion of the demand for chicken in this sector. The food service sector is a major purchaser of chicken. For instance, 26% of Canadian chicken consumption was accounted for by the fast food sector in 1993 (CCMA, 1994).

Empirical evidence supports the idea that profit margins may exist in all sectors of supply management industries. Positive quota values are evidence that rents are being generated in the farm sector. For the processing sector, Coffin, Romain and Douglas found, when comparing the periods 1974–79 and 1965–72, that the processor/farm markups in the Canadian chicken industry increased relative to those in the United States. That is the introduction of supply management coincided with an increase in the processors' markup (see Table 18.1).[3] For the retail sector, Coffin, Romain and Douglas provided evidence that, although retail margins have been declining in the United States since the 1960s, retail margins in Canada have been rising.

The existence of supply management has created rents that are distributed among the various sectors of the industry according to the bargaining or market power of the sectors. The existence of rents in all segments of the market chain provides one explanation as to why supply management has received substantial industry support over the years.

Prior to supply management, competition from US imports effectively limited the ability of all sectors — farm, processing, retail — to earn rents. In contrast, the introduction of supply management created an opportunity where virtually all sectors could earn rents. Thus, support for supply management has been reasonably widespread throughout the marketing chain.

Figure 18.2 illustrates the effect of introducing supply management in the farm, processing and retail sectors. Prior to the introduction of supply management, Canada was essentially self-sufficient in chicken.[4] The retail supply curve, S_{Retail}, is drawn so that it intersects the Canadian demand curve, D, at quantity x. The supply curves at the processing and retail level are constructed by adding the price of the processing and retailing services to S_{Farm} and $S_{Processor}$, respectively. This construction reflects the assumption that both sectors were relatively competitive prior to the introduction of supply management. Given a farm-level supply curve, S_{Farm}, the farm price in Canada was equal to w.

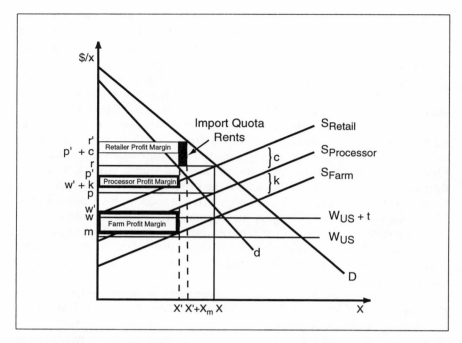

FIGURE 18.2   Impact of the Introduction of Supply Management on the Farm, Processing, and Retail Sectors.

The introduction of supply management in Canada resulted in slower output growth as compared to what it would have been otherwise. As Table 18.1 shows, Canadian production growth was lower than that in the United States from 1974 to 1984. Although the growth dropped in both Canada and the United States in 1979–84 as compared to 1974–79, the drop in Canada was larger. The first full year that supply management was used in the Canadian chicken industry was 1979.

The relative drop in Canadian production can be modeled by reducing the quantity of production from x to x' as in Figure 18.2. That is in the absence of supply management, the level of production would have been greater. Assume that supply management provided the marketing board with power to raise the producer-level price from w to w'. As a result, the farm-level sector increased its rents by the area, "Farm Profit Margin."[5] Quantity x' is processed by the processing sector. Although the price the processing sector paid for the farm product rose, the processing sector was able to more than pass this increase on to the retail sector. The result was that the processing sector was able to earn rents equal to the area, "Processor Profit Margin." The price paid by retailers for the processed product is p'. The retail industry was also able to generate rents, "Retailer Profit Margin," as the result of the introduction of supply management. Finally, import quota holders were able to generate rents as a result of supply management. Assuming import quota holders can purchase imports at price r, they are able to capture the difference between price r' and price r on imports x_m. The rent obtained from holding import quotas is labeled "Import Quota Rents."

TABLE 18.1   Growth in Production in Canada and the United States, 1974–79 and 1979–84

Period	Growth in Production (%)	
	Canada	United States
1974–79	21.9	30.8
1979–84	6.2	16.8

Source: CCMA, 1994

The argument outlined above assumes that supply management had no impact on farm production costs and processing and retailing service costs. However, it is possible that supply management could have resulted in increased costs. It is also possible that different sectors took different intervals of time to capture the rents. Recognition of these factors could easily be added to the model.

The vertical model outlined above considers a single market and no trade between markets. Such a model captures the essence of how prices and quantities were determined under supply management during the 1970s and early 1980s. During this period, retailers were relatively decentralized, and purchasing decisions were largely made on a provincial or regional level. Because of the relative lack of interprovincial trade in processed products, each province or region was essentially a separate market. Thus, once a marketing board established the provincial output, the retail price in the province was essentially set.

In the last five to ten years, however, the retailing industry has changed substantially. Instead of making their purchasing decisions at a provincial or regional level, retailers now make their purchasing decisions centrally. For instance, a centrally located buyer will handle all the bids on chicken delivered to markets in different parts of Canada. The processor that can provide the lowest price in a particular market will obtain the contract to supply chicken to that market. That market may or may not be in the province or region where the processor is located. The result is that processors in one region are now competing with processors from other regions, both in their "home" market and in the other processors' "home" markets. For instance, some of the chicken marketed in Alberta may be supplied by Ontario processors while at the same time some of the chicken marketed in Alberta may be supplied by Ontario processors.

This change in the marketing of processed and further processed chicken was the result of a number of factors. For instance, technological change in chicken processing led to the development of chicken breasts and processed chicken meat. Compared to eviscerated chicken, which was the norm prior to the 1980s, processed chicken has a higher value per weight, thus making transportation costs less of a factor. As well, the introduction of processed chicken led to the development of new market niches such as McDonald's Chicken McNuggets. New marketing arrangements were also associated with these new market niches. For instance, when McDonald's looked for a source of Chicken McNuggets, it looked for one plant that would supply all of Canada. The result of new products, lower transportation costs relative to product value and the introduction of new market arrangements was a substantial increase in the interprovincial chicken trade. Much of this interprovincial trade involved cross-provincial trade, for example, processed product simultaneously moving from Alberta to Ontario and from Ontario to Alberta.

Technological change and new marketing arrangements are only two of the reasons for cross-provincial trade in processed and further processed chicken. Although the existence of differentiated products explains some of the cross-provincial trade, another factor is the market power possessed by the processors.

A Horizontal and Vertical Model of Supply Management

The basis for the horizontal model of supply management developed in this chapter is that the processing and retailing sectors are not perfectly competitive. Cross-provincial trade can be expected as a result of this unbalanced competition. In contrast, under conditions of perfect competition, such cross-provincial trade does not occur. Instead, the product flows in a single direction from the relatively lower cost areas to the relatively higher cost areas. This chapter examines the consequences of this cross-provincial trade in processed and further processed products.

To examine the impact of oligopoly behavior in the context of the regional nature of Canadian supply management, consider a model with two regions.[6] Processors in region 1 and region 2 have costs of $w_1 + k_1$ and $w_2 + k_2$, respectively. If processors have no market power, then the price will be set at $p_i = w_i + k_i$ ($i = 1,2$). In terms of product flow, two alternatives are possible: 1) If the difference between p_1 and p_2 is less than the transportation cost between region 1 and region 2, then processors supply only the market in their region; or 2) If the price differential is equal to the transportation cost, then the product is exported from the less expensive to the more expensive region.

If processors have market power, then the outcomes are entirely different. First, $p_i > w_i + k_i$ ($i = 1,2$). Second, although it is possible that the product could flow in only one direction, it is more likely that the processed product will flow from region 1 to region 2 and vice-versa. That is, the product will be cross-shipped between the regions. Because price is not equal to marginal cost, firms can earn positive profits. Consequently, firms with different cost structures can coexist in the same industry. Although low-cost firms have a higher market share and a greater markup over cost, high-cost firms can still have a positive market share and earn a positive profit (Brander and Krugman, 1983). When processors sell into a market in another region, the presence of transportation costs effectively acts to increase marginal costs for the processor who is shipping the product. Consequently, although a processor's share of the "other" market is positive, it is less than his market share in his "home" market.[7]

The consequence of this cross-provincial trade is that the quantity produced in a province or region is no longer the quantity consumed in that region. From the perspective of price determination, the implication is that the quantity established by marketing boards in a province no longer determines the retail price in that province. Although the live product is usually processed in the region in which it is produced, the processed product can be and often is transported to other retail markets.[8] Thus, the retail price in a region is no longer determined by the product produced in that region. Instead, the retail price in the province is dependent on the quota decisions made by all marketing boards across Canada as well as on the decisions made by processors and retailers across Canada.

This change in the marketing system means the model outlined earlier is not appropriate for understanding how prices and quantities are determined currently under supply management. With trade and cross-trade in processed product, price determination is fundamentally altered. Processor prices are now determined much more centrally with retailers choosing processors that have the most competitive bids. To be competitive, processors are putting pressure on the marketing board in their region to lower the farm price. Indeed, processors realize that if they can obtain a lower farm price, they will be able to expand their market share with retailers across Canada. Thus, processors are offering marketing boards the opportunity of substantially increased production as a reward for lower farm prices.

The consequence of this structural change is that provincial marketing boards no longer have the ability to negotiate with their processors a price that is independent of the quantity the board wishes to produce. The price and the quantity in each region are now linked. Provincial marketing boards can no longer act unilaterally, and provincial rivalries become more important. For instance, consider the two-region model outlined above. The marketing board in region 1 can expand its production by reducing the farm price in its region. The reduction in price makes the processors in region 1 more competitive in both regions (through a lower processor price) thus enabling them to sell more product.

As a consequence, however, the marketing board in region 2 will be pressured by the processors in region 2 to reduce the farm level price so that the processors in that region can compete with processors in region 1. If this rivalry becomes too intense, both marketing boards can find themselves in a much worse position than the one they were in before they started their price cuts. Formally, the linking of the regional markets creates an opportunity for marketing boards to behave in a noncooperative fashion vis-à-vis each other.

The structural change outlined above is also likely to have an additional consequence. Increased inter- and cross-provincial trade leaves processors increasingly vulnerable to the bargain they can reach with the marketing board in their province. For instance, processors in one region could be severely disadvantaged vis-à-vis processors in other regions if the marketing board with which they are dealing is not able to respond quickly to changing market conditions or is unwilling to offer output or price guarantees. As a result, increased vertical integration is likely to occur in the chicken industry as processors try to lock in production amounts at known prices.

Institutional History of the Industry

The recent institutional history and current politics of the chicken industry closely reflect the changes in market structure described above and the resulting alterations to pricing and allocation decisions. Since May, 1994, the CCMA has allocated production quota according to a procedure substantively different than that followed for the preceding 15 years. This new procedure is also substantively different from that currently followed by the other three feather agencies (eggs, turkey and broiler hatching eggs). The market changes described above for the chicken industry have not occurred to the same extent in these other industries.

The new production-allocation procedure is based on a bottom-up approach in which the provincial producer price and the provincial production quota are simultaneously determined in every province by explicit negotiations between the provincial marketing board and the provincial processors or the provincial processors' association. Previously, the provincial production quota and the producer price were determined independently and sequentially. The recent changes to the pricing and allocation mechanism closely match the results of the theoretical model presented in the last section of this chapter.

The change in the allocation procedure resulted from Ontario and Québec seeking increases in the production quota allocated to them. These increases were sought, in part, because of the changing market conditions outlined earlier in this chapter. Growing populations in Ontario and Québec, the concentration of further processing facilities in these provinces and the competitive edge that the large processors in these provinces have in chicken production because of scale and scope economies were also contributing factors. The national agency refused Ontario and Québec's request for quota increases, adhering instead to the standard, more or less fixed proportional distribution of national production quota. The fixed-proportion allocation mechanism was established as the principle of production allocation and was embodied in the federal-provincial agreement that established CCMA's marketing plan in 1978.

The General Agreement on Tariffs and Trade (GATT), in place prior to the conclusion of the Uruguay Round, also precluded any significant changes to the structure of the domestic industry. The pre-Uruguay Round agreements contained an exemption under Article XI.2.c that allowed countries to impose import restrictions if they also put in place measures to restrict domestic production. If a country would not act to limit domestic supply, Article XI would not apply, and the country would be forced to admit imports. As a consequence, there were substantial costs to any province or region that attempted to unilaterally expand production.

The December, 1993 GATT accord replaced import restrictions with tariffs, which are fully delinked from internal trade regulations. As a result, Ontario was able to take steps which under the pre-Uruguay Round agreements would have resulted in disruption of domestic markets and loss of protection under Article XI.2.c. In particular, Ontario set its production quota levels independently of the production quota allocations of other provinces and without regard for its historical share of the national market.

These steps were taken, in part, as a response to the changes in market conditions described above. New products, lower transportation costs and the introduction of new market arrangements resulted in increased interprovincial trade and an increased ability for processors in one region to capture markets in another region. The change in the international environment, defined by revisions to GATT, facilitated these actions. Ontario's initiatives were not taken solely with the view of maximizing returns from new market opportunities; they were also motivated by a need to respond to the threat of competition from the United States. This competition could occur after the next GATT round or sooner if the United States' challenge to the North American Free Trade Agreement (NAFTA) is upheld.

To a large degree the steps taken by Ontario parallel the steps taken by US industry during the past ten years. Concentration of production and processing has progressed quickly in response to rapid market growth for further processed products in the United States, the need for investment in processing facilities and the need for national distribution networks. The events in the United States suggest that the pace of vertical integration will accelerate in Canada as integrated processing and further processing companies seek to lock in production at fixed prices and extend control over the pricing, manufacture and delivery of chicken products, from feed and chick to table.

Less than a year after the introduction of the new system, only one of New Brunswick's two processing plants remains in operation. The management at the remaining plant has taken aggressive steps to introduce an integrated program of feed and chick purchases to the producers delivering to their plant. Management has also implemented much longer payment terms (30 days versus the previous standard of seven days) that will effectively bind producers to this one plant. Management has indicated that the only alternative to these changes is the closing of the plant.

The market changes over the past 10 years suggest that a new institutional structure will have to be developed for supply management in Canada. The theory presented above predicts that provincial marketing boards will no longer be able to act unilaterally and that interprovincial rivalries will become more important. The logical implication is that considerable strain and antagonism can be expected to characterize the political process that will modify supply management's institutional structure. Such is in fact the case.

For instance, there is considerable apprehension about the bottom-up approach, particularly by producers in Atlantic Canada and Québec, and by many provincial governments. Specific concerns are that the system could result in price-depressing surpluses and intensified competition between provinces for market share and that the system provides no safeguards or redress to assure market stability and discipline.

To meet these concerns, federal and provincial ministers of agriculture at their December, 1994 meeting in Toronto directed CCMA to pursue the need for the introduction of additional safeguards and discipline into the system as a matter of priority and to report their findings to the ministers by March 1, 1995. The question of safeguards to assure market stability and discipline has been a matter of considerable debate in the industry since the inception of the new bottom-up system. Needless to say, as predicted by the theory, no practical proposal for the workings of such a safeguard mechanism has been suggested by any stakeholder nor is such a proposal expected.

Conclusion

The prices and output determined under supply management depend on the behavior of all the sectors within these industries. Included among these sectors are the retailing, processing and farm sectors. Although the role of the farm sector in determining prices and quantities has been studied extensively, little is known about the role of the processing and retail sectors (Schmitz, 1993). This chapter develops a conceptual model that illustrates how prices and output are determined and explicitly incorporates the role of all sectors. Since there is substantial evidence that each of the sectors has either bargaining or market power, the model also incorporates this aspect.

The vertical relationships that exist between the various sectors are not the only factors influencing output and price determination. The regional (or horizontal) structure of the Canadian market also has a substantial impact on prices, quantities and the location of production. The impact of this horizontal structure has been increasingly important during the last five to ten years. During this period, there has been a substantial increase in interprovincial trade in chicken. This increase is the result of new products, lower transportation costs relative to product value and the introduction of new market arrangements.

These changing market conditions have profound implications for the manner in which prices and output are determined under supply management. Marketing boards have been accustomed historically to setting price and quantity separately. Decisions regarding quantity often reflect political considerations, for example, provincial self-sufficiency, historical production patterns and provincial balancing, while pricing decisions are made on the basis of the bargaining power that the marketing board has relative to the processors.

When the recent changes to the structure of the Canadian chicken market are taken into account, the separation of price and output decisions is no longer possible. The price at which farm output can be sold in a province is effectively determined when total Canadian output, not provincial output, is determined. Decisions about producer prices and provincial production quotas can no longer be made independently of each other but must be made simultaneously. The prices and output set in one province affect the prices and output in other provinces.

The result is greater provincial rivalry among marketing boards and among processors as different companies and regions seek to expand their production base. Increased pressure for vertical integration is also likely as processors attempt to provide themselves with guarantees on production and prices. Considerable strain and antagonism are also apparent in the political decisions regarding the future direction of supply management.

Notes

1. The views expressed in this chapter are those of the authors and do not necessarily reflect those of Agriculture and Agri-Food Canada.

2. The slope of d relative to D indicates the degree of market power possessed by retailers. Under perfect competition, the derived demand curve is simply the retail demand shifted down by the cost of retailing service, c. As the degree of market power increases, the slope of d becomes steeper relative to that of D. Thus, as the retailers' market power increases, the derived demand becomes more inelastic, and the price reduction required to get retailers to purchase a given additional amount of output becomes larger. (See Karp and Perloff, 1993). Since the vertical distance between D and d is greater than c, supranormal profits are earned in the retailing sector.

3. In this chapter, the term markup refers to the difference between the price at different levels of the vertical system, for example, r - w, while the term margin will refer to the profit margin, for example, r-(w+k).

4. In 1974, Canada imported 3.369 million kg of chicken, just over 1% of production (CCMA, 1994). By 1976, Canada was importing 23.887 million kg of chicken, about 7% of production. During this period, exports fell from 9.291 million kg to 0.253 million kg.

5. The area below w and above the S_{Farm} curve represents the returns to the owners of the inputs used in the production of x. As a result of the reduction in output, the input owners lost the area between w and m, to the left of supply curve S_{Farm}.

6. A two-region model is the simplest model through which regional relationships can be examined.

7. Cross-shipments only occur when the processors act in a noncompetitive and noncollusive fashion. If the processors were to behave perfectly competitively, then $p_i = w_i + k_i$ (i = 1,2), and one of the two competitive outcomes would occur. If the processors were to act in a collusive fashion, the same two competitive outcomes would occur. Because the processors are acting in a collusive fashion, they will choose the lowest cost method of producing the output; the competitive outcome is the lowest cost method of allocating production.

8. Although there is some movement of live chicken between Ontario and Québec and between Québec and the Maritimes, the amount is less than 4% of total production (CCMA, 1994).

References

Brander, J., and P. Krugman. 1983. "A Reciprocal Dumping Model of International Trade." *Journal of International Economics* 15: 313–321.

Canadian Chicken Marketing Agency. 1994. *Data Handbook 1994.* Ottawa, Ontario, Canada.

Coffin, G. H., R. F. Romain, and M. Douglas. 1989. "Performance of the Canadian Poultry System Under Supply Management." Department of Agricultural Economics, Macdonald College of McGill University, Ste-Anne-de-Bellevue, Québec, Canada and le Département d' Economie Rurale, Université Laval, Québec City, Québec, Canada.

Ferguson, R. 1992. *Compare the Share Phase II: The Comparisons Continue.* Brochure prepared by the Office of Hon. Ralph Ferguson, MP. Lambton-Middlesex, Québec, Canada (May).

Karp, L. S., and J. M. Perloff. 1993. "A Dynamic Model of Oligopoly in the Coffee Export Market." *American Journal of Agricultural Economics* 75 (2): 448–457.

Schmitz, A. 1993. "Supply Management in the 21st Century." Paper presented at the Supply Management Conference, Ottawa, Ontario, Canada (14–15 June).

19

Will the Supply Management Cartel Stand?

T. G. Schmitz and J. J. Skinner

Abstract

This chapter analyzes the economic implications of a major production increase by Ontario chicken producers on the Canadian broiler industry. An economic, spatial oligopoly model of Canada is developed that estimates trade flows and the economic benefits accruing to provincial producers as well as to consumers. The model shows that a major increase in the supply of broilers from Ontario would cause economic losses to accrue to producers in all provinces if the current supply management system remains intact. On the other hand, this policy could force retaliatory action on the part of other provinces that would cause a breakup of the national supply management cartel. Under a new regime in which marketing boards from each individual province compete with each other for a share of the Canadian market, Ontario producers could accrue additional economic benefits if they could act as a leader in the industry. In this scenario, producers in the prairie provinces would also accrue economic benefits while producers in British Columbia, Québec and the Maritime provinces would suffer losses.

Introduction

The elimination of Article XI of the General Agreement on Tariffs and Trade (GATT) and the replacement of nontrade barriers with tariff-rate quotas[1] has caused concern for the future of supply management. However, as Schmitz, de Gorter and Schmitz (1996) discuss elsewhere in this volume, Canadian supply-managed industries will not face legal repercussions from international players in the industry in the near future because of the minimum access rule under the new GATT agreement. The agreement resulted from the Uruguay Round of negotiations, which essentially insulates Canadian producers from import levels above the minimum access level. As implied by the tariff-quota equivalents estimated by Moschini and Meilke (1991), the over-quota tariffs for most supply-managed commodities, as specified by the new Uruguay Round Agreement (URA), are prohibitive. Hence, the two-tiered tariff-rate quota rule effective this year will act as an import quota for the entire six years of the URA.

As a result of this international uncertainty, Ontario broiler producers have either threatened to increase production by 20% over their current quota allotment or withdraw altogether from the Canadian Chicken Marketing Agency (CCMA). In retaliation other provinces are considering different strategic policy options.[2] This chapter uses spatial equilibrium in an oligopolistic economic setting for the prediction of current trade flows and the determination of the economic impact of a 20% increase in Ontario's production on different provinces. Two cases are considered. In the first case the other provinces do not react to such a move by Ontario. That is, they do not alter their current quota allocation scheme. In the second case it is assumed that the national structure collapses as provincial boards across Canada compete with each other in the sense that each province acts as an oligopolist within the industry.[3] A comparison of these two scenarios will lend insight that will help answer the question, "Will the Cartel Stand?"

An Economic Model of the Canadian Broiler Industry

In order to model the Canadian broiler industry, each provincial marketing board is treated as a member of a trading oligopoly that is separated by significant transportation costs between regions and faces distributional constraints placed upon it by processors in each region. Because of the central-planning authority of each provincial board, producers are assumed to act as a group within each province. By definition, these individual producer groups form an oligopolistic trading structure across Canada. However, in the Canadian supply management structure producers do not enjoy the full benefits of a traditional oligopolistic structure because the industry is not as vertically integrated as it is in the United States. This places additional distributional constraints upon the various provincial producer groups because of the interaction between producers, processors, retailers and consumers. In particular, because the transportation of live broilers is prohibitively expensive, it is the processor — not the producer — that decides how the product will be distributed. This fact creates additional arbitrage constraints that the producers must consider when maximizing profits.

For simplicity, the following model combines the processors and retailers within a particular province in such a way that each acts in a perfectly competitive fashion. That is, processors and retailers are assumed to be price-takers on both the buying and selling side of the market. Each province has one monopolistic producer group, the provincial board, that must sell to the processor in its province.[4] This processor can then sell to the home retailer, incurring no measurable transportation cost, or to a retailer in any other province at a strictly positive transportation cost. It is assumed that the retailer in each province can sell its product only to consumers in its home province. There is an arbitrage rule that governs the way in which processed product from one province flows toward a retailer in any particular province. This rule states that an amount produced by producer i will only be sold by processor i to retailer j if the price, minus the cost of transportation from region i to region j, is at least as large as the price that processor i could receive from any other retailer. This will happen because both retailers and processors are price-takers, and the processor's only rule is to sell into the market which yields the highest price.

The above market structure can be formalized using the following notation which holds for $\forall\, i, j \in (1, \ldots, n)$:

Q_{ij}	quantity produced in market i that is consumed in market j
$Q_{i\bullet} = \sum_{j=1}^{n} Q_{ij}$	total quantity produced by producer i
$Q_{\bullet i} = \sum_{j=1}^{n} Q_{ji}$	total quantity consumed in market i
$Q = \{Q_{ij}\}\ \forall\, i, j \in (1,\ldots,n)\ t_{ij}$	cost of transportation from market i to market j
$P_i^C(Q_{\bullet i})$	inverse consumer demand function in market i
$P_i^R[P_i^C(Q_{\bullet i})]$	derived inverse retailer demand function in market i
$P_{ij}^{\,S}\,(P_j^R[P_j^C(Q_{\bullet j})],\ t_{ij})$	derived inverse demand for processor i from market j
$\dfrac{\sum_{j=1}^{n} P_{ij}^S\left(P_j^R\left[P_j^C(Q_{\bullet j})\right], t_{ij}\right)\bullet Q_{ij}}{Q_{i\bullet}}$	average eviscerated price received by producer i

The model is depicted in Figure 19.1 for the case of two provinces, i and j. Consider province i in which inverse consumer demand, $P_i^C(Q_{\cdot i})$, is a function of the total quantity sold by each province into market i. Now consider the retailer in market i that can either purchase the product from the processor in its home market i or some other market j. If the retailer purchases from its home market, the retail buying price will simply equal the wholesale selling price in that market, or $P_i^R[P_i^C(Q_{\cdot i})]$. However, if the retailer purchases from market j, it will only pay $P_i^R - t_{ji}$ to the processor, assuming that the retailer must pay for transportation.

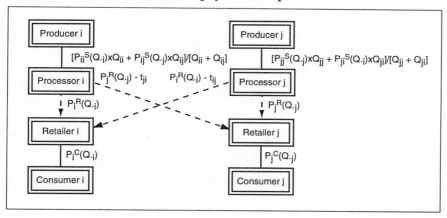

FIGURE 19.1   Spatial Oligopoly Model for Two Provinces

Now consider the processor. Processor i can possibly receive two prices. $P_{ii}^S(P_i^R[P_i^C(Q_{\cdot i})])$ is the price that the processor will receive for the quantity Q_{ii} that it sells into its home market. In addition, the processor will sell into market j if the price it receives from retailer j, determined by $P_j^R[P_j^C(Q_{\cdot j})] - t_{ij}$ is at least as high as the price it could receive from retailer i, $P_i^R[P_i^C(Q_{\cdot i})]$. If the processor does sell any positive amount, Q_{ij}, into market j, it will receive a price determined by $P_{ij}^S(P_j^R[P_j^C(Q_{\cdot j})], t_{ij})$. In addition, producer i will receive a price that is a weighted average of the prices that its home processor receives for selling its product into each market. This weighted, average price for the two markets in question can be written as $[P_{ii}^S \cdot Q_{ii} + P_{ij}^S \cdot Q_{ij}] / Q_{i\cdot}$.

Mathematical Model

The aforementioned market structure can be used to formulate a mathematical model that can be used for further analysis. The following model assumes that within each province producers maximize profits with respect to the quantity sold into each market. Further, it is assumed that producers in each province have Cournot conjectures with respect to producers in other provinces. That is, the producer in each province takes as given the quantity sold by all other provinces into each market. The above behavioral assumptions, combined with aforementioned arbitrage constraints, lead to an equilibrium that is established using the standard Nash equilibrium concept.

The equilibrium trade flows are given as the solution to the following problem:

$$\text{Maximize } \{\, \Pi_i = \sum_{j=1}^{n} TR_{ij} - TC_i \,\} \text{ w.r.t. } \{Q_{ij}, \forall\, j=1,...,n\}$$

subject to:

(1) $\qquad Q_{ij} \geq 0 \qquad\qquad\qquad\qquad \forall\, j=1,...,n$

(2) $\qquad [\, (P_j^R(\bullet)\text{-}t_{ij}) - (P_k^R(\bullet)\text{-}t_{ik})\,] \bullet Q_{ij} \geq 0 \,, \forall\, j,k \in \{1,...,n\}\,, j \neq k$

where n is the number of regions under consideration, $TR_{ij}(Q) = P_{ij}^S(Q_{\bullet j}) \times Q_{ij}$ equals the total revenue received by producer i and $TC_i(Q_{i\bullet}) = C_i(Q_{i\bullet})$ equals producer costs. The method of Lagrangian multipliers leads to the following Kuhn-Tucker conditions that must be satisfied for producers in all i regions:

(1) $\qquad\qquad \dfrac{\partial}{\partial Q_{ij}}[\Psi_i] \leq 0\,, Q_{ij} \geq 0 \text{ and } Q_{ij} \times \dfrac{\partial}{\partial Q_{ij}}[\Psi_i] = 0$

$\qquad\qquad \forall\, j \in \{1, \ldots, n\}$

(2) $\qquad\qquad \dfrac{\partial}{\partial \mu_{jk}}[\Psi_i] \geq 0\,, \mu_{jk} \geq 0 \text{ and } \mu_{jk} \times \dfrac{\partial}{\partial \mu_{jk}}[\Psi_i] = 0$

$\qquad\qquad \forall\, j,k \in \{1, \ldots, n\}\ (j \neq k)$

where:

$$\frac{\partial}{\partial Q_{ij}}[\Psi_i] = \left[P_{ij}^S + P_{ij}^{S'} Q_{ij} - C_i' + \sum_{k \neq j}^{n} \left[P_j^{R'}(\mu_{jk} Q_{ij} - \mu_{kj} Q_{ik}) + \mu_{jk}(P_j^R - P_k^R - T_{j-k}^i) \right] \right]$$

$$\forall\, j \in \{1, \ldots, n\}$$

$$\frac{\partial}{\partial \mu_{jk}}[\Psi_i] = \left[P_j^R - P_k^R - T_{j-k}^i \right] \qquad \forall j,k \in \{1, \ldots, n\}\ (j \neq k)$$

where $T_{k-l}^i = [\, t_{ik} - t_{il}\,]$ is the difference between the cost of transporting eviscerated broilers to region k and the cost of transportation to region l.[5]

Empirical Model Assumptions

The above mathematical approach can be applied to the 10 provinces across Canada. Under the empirical model presented here, the first six provinces from west to east have been considered separately while the maritime provinces east of Québec have been combined under one heading called the Atlantic provinces because their production is relatively small. These regions are depicted in Figure 19.2. They are: British Columbia (BC), Alberta (AB), Saskatchewan (SK), Manitoba (MB), Ontario (ON), Québec (QE) and the Atlantic provinces (AT). Hence, there are a total of n=7 members that form the broiler oligopoly in Canada.

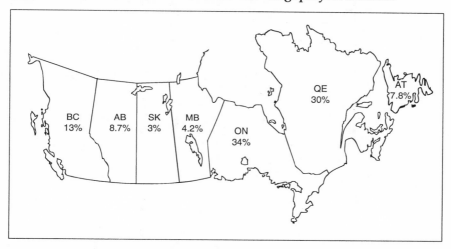

FIGURE 19.2   Share of Broiler Production per Province Across Canada

Table 19.1 gives some raw broiler data for 1992, the year used as the base for this study. For each province, total domestic consumption data are unknown but are estimated using the 1992 per capita consumption rate reported in Statistics Canada (23–202) multiplied by census data on population per province. Global imports are given as millions of kilograms eviscerated and only represent regional landed imports. This model takes imports as exogenous[6] and assumes that all produce is consumed in the province in which it originally landed. The eviscerated producer price is estimated using reported live weight prices adjusted for a 70% dress weight. Table 19.1 also gives 1992 wholesale and retail prices.

Total production per province for 1992 is also shown in Table 19.1. These production levels are taken from Statistics Canada (23–202) data and are in millions of eviscerated kilograms. They are also depicted in Figure 19.2 where the numbers shown represent the percentage of total Canadian production

TABLE 19.1 Canadian Broiler Data for 1992

	BC	AB	SK	MB	ON	QE	AT	Canada
Domestic production	74.760	49.756	17.117	24.192	192.899	170.555	44.707	573.986
Global imports	0.970	0.373	0.232	2.868	40.663	11.9947	0.000	57.053
Consumption	76.175	59.198	22.943	25.336	233.278	159.972	54.137	631.039
Producer price (l)	114.40	115.40	113.40	112.90	115.10	114.10	119.00	114.90
Producer price (e)	163.43	164.86	162.00	161.29	164.43	163.00	170.00	164.14
Wholesale price	225.00	245.30	274.00	271.40	241.00	241.20	260.50	243.13
Retail price	371.00	428.70	388.20	401.20	449.50	399.10	428.00	416.96

Domestic production is in million kilograms eviscerated (Statistics Canada 23–202).
Global Imports are reported as million kilograms eviscerated, and the data is on landing only (Statistics Canada, *Merchandise Trade Data Handbook*).
Consumption is calculated using data and per capita consumption of 23.1 kg/year (includes imports) (Statistics Canada, Canadian Census Bureau).
The producer price is reported as live weight cents/kg (under 2.3 kg) (Agriculture Canada *Poultry Market Review Handbook*, 1992).
The eviscerated price is in cents/kg and is calculated from the live weight using an average dress weight of 70% (Agriculture Canada *Poultry Market Review Handbook*, 1992).
The wholesale price is reported as eviscerated cents/kg (Agriculture Canada *Poultry Market Review Handbook*, 1992).
The retail price is reported as eviscerated cents/kg (Agriculture Canada *Poultry Market Review Handbook*, 1992).

supplied by each region. Table 19.2 shows average cost data, most of which is reported by the CCMA *1990 Cost of Production Study* (1992). However, catching costs are adjusted to reflect the average hourly wage rate differences across provinces. Also, cost of production (COP) data on feed costs are not used. Instead, these costs are calculated using the average 1992 broiler feed price for each province, multiplied by an assumed feed conversion ratio which is estimated at 1.8:1 (Hamilton, Proudfoot, Dewitt and Jansen, 1991). Transportation cost data are reported in Table 19.3 and are based on a .004 cents per kilogram per kilometer rate used by Kindersley Trucking Company in Kindersley, Saskatchewan, Canada.

In terms of functional forms, it is assumed that demand is linear, and a demand elasticity of $\eta = -0.91$[7] is used to calculate intercept and slope parameters. Quadratic cost functions are assumed so that marginal cost is linear. The average cost data reported in Table 19.2 is used as a proxy for a point estimate of marginal cost along with an estimated supply elasticity of $\varepsilon = 2.0$.[8] Further, it is assumed that retailers use a gross markup rule so that the retail price equals a certain percentage markup over the wholesale price in the home province. Similarly, processors use a gross markup rule so that their selling price is equal to a certain percentage markup over the free on board (f.o.b.) price of the home producer.

Empirical Results

The data presented in Table 19.1 and Table 19.2 along with the mathematical model outlined earlier in this chapter can be used to predict trade flows across Canada. In addition, the economic benefits/losses accruing to producers and consumers can be obtained using standard surplus measures.[9]

Estimated Trade Flows Under the Current Structure

Because data on trade flows among provinces are not known, it is necessary to first estimate trade flows using 1992 quota allocation levels so that we have a base model with which we can compare our other results. This type of model assumes that after quotas are allocated, trade flows are predicted using competitive behavior on the part of retailers and processors. The economic model outlined earlier in this chapter, along with the additional assumption that total production in each province equals 1992 levels, is used as a base case. The resulting trade flows under this model are depicted in Figure 19.3, where the numbers represent total production and the arrows depict trade directions.

TABLE 19.2 Average Cost Data for 1992

cents/kg	BC	AB	SK	MB	ON	QE	AT
Catching	2.26	2.10	1.97	1.99	2.18	2.13	1.96
Energy	3.18	4.66	4.21	4.95	5.28	6.25	4.62
Repair	2.20	2.34	2.56	2.09	2.70	2.63	2.15
Labor	13.87	12.83	12.06	12.21	13.32	13.05	11.98
Chicks	25.61	27.64	25.61	26.80	27.75	24.66	27.47
Feed	47.60	42.78	41.77	43.37	44.80	51.22	53.59
MC (live weight)	94.72	92.35	88.18	91.41	96.03	99.94	101.77
MC (eviscerated)	135.31	131.93	125.97	130.59	137.19	142.77	145.39

Most of these costs are taken from average cost data in cents/kg as reported by the CCMA (1993).
Catching costs are adjusted using average catching times and hourly wage rates per province.
Feed costs are adjusted using the average 1992 price broiler feed price in each province times the fed conversion ratio (1.8:1).
In order to get a point estimate of the live weight marginal cost, the average cost is used as a proxy.
The eviscerated marginal cost is calculated using the live weight cost and a 70% dress weight.

TABLE 19.3 Transportation Costs for Eviscerated Broilers across Canada

cents/kg	BC	AB	SK	MB	ON	QE	AT
British Columbia	0.00	6.00	7.90	11.80	21.10	22.33	28.51
Alberta	6.00	0.00	1.90	5.80	15.10	16.33	22.51
Saskatchewan	7.90	1.90	0.00	3.90	13.20	14.43	20.61
Manitoba	11.80	5.80	3.90	0.00	9.30	10.53	16.71
Ontario	21.10	15.10	13.20	9.30	0.00	1.23	7.41
Québec	22.33	16.33	14.43	10.53	1.23	0.00	6.18
Atlantic provinces	28.51	22.51	20.61	16.71	7.41	6.18	0.00

Source: Based on a .004 cents/kg/km rate SPI (Korg trucking) Kindersley Trucking Company, Kindersley, Saskatchewan, Canada.

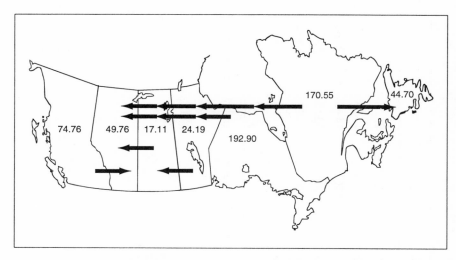

FIGURE 19.3   Estimated Trade Flows under 1992 Production Quota Levels (million kg)

Ontario is the largest broiler producer as it has the largest population. Ontario ships west to Alberta, Saskatchewan and Manitoba although most of its exported product winds up in Manitoba (roughly 4 million kilograms). It also sells 1 million kilograms to Saskatchewan and under 10,000 kilograms to Alberta. Most of Québec's exports (12 million kilograms) move eastward to the Atlantic provinces because of the high costs associated with producing broilers in those areas, and it also ships 2.3 million kilograms west to Saskatchewan and almost 6 million kilograms to Ontario. British Columbia ships 6.9 million kilograms east to Alberta while Saskatchewan sends roughly 22,000 kilograms west to Alberta. Alberta does not trade at all because of its high retail price that results from large processor and retailer markups. Figure 19.4 shows consumer surplus (CS) and producer surplus (PS) levels under the current base scenario that will be used as a comparison when analyzing the results of the models that follow. Total surplus (TS) is also shown.

Ontario Increases Production by 20% with No Retaliation

This model illustrates the effects of a 20% production increase that Ontario threatened to carry out. It is assumed that the other provinces do not retaliate; they stay within the CCMA and do not adjust their quota allocation. The economic model previously outlined, adjusted by this 20% increase, is used for this analysis. This dramatic increase in Ontario supply will disrupt the current pattern of trade flows established under the base scenario. As a result of this increase, a new equilibrium will be established. The resulting trade flows under this scenario are shown in Figure 19.5. The major result of the expanded production level in Ontario is that most provinces absorb the extra production. Ontario not only increases its exports to the western provinces, it creates new markets in the East. Ontario's exports to Alberta, Saskatchewan and Manitoba increase to 9.2, 4.1 and 7.4 million kilograms, respectively. In addition, it exports 3.7 and 1.2 million kilograms into Québec and the Atlantic provinces, respectively. Notice that this has resulted in a complete trade flow reversal. Québec previously exported to Ontario, but now the direction of trade has been reversed. Québec no longer sells to the West as Ontario floods those markets, and as a result, Québec's home sales increase by even more than the previous exported quantities. There are also some unpredictable side effects as a result of the spatial equilibrium outcome under this scenario in that British Columbia reduces its Alberta exports from almost 6 million kilograms to under 1 million kilograms, and Manitoba now sends roughly one million kilograms into Alberta, whereas under the base case, it exported none.

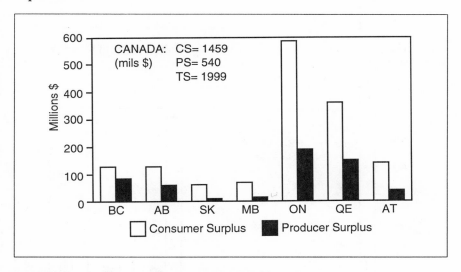

FIGURE 19.4   Surplus under the Current Base Model

The effect of a 20% Ontario production increase on producer and consumer surplus is depicted in Figure 19.6. Every province — including Ontario — suffers losses in producer surplus and gains in consumer surplus. However, Ontario suffers only a 7.8% loss in producer surplus while all other provinces experience between 12 and 13% losses. Obviously, consumers would welcome such a move because it results in a net increase in consumer surplus. Under this model, giving equal weight to the interests of both consumers and producers together, Canada would see an increase in total economic surplus of 7.4% over the base model.

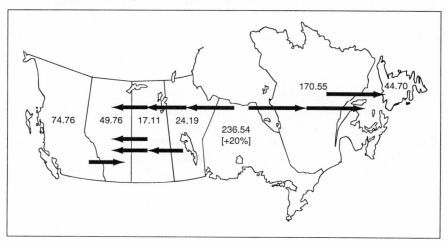

FIGURE 19.5   Estimated Trade Flows under a 20% Production Increase by Ontario with No Retaliation (millions kg)

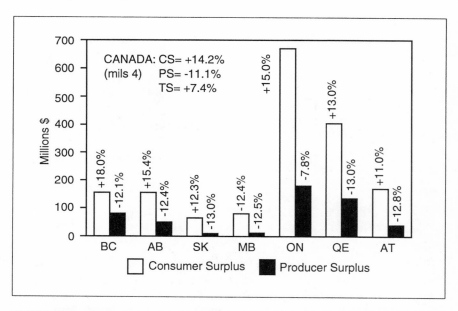

FIGURE 19.6   Change in Surplus if Ontario Increases Production by 20% (no retaliation)

Ontario Increases Production by 20% and Other Provinces Retaliate

This section also models the effects of a 20% production increase by Ontario. Here we assume that, as a result of this move, the national CCMA structure collapses, and the individual provincial boards adjust their quantities in order to reach a new quota-allocation scheme within their provinces. This would be consistent with a spatial equilibrium that optimizes each province's profits accepting the others' quantities as given, that is Cournot-Nash. The empirical model discussed earlier in this chapter is used to analyze this situation by simply adding a constraint that sets Ontario's supply at a level of 20% over its current base value.

Because of the significant computational complexities inherent in this type of modeling exercise, two additional assumptions are made in order to make the problem tractable. First, it is assumed that because of the large increase in Ontario's level of production, Ontario ships its product east into Québec and the Atlantic provinces. This implies that Québec can only sell into its own market and possibly sell into the Atlantic provinces — it cannot ship any product to the West. Second, it is assumed that all provinces west of Ontario can sell into British Columbia. This additional assumption implies that British Columbia cannot trade at all. That is, all product that is produced in British Columbia must be consumed at home.

As expected, the scenario outlined above causes a dramatic change in the industry structure in terms of trade-flow directions and total supply across the country as shown in Figure 19.7. Regional production levels fall dramatically in most other provinces, except in the smaller, low-cost regions of Manitoba and Saskatchewan. While production levels in Alberta go down by 7.2%, those in Saskatchewan and Manitoba increase by 83% and 54%, respectively. Trade flows become regionalized as the spatial oligopoly equilibrium eliminates any inefficiencies associated with hauling long distances that were present under the national supply management scheme. Manitoba, Saskatchewan and Alberta export to all of the western provinces, while British Columbia now sells only into its home market. Ontario and Québec sell only to the provinces east of themselves. Ontario floods the markets of Québec and the Atlantic provinces with huge levels of exports of 46 and 26 million kilograms, respectively. This drives Québec's production down by almost 50% so that they sell only 3.6 million kilograms to the Atlantic provinces.

What are the surplus effects of this change in the industry structure? Because the other provinces have been able to reoptimize by adjusting production levels, the producer effects are not as negative as one would think, and consumers actually lose. Figure 19.8 shows that consumers lose anywhere between 11% and 26% as a result of this restructuring. For producers, the results are mixed. British Columbia and the Atlantic provinces lose 26% and 29%, respectively, in producer surplus, and Québec loses 25% of its surplus. However, Alberta producers actually gain 6.6% while Saskatchewan and Manitoba gain 123% and 61%, respectively, because they are low-cost producers with potential for expansion. As a result of the 20% increase in its own production, Ontario producers would gain 32.6% in producer surplus.

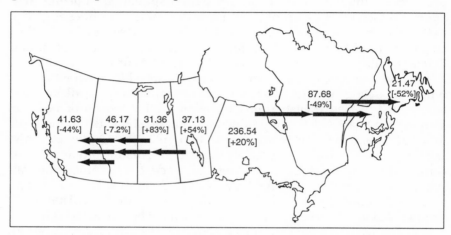

FIGURE 19.7   Estimated Trade Flows under a 20% Production Increase by Ontario with Retaliation (millions kg)

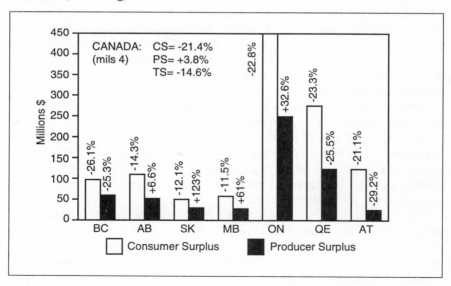

FIGURE 19.8   Change in Surplus if Ontario Increases Production by 20% (with retaliation)

This scenario loosely illustrates the concept of the benefits that accrue to a producer who can act as an industry leader. Ontario is essentially acting as a leader by announcing a 20% increase in production.[10] If the other provinces adjust in a noncooperative Cournot-Nash fashion, Canadian consumers will lose 21.4% of their surplus while Canadian producers as a whole will see a slight 3.8% gain in surplus. However, non-prairie producers outside of Ontario, for example, Québec, are big losers in terms of both consumer and producer surplus.

Conclusions

We have analyzed the economic impacts of a 20% production increase by Ontario on both producers and consumers using two different scenarios. In the first scenario, if no other province responds to the increased production, the producers in all provinces will lose. Why, then, would Ontario consider such a move? It may be that because Ontario has such a large domestic market share, it would be willing to accrue short-term losses if those losses enabled it to run other provinces out of the industry. This bluffing strategy could be used to keep the other provinces in line. The strategy could be perceived as credible by the other provinces because the losses to Ontario producers would be relatively fewer than the losses that would accrue to other producers. Another reason for this strategy on the part of Ontario producers is that Ontario has a large portion of the further processing sector in Canada for which it cannot fully provide, given its current quota allotment.[11] Finally, an increase in production by Ontario *ceteris paribus* would be better for consumers and hence the country as a whole. This may lead to increasing support of such supply expansionist strategies from other areas of government as well as from various consumer groups.

In the second scenario, the collapse of the so-called chicken "cartel" would have varied effects on different regions. With Ontario as a leader, assuming they stick to their high-production strategy even after the national structure collapses, the above model shows that Ontario producers would gain 32.6% in producer surplus.[12] Producers in the prairie provinces would also be better off — 6.6% increase for Alberta, 123% increase in surplus for Saskatchewan and a 61% increase for Manitoba producers.[13] In contrast, Québec, British Columbia and the Atlantic provinces would definitely lose in both scenarios. This lends evidence to support the theory that, in terms of comparative advantage, the current quota distribution scheme is not satisfactory. If this industry is to remain more efficient in the future, production should move from high-cost areas into low-cost centers with a low-population base.

Notes

1. This author defines a tariff-rate quota in the context of GATT as a policy in which the country must allow a certain quantity of imports at the world market price, up to the rule defined by minimum access. Any extra amount must be sold at the tariff markup price.

2. For a more comprehensive summary of the issues surrounding supply management, see Schmitz and Schmitz (1994).

3. This model is entirely different from the previous model developed by Schmitz and Skinner (1996) because in that model producers were allowed to maximize profits with no additional constraints placed on them by the processing and retail sectors. The model presented here is a preliminary version of a model that will eventually be developed for a Ph.D. dissertation by T. G. Schmitz to be completed at the University of California at Berkeley.

4. It is assumed that the transportation of live birds from a producer within a province to a processor in that same province is negligible.

5. Note that μ_{kl} is the lagrangian multiplier for the klth constraint.

6. In actuality, imports should be endogenous because the URA rules imply that an amount equal to 7.5% of domestic production will be allowed to enter the country. This question is beyond the scope of this chapter and will be resolved in a forthcoming dissertation by T. G. Schmitz.

7. Most studies report the demand elasticity anywhere between -0.75 and -1.0.

8. Most would agree that the marginal cost curve of the broiler industry in Canada is pretty flat. If anything, this estimate may be too low which would imply that changes in trade flows resulting from this model are smaller than those under a higher supply elasticity.

9. For a theoretical discussion on the use of economic surplus as a welfare measure, see Just, Hueth and Schmitz (1982).

10. This is in no way meant to indicate that they act as a Stackelberg leader in the industry. However, it would be interesting to see what the Stackelberg outcome might look like.

11. Very little in terms of economics literature has been done to analyze the impacts of the growing further processing industry on quota allocation across Canada. Further research in this area is needed.

12. Again, it must be emphasized that this result will only hold if Ontario sticks to its high production levels. However, it may be time-inconsistent for it to do so since once the national structure has collapsed with the provincial structure still in place, it would be better off to greatly reduce production.

13. It should be noted that this result tends to support the hypothesis that even under the current CCMA structure, the prairie provinces should be allowed to expand production while the other provinces contract.

References

Agriculture Canada. 1992. *Poultry Market Review Handbook*. Ottawa, Ontario, Canada.
Canadian Census Bureau. 1992. *1991 Census Report*. Ottawa, Ontario, Canada.
CCMA. 1993. *1990 Cost of Production Study Update*. Ottawa, Ontario, Canada.

Hamilton, R., F. Proudfoot, W. Dewitt, and H. Jansen. 1991. *Raising Chicken and Turkey Broilers in Canada.* Agriculture Canada Publication No. 1860/E Ottawa, Ontario, Canada.

Just, R., D. Hueth, and A. Schmitz. 1982. *Applied Welfare Economics and Public Policy.* Prentice Hall Publishing, Inc., Englewood Cliffs, NJ.

Moschini, G., and K. D. Meilke. 1991. "Tariffication with Supply-management: The Case of the US–Canada Chicken Trade." *Canadian Journal of Agricultural Economics* 39: 55–68.

Schmitz, A. 1983. "Supply-management in Canadian Agriculture: An Assessment of the Economic Effects." *Canadian Journal of Agricultural Economics* 30: 135–152.

Schmitz, A., H. de Gorter, and T. G. Schmitz. 1996. "Consequences of Tariffication." This volume.

Schmitz A., and T. G. Schmitz. 1994. "Supply-management: The Past and the Future." *Canadian Journal of Agricultural Economics* 42(2): 125–148.

Schmitz, T. G., and J. Skinner. Forthcoming. "GATT and Structural Change in the Canadian Broiler Industry," in T. Becker, R. Gray, and A. Schmitz, eds., *Impact of GATT on World Agriculture.* Saskatoon, Saskatchewan, Canada: University of Saskatchewan Press.

Statistics Canada. 1992. *Poultry and Egg Production Handbook.* Publication No. 23–202, Ottawa, Ontario, Canada: The Queen's Printer.

_____ . 1992. *Merchandise Trade Data Handbook.*

Sullivan, A. 1981. "A Statistical Summary of Marketing Boards in Canada." *Supply and Services Canada.* Agriculture Canada Publication, Ottawa, Ontario, Canada.

Vercammen, J., and A. Schmitz. 1992. "Supply-Management and Import Concessions." *Canadian Journal of Economics* 4: 957–971.

_____ . 1993. "Deregulating Supply-Managed Industries: The Unexpected Trade Effects." University of British Columbia Working Paper, Vancouver, British Columbia, Canada (7 February).

About the Book

GATT, and to a lesser extent, NAFTA and CUSTA, has significantly affected certain sectors of North American agriculture, while other sectors, protected by special regulation before the free trade agreements, remain essentially untouched. This volume surveys and analyzes the effects of the agreements on different agricultural sectors in the United States and Canada and explores the implications of these changes for the future of U.S.-Canadian trade.

The book opens with an overview of the agricultural trade impacts of the three international trade agreements. Case studies on specific agricultural sectors such as the dairy, tobacco, peanut, and sugar industries in the United States and the dairy and poultry industries in Canada provide greater detail. Adopting Stigler's theory of regulation, the authors show how special-interest politics—for example, in Canadian supply management—has distorted the institutional and structural characteristics of certain branches of agricultural industry, impeding growth in overall economic performance as well as in international trade.

Index